Modern Technologies in Agriculture: Sensing Approaches

Modern Technologies in Agriculture: Sensing Approaches

Edited by Elsie Kennedy

SYRAWOOD
PUBLISHING HOUSE
New York

Published by Syrawood Publishing House,
750 Third Avenue, 9th Floor,
New York, NY 10017, USA
www.syrawoodpublishinghouse.com

Modern Technologies in Agriculture: Sensing Approaches
Edited by Elsie Kennedy

© 2023 Syrawood Publishing House

International Standard Book Number: 978-1-64740-422-2 (Hardback)

This book contains information obtained from authentic and highly regarded sources. Copyright for all individual chapters remain with the respective authors as indicated. All chapters are published with permission under the Creative Commons Attribution License or equivalent. A wide variety of references are listed. Permission and sources are indicated; for detailed attributions, please refer to the permissions page and list of contributors. Reasonable efforts have been made to publish reliable data and information, but the authors, editors and publisher cannot assume any responsibility for the validity of all materials or the consequences of their use.

Trademark Notice: Registered trademark of products or corporate names are used only for explanation and identification without intent to infringe.

Cataloging-in-publication Data

Modern technologies in agriculture : sensing approaches / edited by Elsie Kennedy.
   p. cm.
Includes bibliographical references and index.
ISBN 978-1-64740-422-2
1. Agricultural innovations. 2. Agriculture--Technology transfer. 3. Technological innovations.
4. Agriculture. I. Kennedy, Elsie.
S494.5.I5 M63 2023
338.16--dc23

TABLE OF CONTENTS

Preface VII

Chapter 1   **Management of Plant Growth Regulators in Cotton using Active Crop Canopy Sensors** 1
Rodrigo Gonçalves Trevisan, Natanael Santana Vilanova Júnior, Mateus Tonini Eitelwein and José Paulo Molin

Chapter 2   **Crop Sensor based Non-Destructive Estimation of Nitrogen Nutritional Status, Yield and Grain Protein Content in Wheat** 17
Marta Aranguren, Ander Castellón and Ana Aizpurua

Chapter 3   **Development of an Autonomous Electric Robot Implement for Intra-Row Weeding in Vineyards** 39
David Reiser, El-Sayed Sehsah, Oliver Bumann, Jörg Morhard and Hans W. Griepentrog

Chapter 4   **Monitoring of the Pesticide Droplet Deposition with a Novel Capacitance Sensor** 51
Pei Wang, Wei Yu, Mingxiong Ou, Chen Gong and Weidong Jia

Chapter 5   **Contribution to Trees Health Assessment using Infrared Thermography** 59
Rui Pitarma, João Crisóstomo and Maria Eduarda Ferreira

Chapter 6   **A Non-Invasive Method based on Computer Vision for Grapevine Cluster Compactness Assessment using a Mobile Sensing Platform under Field Conditions** 73
Fernando Palacios, Maria P. Diago and Javier Tardaguila

Chapter 7   **Utilisation of Ground and Airborne Optical Sensors for Nitrogen Level Identification and Yield Prediction in Wheat** 92
Christoph W. Zecha, Gerassimos G. Peteinatos, Johanna Link and Wilhelm Claupein

Chapter 8   **A Comprehensive Study of the Potential Application of Flying Ethylene-Sensitive Sensors for Ripeness Detection in Apple Orchards** 105
João Valente, Rodrigo Almeida and Lammert Kooistra

Chapter 9   **Acquisition of Sorption and Drying Data with Embedded Devices: Improving Standard Models for High Oleic Sunflower Seeds by Continuous Measurements in Dynamic Systems** 122
Simon Munder, Dimitrios Argyropoulos and Joachim Müller

Chapter 10   **UAV-Borne Dual-Band Sensor Method for Monitoring Physiological Crop Status** 136
Lili Yao, Qing Wang, Jinbo Yang, Yu Zhang, Yan Zhu, Weixing Cao and Jun Ni

Chapter 11 **Development and Field Evaluation of a Spray Drift Risk Assessment Tool for Vineyard Spraying Application**...155
Georgios Bourodimos, Michael Koutsiaras, Vasilios Psiroukis,
Athanasios Balafoutis and Spyros Fountas

Chapter 12 **Non-Invasive Tools to Detect Smoke Contamination in Grapevine Canopies, Berries and Wine: A Remote Sensing and Machine Learning Modeling Approach**.............................175
Sigfredo Fuentes, Eden Jane Tongson, Roberta De Bei, Claudia Gonzalez Viejo,
Renata Ristic, Stephen Tyerman and Kerry Wilkinson

Chapter 13 **Assessing Topsoil Movement in Rotary Harrowing Process by RFID (Radio-Frequency Identification) Technique**...191
Ahmed Kayad, Riccardo Rainato, Lorenzo Picco, Luigi Sartori
and Francesco Marinello

Chapter 14 **Predicting Forage Quality of Warm-Season Legumes by Near Infrared Spectroscopy Coupled with Machine Learning Techniques**...200
Gurjinder S. Baath, Harpinder K. Baath, Prasanna H. Gowda, Johnson P. Thomas,
Brian K. Northup, Srinivas C. Rao and Hardeep Singh

Chapter 15 **Calibration and Validation of a Low-Cost Capacitive Moisture Sensor to Integrate the Automated Soil Moisture Monitoring System**...215
Ekanayaka Achchillage Ayesha Dilrukshi Nagahage,
Isura Sumeda Priyadarshana Nagahage and Takeshi Fujino

Permissions

List of Contributors

Index

PREFACE

This book was inspired by the evolution of our times; to answer the curiosity of inquisitive minds. Many developments have occurred across the globe in the recent past which has transformed the progress in the field.

A sensor is a device that detects and responds to some type of input from the physical environment such as light, heat, motion, moisture and pressure. Agriculture sensors are sensors used in smart farming that help in heat detection, animal identification and health monitoring. There are several types of agricultural sensors including electronic sensors, location sensors, mechanical sensors, optical sensors, airflow sensors, and dielectric soil-moisture sensors. The use of remote sensing techniques to schedule nitrogen fertilization can assist in reducing the risk of harvest loss by providing a suitable rate of nitrogen at the time and place where it is needed by the crops. Precision agriculture entails the use of sensor technologies for nutrient and pesticide application, yield mapping and prediction, irrigation control, and soil sensing. This book contains some path-breaking studies on the application of sensors in agriculture. It will serve as a valuable source of reference for graduate and postgraduate students.

This book was developed from a mere concept to drafts to chapters and finally compiled together as a complete text to benefit the readers across all nations. To ensure the quality of the content we instilled two significant steps in our procedure. The first was to appoint an editorial team that would verify the data and statistics provided in the book and also select the most appropriate and valuable contributions from the plentiful contributions we received from authors worldwide. The next step was to appoint an expert of the topic as the Editor-in-Chief, who would head the project and finally make the necessary amendments and modifications to make the text reader-friendly. I was then commissioned to examine all the material to present the topics in the most comprehensible and productive format.

I would like to take this opportunity to thank all the contributing authors who were supportive enough to contribute their time and knowledge to this project. I also wish to convey my regards to my family who have been extremely supportive during the entire project.

Editor

Management of Plant Growth Regulators in Cotton using Active Crop Canopy Sensors

Rodrigo Gonçalves Trevisan *, Natanael Santana Vilanova Júnior, Mateus Tonini Eitelwein and José Paulo Molin

Precision Agriculture Laboratory, Department of Biosystems Engineering, Luiz de Queiroz College of Agriculture, University of São Paulo, Piracicaba 13418-900, São Paulo, Brazil; natanael.vilanova@usp.br (N.S.V.J.); mateu@smart.agr.br (M.T.E.); jpmolin@usp.br (J.P.M.)
* Correspondence: rodrigo.trevisan@usp.br

Abstract: Factors affecting cotton development present spatial and temporal variability. Plant growth regulators (PGR) are used to control vegetative growth, promote higher yields, better fiber quality, and facilitate mechanical harvest. The optimal rate of PGR application depends on crop height, biomass, and growth rate. Thus, the objective of this study was to evaluate optical and ultrasonic crop canopy sensors to detect the crop spatial variability in cotton fields, and to develop strategies for using this information to perform variable rate application (VRA) of PGR in cotton. Field trials were conducted in Midwest Brazil during the 2013/2014 and 2014/2015 crop seasons. Two optical and two ultrasonic active crop canopy sensors were evaluated as tools to detect crop variability. On-farm trials were used to develop and validate algorithms for VRA based on within-field variations in crop response to PGR applications. The overall performance of the sensors to predict crop height and the accumulation of biomass in cotton was satisfactory. Short distance variability was predominant in some fields, reducing the performance of the sensors while making current technology for variable rate application of PGR inadequate. In areas with large scale variability, the VRA led to 17% savings in PGR products and no significant effect on yield was observed. Ultrasonic sensors present can be a low-cost alternative to implement variable rate application of PGR in real time.

Keywords: ultrasonic sensors; spatial variability; variable rate application; *Gossypium hirsutum* L.

---

1. Introduction

Cotton (*Gossypium* ssp.) is among the most important fiber crops, with approximately 35 million hectares grown throughout the world. Global demand has gradually increased since the 1950s, with an average annual growth of 2%. About 350 million people around the world carry economic activities related to cotton, making it one of the 20 most important commodities in the world market in terms of its value [1].

Brazil produces about 1.7 million tons of cotton lint per year, being placed among the top five global producers, after China, India, Pakistan, and the United States. Brazil is also the fourth largest exporter in the world, with an area planted with cotton in the last five crop seasons averaging 1 million hectares, with some variations due to the market scenarios for the sector and production costs [2]. The national average cottonseed yield surpassed the 4000 kg ha^{-1} mark in the 2014/2015 season, which makes Brazil the country with the highest yields of dryland cotton cultivated in the world. Although this average is reasonable, many producers have focused on improving techniques to increase crop productivity by adopting higher levels of technology.

The cotton height must be periodically managed to decrease the abortion of reproductive structures by carbohydrates competing with the vegetative growth [3]. Cotton growth and its response to application of plant growth regulators (PGR) highly depends on temperature and water availability. The decisions about PGR rate, timing, and the general crop management remains a challenge for growers, especially due to spatial variability of soil conditions, which demands the use of variable rate application (VRA) of nitrogen and PGR.

Crop canopy vegetation indexes have been used to detect infield variability. Active and passive optical reflectance sensors mounted in various platforms have shown good performance to predict biomass, plant height, height-to-node ratio, nitrogen nutrition status, crop maturity, and lint yield [4–8]. The performance of sensors can vary throughout the crop season, due to crop canopy changes and management practices. Good results are usually found in early season (pin-head square to early bloom), but poor performance may occur due to the known effect of signal saturation when used later in the season, due to dense canopy and changes in the spectral signature of the plants.

Normalized difference vegetation index (NDVI) and plant height have a good correlation with stem mass and leaf mass during the early and late crop stages, but not during mid-season of cotton cycle [9]. In mid-season, more than 90% of NDVI values exceeded 0.8, and nearly half were above 0.9, which showed saturation of optical sensor. PGR applications promote increased chlorophyll concentration and reduced leaf area, which can also affect the sensor readings [10]. Unlike optical readings, crop heights have shown a good correlation with crop biomass through the entire season, even after peak bloom, which could be used to improve the detection of infield variability. This could allow farmers to make better decisions and assist in variable rate application of PGR.

Plant height is an important deciding factor for application of PGR. Nevertheless, other crop parameters can also be used to guide PGR application decisions, and may even be better than crop height growth in some circumstances. The height-to-node ratio (HNR) is calculated by dividing plant height by the total number of main stem nodes, and is equal to the average internode length. The height of the top five nodes is an estimate of current crop growth rate, and can be a good indicator of rank growth and the need to apply higher rates of PGR. Plant growth rate was shown to be inversely correlated to its leaf PGR concentration. Therefore, the total dry biomass should also be accounted when deciding PGR application rates [3].

Cotton production systems can be improved with better understanding of the interactions between soil and crop spatial variability and its influences on the best management strategy regarding PGR application rates. Thus, the objective of this study was to evaluate optical and ultrasonic crop canopy sensors to detect crop spatial variability in cotton fields, and to develop strategies for using this information to perform variable rate application of plant growth regulators in cotton.

2. Materials and Methods

Fields trials were located in the states of Goiás (2014 crop season—GO) and Mato Grosso (2015 crop season—MT), Midwest Brazil. Five fields were studied for all research protocols. The soil in the experimental fields ranged from clay Oxisols to sandy loam Quartzipsamments, with different levels of spatial variability in soil clay content. More details of the experimental fields are presented on Table 1.

Table 1. Field and crop characteristics of areas used for data collection.

Field *	State	Crop Season	Area (ha)	Row Spacing (m)	Variety	Emergence Date	Seed Density (Seed ha^{-1})
CF1	GO	2013/2014	93.8	0.80	FM 975 WS	8 January 2014	100,000
NF1	GO	2013/2014	88.5	0.45	FM 975 WS	23 January 2014	190,000
CF2	MT	2014/2015	139.7	0.76	DP 1243 B2RF	7 January 2015	95,000
NF2	MT	2014/2015	203.4	0.45	TMG 81 WS	22 January 2015	205,000
CF3	MT	2014/2015	142.3	0.76	TMG 81 WS	17 January 2015	100,000

* CF: conventional row spacing cotton; NF: narrow row cotton; GO: Goias State; MT: Mato Grosso State.

2.1. Sensor Performance to Predict Crop Parameters

Two types of optical sensor systems were used (Table 2). The first optical sensor system—OPS1 (N-Sensor™ ALS, Yara International ASA, Dülmen, Germany), was installed above the vehicle cabin and the readings were taken from an oblique position. The OPS1 data acquisition was made in Yara N-Sensor software running on Microsoft Windows operating system. The sensors measure canopy reflectance in red edge (730 nm) and near infrared (760 nm), and a scaled logarithmic difference of reflectance at the two wavelengths was used as the vegetation index in all comparisons [11].

Table 2. Technical specifications of crop canopy optical sensors evaluated in cotton fields.

Sensor System	OPS1	OPS2
Model	N-Sensor ALS	Crop Circle ACS 430
Light source	Xenon	Polychromatic modulated LED
Spectral bands	730 nm (RedEdge) 760 nm (NIR)	670 nm (RED) 730 nm (RedEdge) 780 nm (NIR)
Vegetation index	$100 \times (\ln(R760) - \ln(R730))$	$\left(\frac{R780 - R730}{R780 + R730}\right)$
Acquisition frequency	1 Hz	5 Hz
Mounting height	2.0–4.0 m	0.6–1.2 m
Field of view	40°–55°	45/10°
Sensor footprint	2.0–4.0 m	0.5–1.0 m

The second optical sensor system, OPS2 (Crop Circle ACS-430, Holland Scientific, Lincoln, NE, USA), was mounted to take readings directly above the crop canopy. The OPS2 was integrated with the GEOSCOUT GLS-420 mapping system (Holland Scientific, Lincoln, NE, USA) for data acquisition. The sensors measure canopy reflectance at three wavelengths, but only the red edge (730 nm) and near-infrared (780 nm) were used to calculate the Normalized Difference Red Edge Index (NDRE), which was used in all comparisons.

In order to evaluate the optical sensors' saturation problem, described above, and to test simpler and less cost-intensive technologies, ultrasonic sensors systems were developed and tested using the same conditions. The objective of these systems was to calculate plant height using the time of flight principle to measure the distance between the top of the canopy and the sensor (Table 3). The first system was US1 (Polaroid 6500, Minnetonka, MN, USA), with a working frequency of 49.4 kHz, using custom software for data acquisition by a data logger (CR 1000 Campbell Scientific, Logan, UT, USA). This system was used for 2013/2014 crop season, and one sensor presented malfunction problems in the last acquisition. Due to the difficulties of finding replacement parts, a new system was developed for the next season. The second system, US2 (HC-SR04, generic sensor), with a working frequency of 40 kHz, uses low-cost hardware commonly used in automation projects and data acquisition based on an Arduino® Mega 2560 (Arduino, Ivrea, Italy).

Table 3. Technical specifications of crop canopy ultrasonic sensors evaluated in cotton fields.

Sensor System	US1	US2
Model	Polaroid 6500	HC-SR04
Working frequency	49.4 kHz	40.0 kHz
Acquisition frequency	1 Hz	5 Hz
Mounting height	0.6–1.2 m	0.6–1.2 m
Field of view	30°	20°
Sensor footprint	0.3–0.6 m	0.2–0.4 m

All sensors were installed in a high-clearance vehicle operating with a swath width of 30 m, at a maximum travel speed of 23 km h^{-1}. The measurements were georeferenced using Global Navigation Satellite System (GNSS) receivers. In 2013/2014 crop season, an L1/L2 receiver without differential correction was used. The OPS2 and US1 were mounted on brackets 1 m externally of tire tracks in each side of the machine, and the OPS1 was mounted above the main cabin, with one sensor head facing each side of the machine (Figure 1). For the 2014/2015 crop season, real-time kinematic (RTK) receivers were used for data collection and referencing sample points. The US2 replaced the US1 used in the previous year. The sensor mounting positions were kept the same and one more sensing units of each OPS2 and US2 were added at each side of the machine, mounted on brackets 1.7 m externally of tire tracks.

Figure 1. Approximate view from each sensor position and mounting points in the high clearance vehicle used to move the sensors through the field. Pictures from 2014/2015 crop season of narrow row cotton cultivated in a sandy loam Quartzipsamment in Mato Grosso state.

All sensor data acquisition was made simultaneously in a single machine pass in each field. After data acquisition, sensor readings were averaged in polygons with 5 × 15 m to produce the variability maps. These dimensions were chosen based on the average distance between sensor readings and half the swath width. Based on the maps, 30 validation locations were selected in each field map in order to represent the entire range of crop variability. In each validation sample point, plant height was hand measured, and destructive plant samples of the aboveground biomass were taken by manually cutting and weighing a 1.0 m subplot consisting of three rows. Although the sampled area was different on narrow row and conventional row cotton, the number of rows was kept constant because the number of plants per sample was similar, due to the differences in plant population. During the 2013/2014 crop season, data acquisition and field sampling were repeated at five important phenological stages of cotton, representing the stages 51, 61, 71, 81, and 95 in BBCH scale [12]. One of the objectives of 2014/2015 crop season was to better understand how crop parameters useful for PGR recommendations interact with each other and with the sensor readings in different scenarios. For this reason, in addition to crop height and biomass, dry biomass, height of top five nodes, and the height-to-node ratio were evaluated at BBCH 71 phenological stage. The number of sampling points was also increased to 54 in each field.

Data obtained in the sampling points was submitted to descriptive statistical analysis and linear regression analysis, comparing each sensor performance for individual fields and sampling dates. Values of fresh biomass for the 2013/2014 crop season were transformed to squared roots before regression in order to meet the assumptions of data normality and homoscedasticity of variances.

Results for this variable are presented after back transformation to original scale. All statistical procedures were performed using the R software, version 3.4.0 [13].

2.2. Variable Rate Application of PGR

The variable rate prescription maps of PGR and fruit ripener were created based on the crop variability detected by crop sensors, relations of crop parameters and sensor readings from this and previous works, and field evaluation of representative points before each application by crop advisors. The methodology used is similar to what was presented in [14], but instead of dividing the field into a few application zones, regression analysis was used to predict optimal PGR rate for every pixel, allowing rates to vary between 80 and 120% of the field average rate for each application. When the optimal rate was less than 50% of field average, the rate was set to zero. Rates between 50 and 80% were set to 80%, and rates greater than 120% were set to 120%, due to the sprayer variable rate limitations.

The experimental layout consisted of on-farm trials with 72 m wide swaths, and variable length according to field position. A random block design was used, with five blocks on narrow row field NF2, and 12 blocks on conventional row spacing field CF3. Each block had three sprayer passes made in variable rate, and another three with fixed rate. The block locations were chosen to represent the field variability, while keeping both treatments in similar conditions. Some parts of the fields were used for nitrogen, and variety trials and were not included in analysis. Prescription maps were transferred into the JD GreenStar 3 controller on a JD 4630 self-propelled sprayer (Figure 2). PGR application was done using nozzles Teejet DG 8003 VS Drift Guard Flat Spray Tip (Spraying Systems Co., Wheaton, IL, USA), spaced at 51 cm, using an average tank mixture rate of 60 L ha^{-1} at 20 km ha^{-1}.

Figure 2. Variable rate application of plant growth regulator in narrow-row cotton using an electronic flow controller.

Cotton was harvested using JD 7760 cotton picker (John Deere, Moline, IL, USA), with Harvest Doc™ yield monitoring system. For the narrow row cotton, a variable row spacing (VRS) kit was used. Yield data was georeferenced using RTK positioning with an acquisition rate of 1 Hz and average speed of 2 m s^{-1}, producing about one yield point per 10 m^2. Yield data was filtered to remove locally extreme values, according to the spatial filtering methodology presented by [15], and then averaged to each plot to further statistical analysis. Variable rate prescription maps were prepared using QGIS version 2.14 [16]. All statistical procedures were performed using the R software, version 3.4.0 [13].

3. Results and Discussion

The proposed methodology allowed the successful data acquisition and performance evaluation for most of the comparisons (Table S1). All results are presented for single crop stage evaluation, focusing on the information most useful for decision making at crop management level. Data logger connection presented some problems and generated missing data points for the OPS2 on CF1 in the BBCH 51. No data acquisition for US1 at BBCH 95 was caused by problems with the sensors.

All problems presented on 2013/2014 crop season were fixed, and no data was lost on 2014/2015 crop season.

3.1. Sensor Performance to Predict Height and Biomass in Different Crop Stages

The two fields used for data acquisition in 2013/2014 presented different scenarios of spatial variability. The NF1 was overall homogeneous, with short scale variability caused by plant population variability, due to excess soil moisture during seeding (Figure 3). The CF1 presented more large-scale variability caused by differences in soil clay content and water retention potential. The range of variability on CF1 was about three times larger than on NF1. Therefore, the differences in sensor performance are related to the differences in spatial variability of these particular fields, rather than row spacing and crop system.

Figure 3. Uneven crop emergence and short scale variability at BBCH 51 in field NF1 during the 2013/2014 crop season and BBCH 51 in field CF3 during the 2014/2015 crop season.

The relation of biomass and crop height was dependent of the crop system and phenological stage (Figure 4). A greater increase in biomass was observed for each unit of increased height in NF1 than in CF1, which is due to the large plant population used in narrow-row cotton, allowing a fast accumulation of biomass. This relation forms the basis of using optical sensors to predict crop height. The OPS1 showed the best overall performance (Table 4, Figures A1 and A2 in Appendix A). This sensor had a higher advantage due to its large footprint, being less affected by row-to-row variability of the fields. This is the main reason for the large differences in performance observed between OPS1 and OPS2. The small footprint of OPS2 generated more noisy data, capturing the variations between plants, in contrast to the overall field variability. The same applies to US1, which presented regular performance on CF1 and insufficient performance on NF1, due to this field's short-scale variability.

Table 4. Performance of crop sensors measured by coefficient of determination and root-mean-squared error to predict crop height and aboveground biomass in five crop stages.

Crop Parameter	Field	Sensor	BBCH * 51	61	71	81	95
Height	NF1	OPS1	0.70 (0.02)	0.64 (0.03)	0.68 (0.05)	0.62 (0.03)	0.67 (0.04)
		OPS2	0.18 (0.03)	0.38 (0.04)	0.04 (0.08)	0.37 (0.04)	0.23 (0.06)
		US1	0.02 (0.04)	0.11 (0.05)	0.00 (0.08)	0.29 (0.05)	
	CF1	OPS1	0.83 (0.02)	0.94 (0.03)	0.90 (0.07)	0.32 (0.12)	0.55 (0.10)
		OPS2		0.52 (0.08)	0.80 (0.10)	0.11 (0.13)	0.48 (0.11)
		US1	0.01 (0.05)	0.75 (0.06)	0.85 (0.08)	0.56 (0.09)	
Fresh Biomass	NF1	OPS1	0.78 (0.50)	0.68 (1.12)	0.54 (2.73)	0.27 (2.75)	0.64 (3.49)
		OPS2	0.43 (0.81)	0.56 (1.27)	0.02 (4.00)	0.28 (2.74)	0.31 (4.81)
		US1	0.04 (1.05)	0.21 (1.71)	0.00 (4.04)	0.00 (2.90)	
	CF1	OPS1	0.67 (0.48)	0.96 (0.76)	0.90 (2.77)	0.17 (5.45)	0.55 (5.08)
		OPS2		0.57 (2.45)	0.80 (3.98)	0.14 (5.55)	0.48 (5.44)
		US1	0.01 (0.71)	0.70 (1.94)	0.71 (4.82)	0.39 (4.67)	

* BBCH: crop stage according to BBCH scale for cotton; OPS1: N-Sensor™ ALS; OPS2: Crop Circle ACS 430; US1: Polaroid 6500; First value in each cell: coefficient of determination (R^2), value in parenthesis: root-mean-squared error (RMSE).

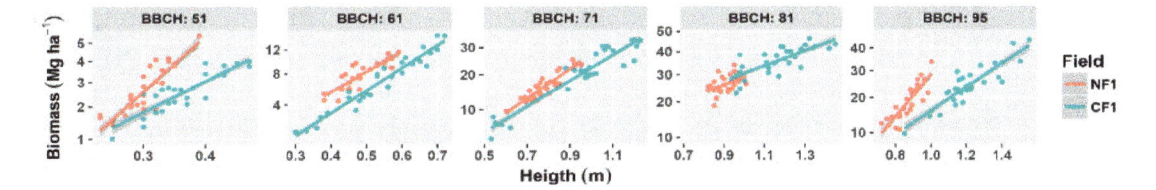

Figure 4. Relation of crop aboveground fresh biomass and crop height in five crop stages (BBCH) for cotton cultivated in two systems: narrow row (NF1) and conventional row spacing (CF1).

The effect of the crop variability magnitude can be better observed when relating the coefficient of determination (R^2) with the root-mean-squared error (RMSE) in each crop stage. Even with large differences in R^2, the RMSE values are consistently similar among sensors in each evaluation stage. The maximum RMSE observed for crop height were in OPS2, with 0.08 m in NF1 and 0.13 m in CF1, which means that the sensor can provide good auxiliary information when large variability is present.

The worst performance of optical sensors was observed at BBCH 81, which agrees with the results obtained by [9]. The high volumes of biomass accumulated at this stage induce a saturation effect on vegetation indexes; moreover, there are also plants in different phenological stages induced by stresses due to water limitation. The US1 presented better performance than the optical sensors at this stage on CF1, however, the best obtained R^2 of 0.56 was lower than the values obtained in the evaluation at the previous BBCH, which were above 0.80 for all sensors.

3.2. Sensor Performance to Predict General Crop Parameters

The relation among crop parameters depends on the crop system and the field (Figures A3 and A4). Some parameters may present better stability to be used for general recommendations across different fields or even crop seasons. Nevertheless, when we consider a single field and crop stage, the parameters are correlated, and the characterization of field variability will be similar if based in any of these parameters.

Better control of data acquisition and sample locations was implemented using real-time kinematic (RTK) technology to avoid the effect of small-scale variability observed on 2013/2014 crop season. With this methodology, more stable results were obtained (Table 5 and Figure 5). Although this was important to better evaluate sensor performance while avoiding other effects, it does raise the question of whether the considered spatial resolution is enough to manage spatial variability effectively.

Table 5. Performance of crop sensors measured by coefficient of determination and root-mean-squared error to predict crop parameters of cotton cultivated in conventional and narrow row spacing.

Field	Sensor *	Dry B.	Fresh B.	H. Five	Height	HNR
CF2	OPS1	0.46 (0.15)	0.66 (0.31)	0.38 (0.02)	0.77 (0.07)	0.50 (0.51)
	OPS2	0.65 (0.14)	0.74 (0.29)	0.51 (0.02)	0.81 (0.07)	0.69 (0.47)
	US2	0.52 (0.15)	0.73 (0.29)	0.41 (0.02)	0.84 (0.06)	0.61 (0.49)
CF3	OPS1	0.65 (0.14)	0.81 (0.21)	0.50 (0.02)	0.72 (0.06)	0.44 (0.37)
	OPS2	0.58 (0.14)	0.71 (0.24)	0.38 (0.02)	0.55 (0.07)	0.37 (0.36)
	US2	0.61 (0.14)	0.75 (0.23)	0.53 (0.02)	0.78 (0.05)	0.43 (0.37)
NF2	OPS1	0.71 (0.19)	0.85 (0.33)	0.66 (0.04)	0.82 (0.08)	0.73 (0.50)
	OPS2	0.74 (0.19)	0.91 (0.27)	0.80 (0.03)	0.89 (0.07)	0.78 (0.47)
	US2	0.71 (0.19)	0.90 (0.27)	0.79 (0.03)	0.92 (0.06)	0.81 (0.45)

* OPS1: N-Sensor™ ALS; OPS2: Crop Circle ACS 430; US2: HC-SR04; Dry B.: dry biomass; Fresh B.: fresh biomass; H. Five: height of top five nodes; Height: hand measured crop height; HNR: height-to-node ratio; First value in each cell: coefficient of determination (R^2), value in parenthesis: root-mean-squared error (RMSE).

The consistency of RMSE values for different sensors and fields is to be highlighted again. The estimation of crop height using any of the sensors presented an RMSE inferior to 0.08 m in all fields. The US2 presented the best performance in estimating crop height, and very similar results to the optical sensors for the other parameters. The range of variations that can be observed in Figure 5 contributed to the good performance observed on these fields. The points representing field NF2 are well distributed over the entire range of values, and this was the field with the best results.

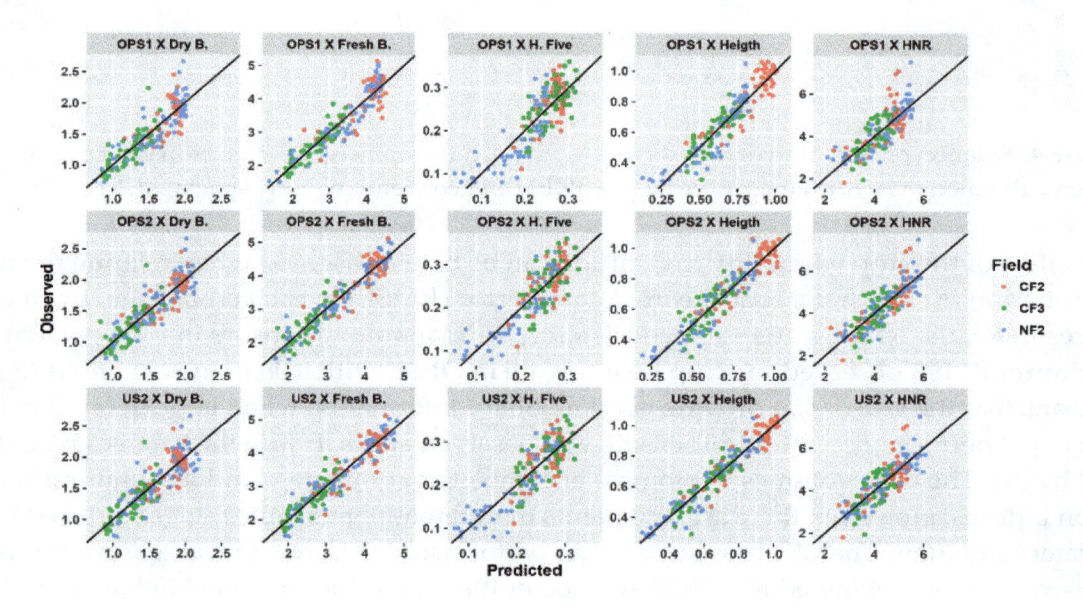

Figure 5. Sensor performance to predict crop parameters: dry biomass (Dry B.), fresh biomass (Fresh B.), height of top five nodes (H. Five), crop height (Height) and height-to-node ratio (HNR) with three crop sensors: OPS1: N-Sensor™ ALS; OPS2: Crop Circle ACS 430; US2: HC-SR04; for cotton cultivated in two systems: narrow row (NF2) and conventional row spacing (CF2 and CF3).

3.3. Crop Response to Variable Rate PGR

Crop height was not statistically affected by the use of fixed rates or VRA of PGR in the two experimental fields. The variations in PGR rates were not sufficient to reduce the variations in crop height when compared to fixed rate application (Figure 6). One of the reasons for this is the machine limitations, due to pressure variation constraints to control water flow [17]. This limitation could be overcome with other variable-rate spraying technology, such as pulse wide modulation at individual nozzles. A direct injection system could also allow for greater variations, but the difficulties with short scale variations would still need to be addressed in further studies.

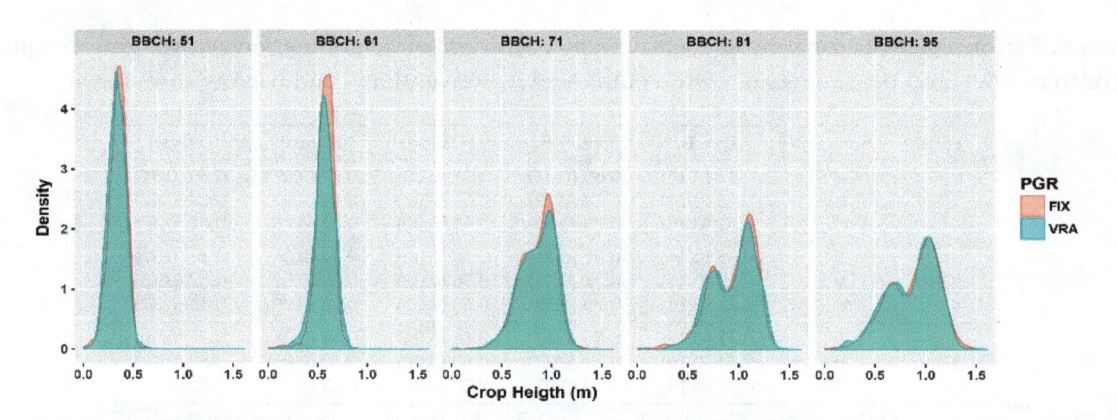

Figure 6. Frequency distribution of crop height in field NF2 under fixed-rate and variable-rate application of plant growth regulators.

Another important reason for the lack of response may be the timing of decisions. In these fields, the most important factor limiting crop development and yields was the soil water retention potential. The crops were rainfed, and enough rain, totaling more than 800 mm, was well distributed from

January to April, until the crop reached BBCH 61. After that, the rain stopped, and the crop had only the water present in the soil to finish its cycle.

The bimodal distribution of plant height, which can be seen in Figure 6, was caused by the large differences in soil characteristics of NF2 (Figure A5). In the soils with low water retention potential, the crop stopped growing due to water limitation, even with the zero-rate of PGR applied. This contributed to PGR savings, but the crop uniformity was not improved. In the regions with high water retention potential, the opposite was observed. Even with 20% increase in PGR rates, the crop presented rank growth in some regions of the field (Figure 7). In these regions, the final height was 30% higher than the maximum desired for harvest using cotton strippers, which was 1.0 m. In this scenario, higher rates of PGR should be applied earlier in the season. The crop sensors used in this study are good tools to estimate current variability of crop parameters; although, the crop management decisions and rate of inputs can be more accurate if spatial variability from previous years and agro-climatic crop development models are used together.

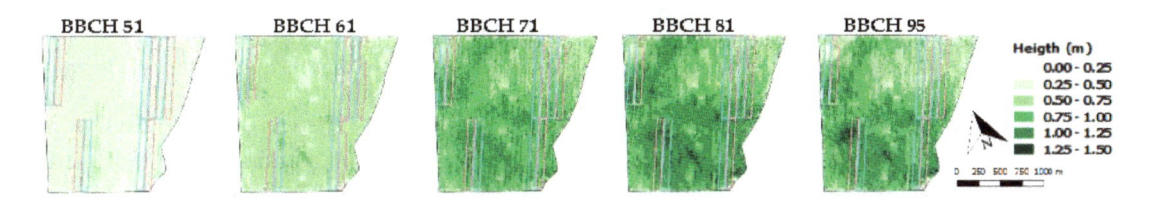

Figure 7. Spatial distribution of crop height in five phenological stages of narrow-row cotton in field NF2.

The results in field CF3 were mostly affected by crop emergence problems. The example picture shown before in Figure 3 represents the problem. The variability of crop height was not reduced by variable rate application (Figure 8). Although there was a large range of plant height and crop development variability, the spraying technology used was not suited for the short-range variations present in this field. The final height in most of this field was less than the desired for harvest with cotton pickers, which is between 0.8 and 1.3 m.

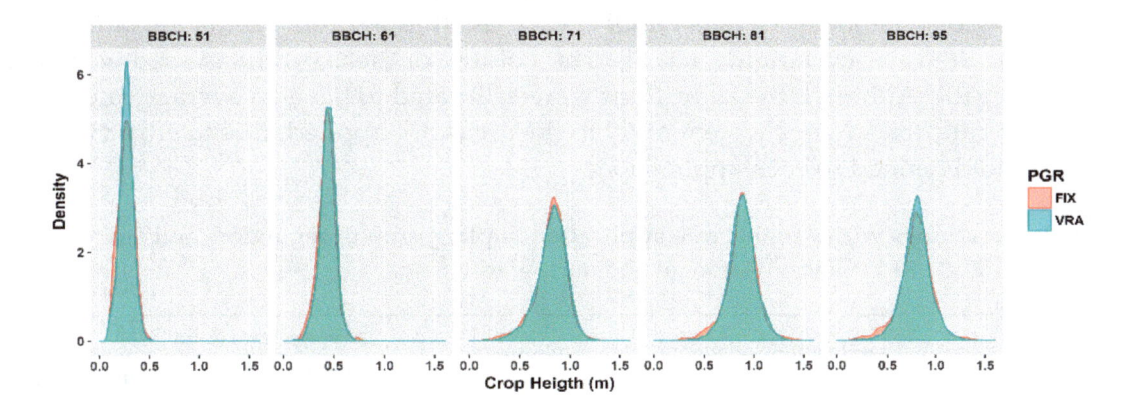

Figure 8. Frequency distribution of crop height in field CF3 under fixed rate and variable rate application of plant growth regulators.

The same weather pattern described for field NF2 occurred in this field. In the area with better soil characteristics, in the northeast part of the field, the 20% increase in all PGR applications was insufficient to reduce crop height (Figure 9). The crop in these places was severely affected by the disease boll rot of cotton, which was accentuated by the higher plant population. These results are a good indicator that previous problems in crop establishment and other management practices have a greater impact on crop performance; therefore, the PGR management must take into account more factors than only crop development. Furthermore, the variable rate application of PGR alone is not enough to manage spatial variability when large soil differences are present. The use of variable rate seeding, variable rate nitrogen, and fungicides applications to match the spatial variability of the crop, needs to be considered. In this context, crop modeling tools well calibrated with on-farm trials representing the local conditions are necessary to anticipate decision making. Crop sensors are a great complimentary tool to validate the decision and make in-season adjustments.

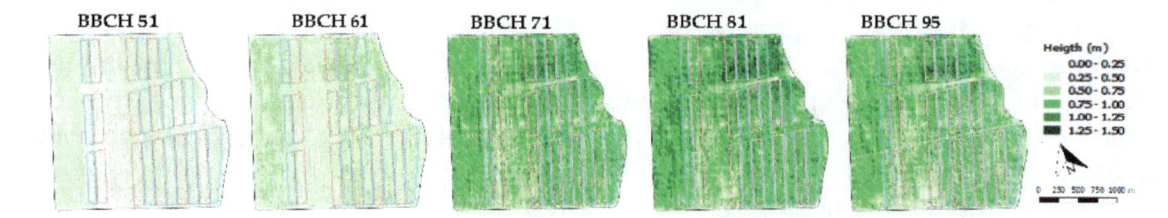

Figure 9. Spatial distribution of crop height in five phenological stages of conventional row spacing cotton in field CF3.

The results obtained by other authors in similar conditions [14] were more conclusive in showing the effect of reduced plant height variability between application zones using VRA with an electronic flow controller in the same machine model used in this study. The authors also found a reduction in the coefficient of variation of the percentage of opened bolls when using the VRA of the fruit ripener between the application zone classes. Although there are differences in the methodologies used in the experiments, the results can be used to provide some general guidelines of implementing VRA of PGR in cotton. When the field can be divided into management zones and long-range variability is predominant, low-cost systems of VRA can produce good results.

3.4. Input Savings and Yield Response to Variable Rate of PGR

Throughout the crop cycle five PGR applications were performed on NF2 and three on CF3 (Table 6). The commercial product PIX HC (mepiquat chloride, 25% ai (active ingredient)) was the PGR used in all applications. Before harvest, the crop received an additional application of defoliant and fruit ripener using commercial products DROPP ULTRA (thidiazuron, 12% ai + diurom, 6% ai) and FINISH (etephon, 48% ai + cyclalinide, 6% ai). The volume of these products used was, on average, 17% lower in the VRA. Although the algorithms were calibrated to have an average rate similar to the fixed application rate, the savings were mainly to the areas not applied, due to crop vigor below the minimum threshold defined to each application.

Table 6. Input savings with variable rate application of plant growth regulators and fruit ripener in cotton variety TMG 81 WS in conventional and narrow row fields.

Field	Application	BBCH	Product	Fixed Rate	Variable Rate
NF2	1st PGR	51	PIX HC	0.030	0.026
	2nd PGR	61	PIX HC	0.050	0.043
	3rd PGR	71	PIX HC	0.080	0.069
	4th PGR	76	PIX HC	0.120	0.104
	5th PGR	81	PIX HC	0.150	0.131
	Defoliant	98	DROPP ULTRA	0.400	0.312
	Boll Opener	98	FINISH	2.000	1.560
CF3	1st PGR	71	PIX HC	0.050	0.046
	2nd PGR	76	PIX HC	0.080	0.053
	3rd PGR	81	PIX HC	0.120	0.080
	Defoliant	98	DROPP ULTRA	0.500	0.458
	Boll Opener	98	FINISH	2.500	2.292

Average Rate (L ha^{-1})

Yield was not statistically affected by the use of fixed rates or VRA of PGR in the two experimental fields (Figures 10 and 11). The automation of all data collection and treatment application made it possible to implement such large trials. The adopted experimental design was adequate to test the technology under field conditions, but the natural field variation also impacted the statistical power of the analysis. The measurement of other information, such as the plant population variability and its use as auxiliary variables, may contribute to better estimates of treatment effects.

The uniformity of the cotton height and opened fruits contribute to a similar yield in the different application zones, once uniform plant height benefits cotton harvest, while the ripener helps to ensure all the cotton is ready to be harvested at the same time. Besides interfering positively with mechanized harvesting procedure, the plant height control provides benefits regarding the harvested cotton, particularly the fiber quality [14]. Fiber quality parameters were not measured in this work, but some field observations confirmed this statement.

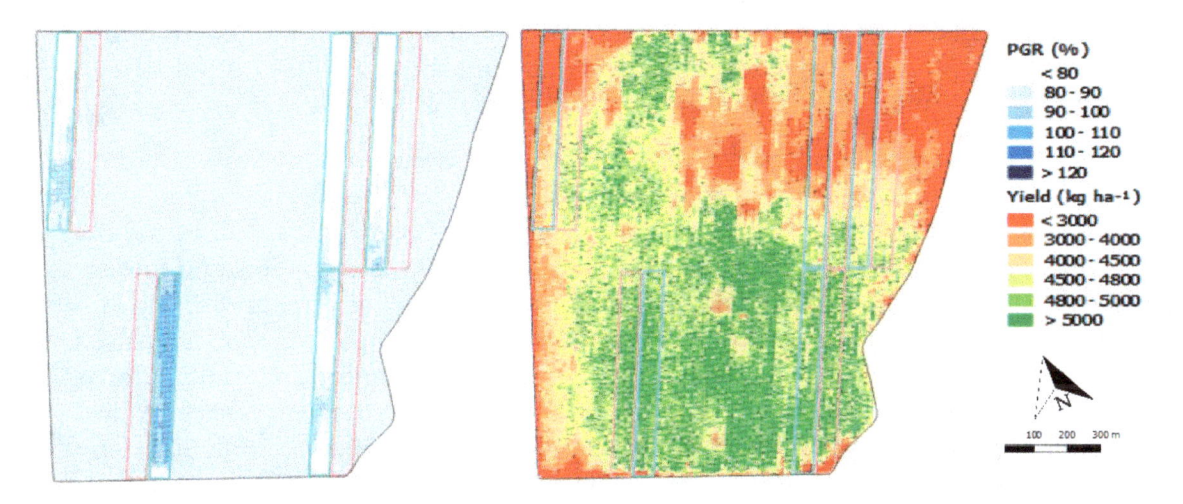

Figure 10. Spatial distribution of applied plant growth regulators and cottonseed yield in narrow row cotton in field NF2.

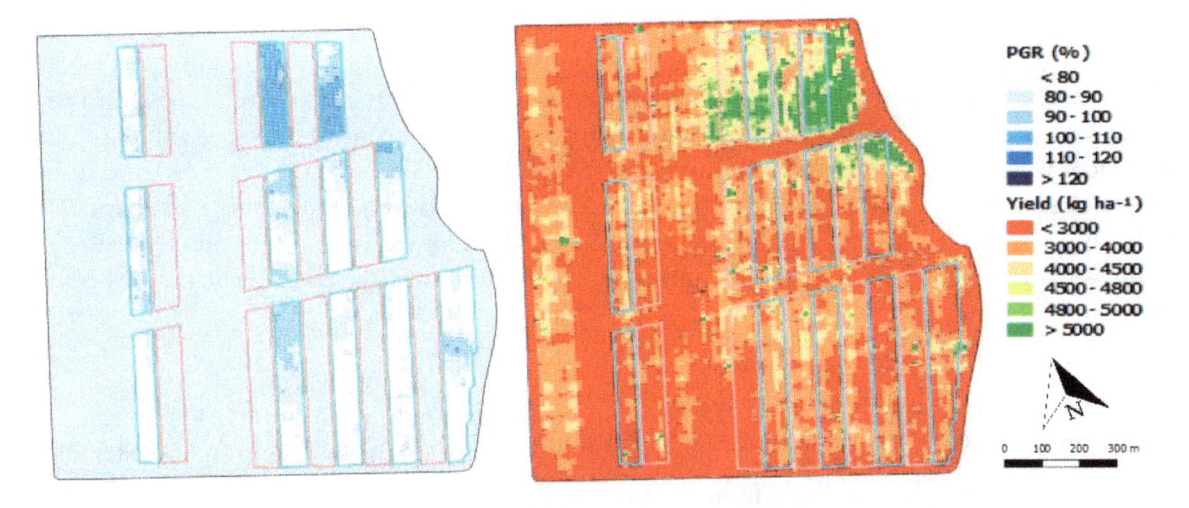

Figure 11. Spatial distribution of applied plant growth regulators and cottonseed yield in conventional row spacing cotton in field CF3.

In order to be profitable, precision agriculture technologies need to be selected by matching the spatial variability resolution of the crop, the diagnostic tool and the management tool [18]. The sprayer used in this work, which is representative of most current sprayers present in Brazilian farms, can only vary the rate across the whole boom, which is usually between 24 and 36 m. In fields with predominant short distance variations, this spatial resolution is not sufficient to apply the required rates. Automatic boom sections, shut off, may help when there are small areas of low crop vigor where no PGR is needed, such as when the field is infested with patches or spots of insects and nematodes. VRA using pressure control is also limited to 20% variation in rates. Based on the final crop height of areas applied using VRA and in the official recommendations [3], this variation is insufficient to compensate crop variations in fields with large differences of crop development.

Most PGRs are applied at the same time as the fungicides and insecticides. In this scenario, variable-rate spraying application can be performed by a sprayer equipped with a direct injection system [17], which has a higher cost when compared to a conventional sprayer equipped with a simple electronic controller of the flow rate. Although this system can apply PGR at variable rates, while keeping other products at fixed rates, the spatial resolution affected. In common systems, the products are injected in the main line, and there is a delay of some seconds between injection and application, which makes the system inadequate for managing short range variability. There are prototype systems with direct injection in each nozzle, but these are still not widely commercially available, and the costs may be prohibitive.

The short scale variability can be minimized, to some extent, with improved operational quality. It was clear in experimental fields that previous crop straw distribution had a great impact on crop emergence, as well as the adjustments on seeding depth. Nevertheless, soil-borne diseases are a problem, usually related to uneven crop emergence and earlier stand reduction, and these will continue to happen when climatic conditions are favorable. To compensate for this short distance variability in crop development, real-time sensors could be coupled with weather and crop models, and nozzles with pulse width modification (PWM) technology to treat each plant individually. The technical and economic viability of this technology still needs to be validated, but the high costs of cotton production may justify this investment. Even using PWM technology, the complexity of tank mixtures used would still be a problem. One alternative may be to consider the joint variation of PGR and the other products, because the performance of these products may also be related to crop biomass and leaf area. More research need to be done before any practical recommendation can be made for these new technologies.

4. Conclusions

The overall performance of the sensors to predict crop height and the accumulation of biomass in cotton was satisfactory. Short distance variability was predominant in some fields, reducing the performance of the sensors while making current technology for variable rate application of plant growth regulators inadequate. In areas with large scale variability, the variable rate application led to 17% savings in plant growth regulators products, and no significant effect on yield was observed. Ultrasonic sensors present a low-cost alternative to implement variable rate application of plant growth regulators in real time.

Author Contributions: Data curation, R.G.T., N.S.V.J. and M.T.E.; Formal analysis, R.G.T.; Funding acquisition, J.P.M.; Investigation, R.G.T., N.S.V.J. and M.T.E.; Methodology, R.G.T., N.S.V.J. and J.P.M.; Project administration, J.P.M.; Software, R.G.T.; Supervision, J.P.M.; Visualization, R.G.T., N.S.V.J. and M.T.E.; Writing—original draft, R.G.T. and N.S.V.J.; Writing—review & editing, M.T.E. and J.P.M.

Acknowledgments: All this work would not be possible without the collaboration of Wink Farm's Group, Bom Futuro Farm's Group, Stara Agriculture Machinery and Yara Research Center Hanninghof support. We also acknowledge the National Council for Scientific and Technological Development (CNPq) for providing scholarship to the first and second authors.

Appendix A

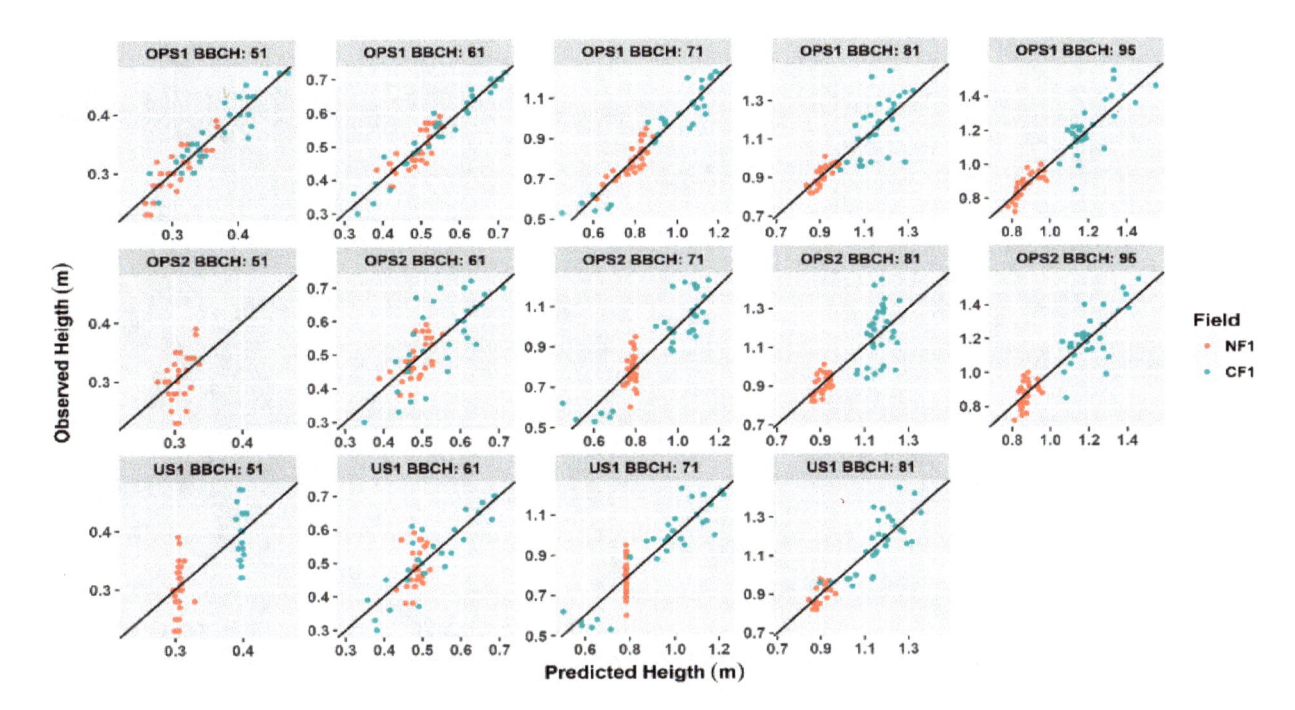

Figure A1. Crop height prediction in five crop stages according to BBCH scale using three crop sensors: OPS1: N-Sensor™ ALS; OPS2: Crop Circle ACS 430; US1: Polaroid 6500, for cotton cultivated in two systems: narrow row (NF1) and conventional row spacing (CF1).

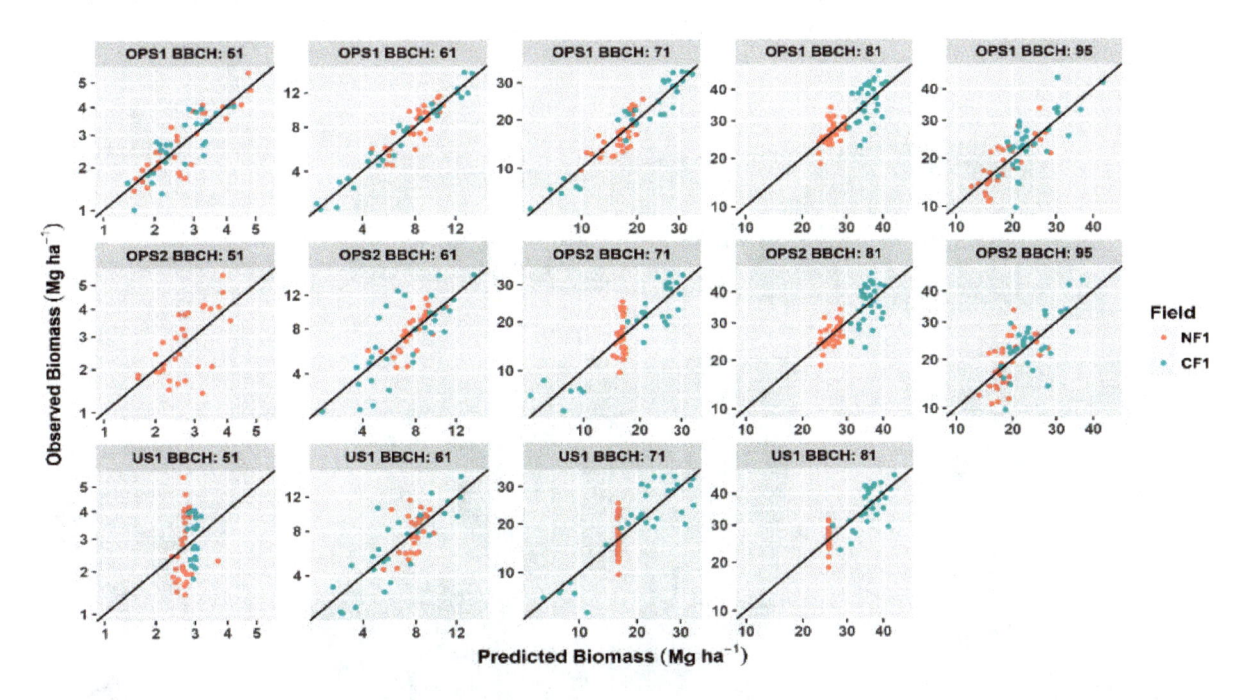

Figure A2. Crop aboveground fresh biomass prediction in five crop stages according to BBCH scale using three crop sensors: optical sensor 1 (OPS1: N-Sensor); optical sensor 2 (OPS2: Crop Circle); ultrasonic sensor 1 (US1: Polaroid), for cotton cultivated in two systems: narrow row (NF1) and conventional row spacing (CF1).

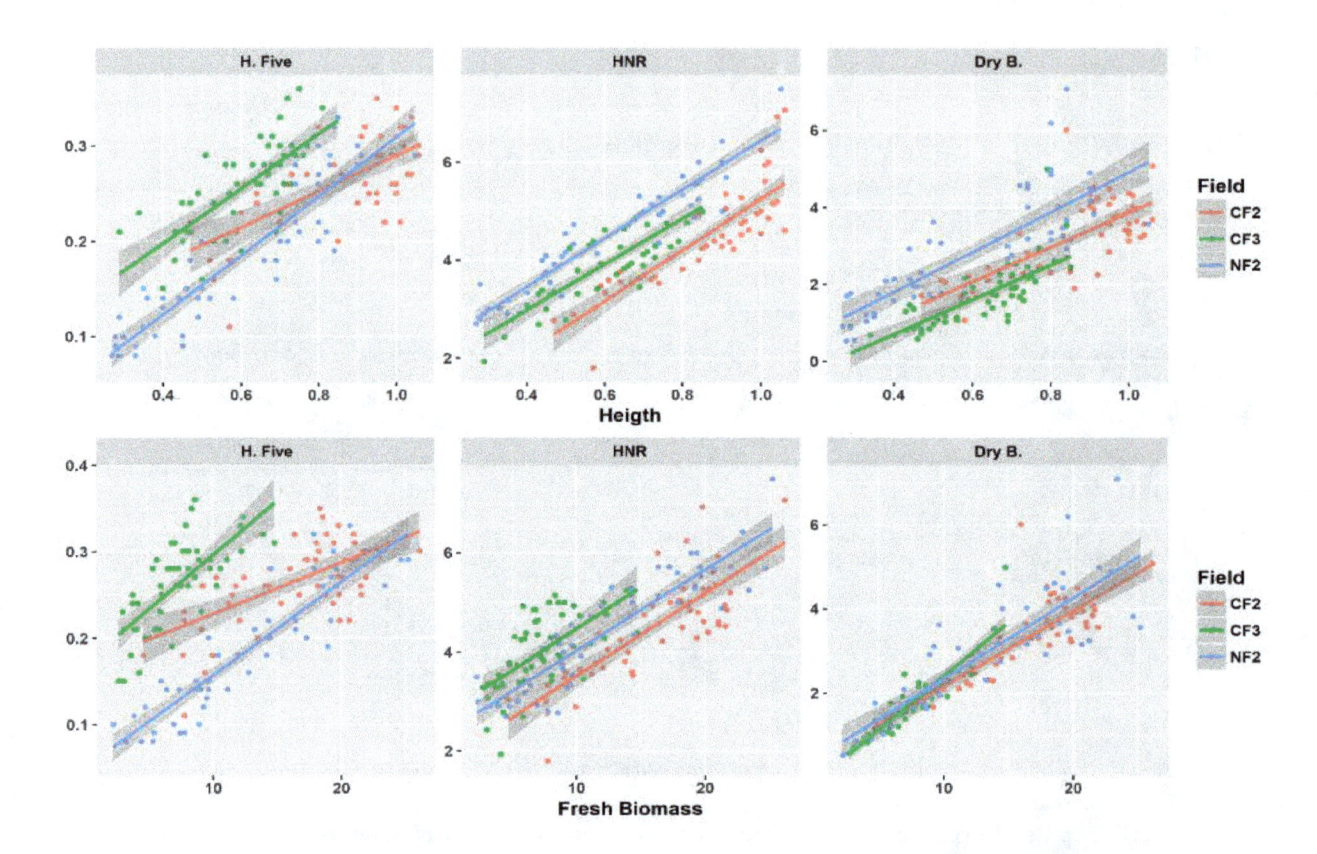

Figure A3. Relation of hand measured crop height and aboveground fresh biomass with height of top five nodes (H. Five), height-to-node ratio (HNR) and dry biomass (Dry B.) for cotton cultivated in two systems: narrow row (NF2) and conventional row spacing (CF2 and CF3).

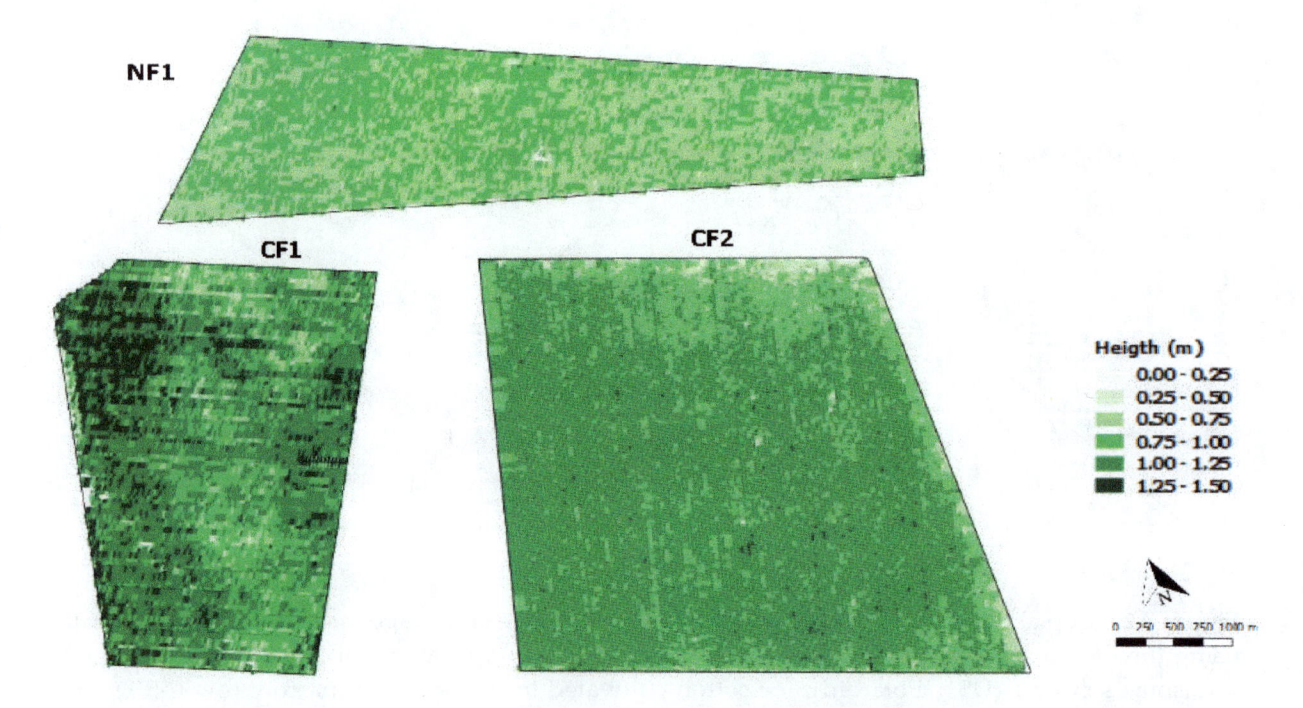

Figure A4. Spatial distribution of crop height measured by ultrasonic sensors before harvest of cotton cultivated in two systems: narrow row (NF1) and conventional row spacing (CF1 and CF2).

Figure A5. Spatial distribution of clay content in field NF2.

References

1. FAO. *South-South Cooperation for Strengthening the Cotton Sector*; FAO: Rome, Italy, 2015.
2. CONAB (Companhia Nacional de Abastecimento). Acompanhamento da safra brasileira. *Décimo Segundo Levant* **2017**, *4*, 1–98.
3. Echer, F.R.; Rosolem, C.A. Plant growth regulation: A method for fine-tuning mepiquat chloride rates in cotton. *Pesqui. Agropecu. Trop.* **2017**, *47*, 286–295. [CrossRef]
4. Leon, C.T.; Shaw, D.R.; Cox, M.S.; Abshire, M.J.; Ward, B.; Wardlaw, M.C.; Watson, C. Utility of Remote Sensing in Predicting Crop and Soil Characteristics. *Precis. Agric.* **2003**, *4*, 359–384. [CrossRef]
5. Sui, R.; Thomasson, J.A. Ground-based sensing system for cotton nitrogen status determination. *Trans. ASABE* **2006**, *49*, 1983–1991. [CrossRef]
6. Gutierrez, M.; Norton, R.; Thorp, K.R.; Wang, G. Association of spectral reflectance indices with plant growth and lint yield in upland cotton. *Crop Sci.* **2012**, *52*, 849–857. [CrossRef]
7. Ballester, C.; Hornbuckle, J.; Brinkhoff, J.; Smith, J.; Quayle, W. Assessment of in-season cotton nitrogen status and lint yield prediction from unmanned aerial system imagery. *Remote Sens.* **2017**, *9*, 1149. [CrossRef]
8. Souza, H.B.; Baio, F.H.R.; Neves, D.C. Using Passive and Active Multispectral Sensors on the Correlation With the Phenological Indices of Cotton. *Eng. Agric.* **2017**, *37*, 782–789. [CrossRef]
9. Vellidis, G.; Savelle, H.; Ritchie, R.G.; Harris, G.; Hill, R.; Henry, H. NDVI response of cotton to nitrogen application rates in Georgia, USA. In Proceedings of the 8th European Conference on Precision Agriculture, Prague, Czech Republic, 11–14 July 2011; pp. 358–368.
10. Foote, W.; Edmisten, K.; Wells, R.; Collins, G.; Roberson, G.; Jordan, D.; Fisher, L. Influence of Nitrogen and Mepiquat Chloride on Cotton Canopy Reflectance Measurements. *J. Cotton Sci.* **2016**, *20*, 1–7.
11. Jasper, J.; Reusch, S.; Link, A. Active sensing of the N status of wheat using optimized wavelength combination: Impact of seed rate, variety and growth stage. In Proceedings of the 7th European Conference on Precision Agriculture, Wageningen, The Netherlands, 6–8 July 2009; Volume 9, pp. 23–30.

12. Bleiholder, H.; Weber, E.; Lancashire, P.D.; Feller, C.; Buhr, L.; Hess, M.; Wicke, H.; Hack, H.; Meier, U.; Klose, R.; et al. Growth stages of mono-and dicotyledonous plants, BBCH monograph. *Fed. Biol. Res. Cent. Agric. For. Berl. Braunschw.* **2001**, *12*, 158. [CrossRef]
13. R Core Team. *R: A Language and Environment for Statistical Computing*; R Core Team: Vienna, Austria, 2017.
14. Baio, F.H.R.; Neves, D.C.; Souza, H.B.; Leal, A.J.F.; Leite, R.C.; Molin, J.P.; Silva, S.P. Variable rate spraying application on cotton using an electronic flow controller. *Precis. Agric.* **2018**, 1–17. [CrossRef]
15. Spekken, M.; Anselmi, A.A.; Molin, J.P. A simple method for filtering spatial data. *Proc. Precis. Agric.* **2013**, *20*, 1531–1538. [CrossRef]
16. QGIS Development Team. *QGIS Geographic Information System*; QGIS Development Team: Las Palmas, Spain, 2017.
17. Antuniassi, U.R.; Baio, F.H.R.; Sharp, T.C. Agricultura de Precisão. In *Algodão no Cerrado do Brasil*; BDPA: Brasília, Brazil, 2015; pp. 767–806.
18. Amaral, L.R.; Trevisan, R.G.; Molin, J.P. Canopy sensor placement for variable-rate nitrogen application in sugarcane fields. *Precis. Agric.* **2017**, 1–14. [CrossRef]

Crop Sensor based Non-Destructive Estimation of Nitrogen Nutritional Status, Yield and Grain Protein Content in Wheat

Marta Aranguren, Ander Castellón and Ana Aizpurua *

NEIKER-Basque Institute for Agricultural Research and Development, Department of Plant Production and Protection, Berreaga 1, 48160 Derio, Biscay, Spain; maranguren@neiker.eus (M.A.); acastellon@neiker.eus (A.C.)
* Correspondence: aaizpurua@neiker.eus

Abstract: Minimum NNI (Nitrogen Nutrition Index) values have been developed for each key growing stage of wheat (*Triticum aestivum*) to achieve high grain yields and grain protein content (GPC). However, the determination of NNI is time-consuming. This study aimed to (i) determine if the NNI can be predicted using the proximal sensing tools RapidScan CS-45 (NDVI (Normalized Difference Vegetation Index) and NDRE (Normalized Difference Red Edge)) and Yara N-Tester™ and if a single model for several growing stages could be used to predict the NNI (or if growing stage-specific models would be necessary); (ii) to determine if yield and GPC can be predicted using both tools; and (iii) to determine if the predictions are improved using normalized values rather than absolute values. Field trials were established for three consecutive growing seasons where different N fertilization doses were applied. The tools were applied during stem elongation, leaf-flag emergence, and mid-flowering. In the same stages, the plant biomass was sampled, N was analyzed, and the NNI was calculated. The NDVI was able to estimate the NNI with a single model for all growing stages ($R^2 = 0.70$). RapidScan indexes were able to predict the yield at leaf-flag emergence with normalized values ($R^2 = 0.70$–0.76). The sensors were not able to predict GPC. Data normalization improved the model for yield but not for NNI prediction.

Keywords: *Triticum aestivum*; RapidScan CS-45; Yara N-Tester™; NNI; precision agriculture; remote sensing

---

1. Introduction

To meet the globally increasing food demand, achieving high grain yields and high-quality grains has become fundamental. For those purposes, N fertilizer is a crucial factor because its application results in an increase in grain quality and grain yield. However, in cereal production, the excessive application of nitrogen fertilisers is common. The optimum management of N fertilization requirements needs a steady monitoring of crop N status throughout the vegetative period [1].

Determining the cereal N status is very important for adjusting the necessary N dose, evaluating crop growth, and estimating yield and grain protein content (GPC) [2–4]. In this sense, the nitrogen nutrition index (NNI) has been commonly utilized to determine the N status of plants during the growing season [5]. The NNI, determines if the N concentration needed to achieve the greatest biomass production is optimum based on a crop's current biomass [1]. The NNI could be helpful to follow N dynamics in a crop canopy and, in this way, identify the deficiencies that suggest a yield decrease. NNI dynamics may be useful under circumstances where cereals are destined to lose their yield. Ravier et al. [1] identified a threshold NNI path for the wheat growing cycle to determine N fertilizer application timing and suggested the minimum NNI values needed for each key growing stage to achieve high yields together with a lower risk of nitrate pollution. Reliable information on crop nutritional status throughout the vegetative period could reveal the need for additional N fertilizer [6,7] and may help develop an innovative method to manage N fertilization. However, to determine NNI, laboratory analytical procedures are needed, thereby making the calculations complicated and time-consuming. To achieve precise N fertilizer management, crop N status should be analysed in-season and at specific sites.

Optical sensing techniques estimate the N content in a plant indirectly, as such techniques cannot measure N content directly. Measurements are rapid, low cost, and can be done intensively over space and time, thereby providing the necessary resolutions required for N fertilizer management [4]. In this sense, non-destructive and instantaneous measurements can be taken for crop blades with chlorophyll meters to use them as estimators of the crop N nutritional status. A previous study showed a good relationship between leaf N concentration and measurements taken with chlorophyll meters [7–9]. Chlorophyll meters and wheat grain yield were related in different studies and used to identify responses to additional fertilizers [9,10] and for recommendations on fertilizer management [11–13]. Moreover, the possibility of using chlorophyll meters to decide if an extra fertilizer dose is required to increase GPC has been studied [7,9]. The readings provided by chlorophyll meters have also been well correlated with the NNI [14,15] and with wheat leaf N concentrations and leaf chlorophyll [7,8].

Active crop canopy sensors have their own light sources, so they are not limited by changeable light conditions, thus making them practical for on-farm management. Plant tissue normally reflects nearly 50% of the near infra-red (NIR) and absorbs nearly 90% of the visible radiation [16]. Information related to crop N status is provided because the ratio of reflectance and absorbance changes with crop N and biomass [17]. Then, vegetation indexes can be calculated with the spectral data collected by the crop sensors. Mistele and Schmidhalter [18] concluded that the NNI can be determined using spectra-based measurements, and Marti et al. [19] positively correlated Normalized Difference Vegetation Index (NDVI) values with wheat yield and biomass. RapidScan CS-45 (Holland Scientific, Lincoln, NE, USA) is a portable ground-based active canopy sensor with a built-in GPS that measures crop reflectance at red (R; 670 nm), red-edge (RE; 730 nm), and near infra-red (NIR; 780 nm) spectra and provides the NDVI and the Normalized Difference Red Edge (NDRE). Previous studies showed that RapidScan CS-45 estimates NNI in rice [20] and allows the fast and precise crop tracking of N status and yield estimations in wheat [21].

Proximal sensor measurements can be repeated several times throughout the wheat growing season, and the information obtained and related to crop N status may be utilized to follow crop N dynamics in real time [8,22–24]. However, remote sensing measurements are usually taken in the middle wheat-growing period to adjust the N fertilizer rate [11,25]. In our area, the highest amount of N is applied at stem elongation (GS30), but the time until harvest is long, and many factors may affect the subsequent N uptake by the crop. However, if the soil is wet [9], it is possible to amend N deficiency until late in the wheat growing season (GS65 [1]; mid-flowering; [26]). In our area, it is possible to use a third application of N fertilizer at leaf-flag emergence (GS37) because there will likely be sufficient rain [9,12] to permit N uptake by the crop. Therefore, it is desirable to follow the crop N status during the vegetative growing season to make decisions related to the optimization of N fertilizer applications [6,12] or to predict the yields and GPC values. In this sense, optical sensing tools could help us understand easily how climate and N rates affect N uptake via crops and, therefore, affect yield and GPC. Ravier et al. [6] developed decision rules for determining N fertilizer application through the wheat growing season as a function of the crop N status or NNI reference values in the key growing stages. However, there are no optical sensing reference values for evaluating wheat crop N status during the vegetative growing period.

The usefulness of the proximal sensing tools for predicting NNI [15,20] and for predicting yield [2,19] and GPC [3,23] has been studied with mixed results. Predictions depend on the agroecosystem environment, and specific correlations for each climate should be developed [23]. The present study was developed under humid Mediterranean conditions to (i) determine if the NNI can be predicted using the proximal sensing tools RapidScan CS-45 and Yara N-Tester™ and if a single or unique model for all growing stages (from GS30 to mid-flowering (GS65)) could be used to predict NNI (or if growing stage-specific models would be necessary); (ii) to determine if grain yield and GPC can be predicted using the RapidScan CS-45 and Yara N-Tester™; and (iii) to determine if the predictions are improved using normalized values rather than absolute values.

2. Materials and Methods

2.1. Study Site

Three field trials were established during three consecutive wheat growing seasons (2014–2015, 2015–2016, and 2016–2017) in Arkaute (Araba, Basque Country, northern Spain) at NEIKER installations (Figure 1) under unirrigated conditions. We refer to the growing seasons as 2015, 2016, and 2017. The climate of the area where the study was carried out was Temperate–Mediterranean [27]. The soil texture was analyzed by the pipette method [28] and classified (0–30 cm, sandy clay loam and 30–60 cm, clay loam) [29]. pH values (8.0–8.5) were high in the soil, which was calcareous [30] and had moderate organic matter content [31] in the upper layer (2%–2.5%). The soil was classified as Typic Calcexeroll [32]. Further experimental details were described by Aranguren et al. [11].

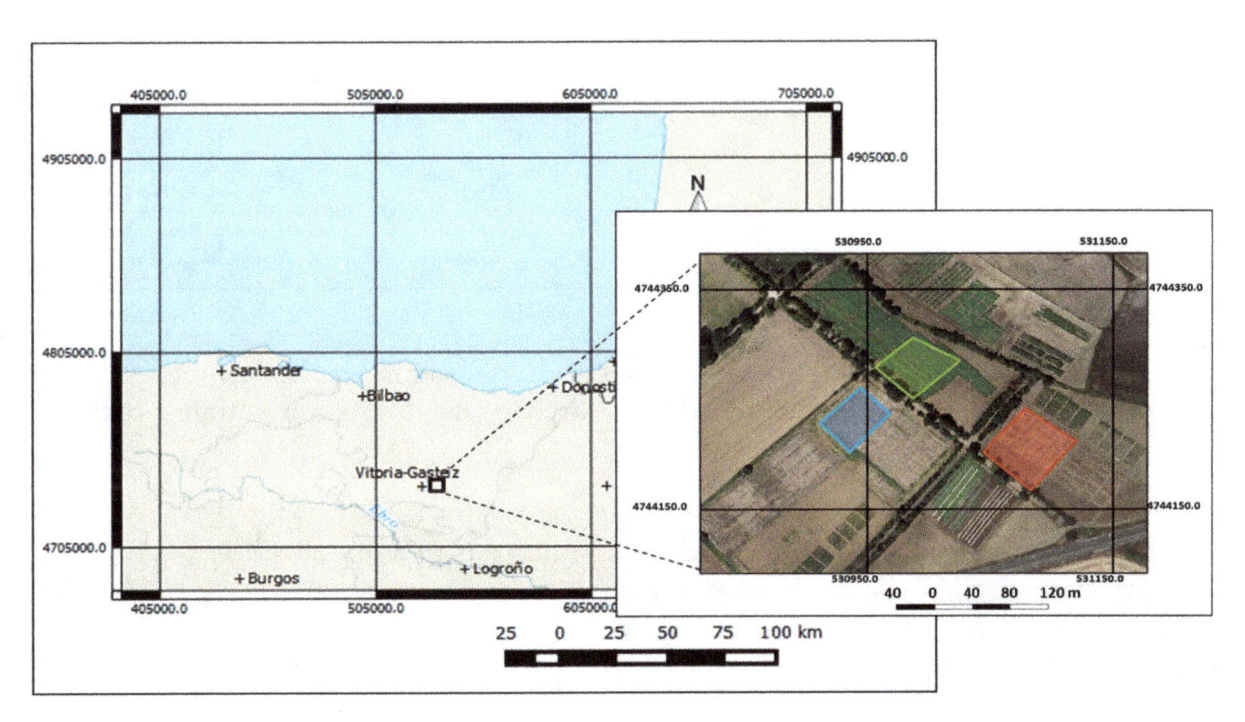

Figure 1. Location of the three field experiments in Arkaute, Araba, Basque Country, northern Spain. The field experiment in 2015 was carried out in the red field. The field experiment in 2016 was carried out in the blue field. The field experiment in 2017 was carried out in the green field.

2.2. Treatments

Three different initial fertilizers were used: dairy slurry (40 t ha^{-1}), sheep farmyard manure (40 t ha^{-1}), and conventional treatment (no basal dressing but 40 kg N ha^{-1}, 18 kg S ha^{-1}, and 45 kg K ha^{-1} at tillering (GS21, [26]). The dairy slurry N content was 192, 144 and 120 kg N ha^{-1} in 2015, 2016, and 2017, respectively. The sheep farmyard manure N content was 336, 592 and 448 kg N ha^{-1} in 2015, 2016, and 2017, respectively. Five N rates, applied at GS30 (0, 40, 80, 120, and 160 kg N ha^{-1}), were combined with the three types of initial fertilization. Regarding mineral fertilization, N was applied as calcium-ammonium-nitrate 27% (NAC) and S and K were applied as potassium sulphate 50%. A control without N (0 N) and an over-fertilized control plot (280 kg N ha^{-1}) were also established. The treatments are shown in Table 1. Organic fertilizers were applied in mid-November each growing season. In the area where this study was carried out, organics are usually applied in combination with mineral fertilizers. The application rate in the experiment (40 t ha^{-1}) is the usual rate applied in the area. Wheat (*Triticum aestivum* var. Cezanne) was sown just after application of the organics. The experiment used a factorial randomized complete block design. The area of each plot was 4 m wide and 8 m long.

Table 1. Fertilization treatments for the field trials (2015, 2016, and 2017) and the N dose applied in each of them. Beginning of tillering (GS21; end of winter [26]) and stem elongation (GS30; [26]).

Initial Fertilization	2015-2016-2017 Topdressing at GS21 (kg N ha^{-1})	Topdressing at GS30 (kg N ha^{-1})	Treatment Identification
Conventional (–)	40	0	40N + 0N
		40	40N + 40N
		80	40N + 80N
		120	40N + 120N
		160	40N + 160N
Dairy Slurry (DS) (40 t ha^{-1})	–	0	DS + 0N
		40	DS + 40N
		80	DS + 80N
		120	DS + 120N
		160	DS + 160N
Sheep manure (SM) (40 t ha^{-1})	–	0	SM + 0N
		40	SM + 40N
		80	SM + 80N
		120	SM + 120N
		160	SM + 160N
Control (–)	–	–	0N
Overfertilized (–)	80	200	280N

(–), no initial fertilization.

Yields were harvested at crop maturity. Total N concentration was determined following the Kjeldhal procedure [33]. GPC was determined by multiplying the total N concentration of the product by 5.7 [34].

2.3. Plant Biomass and Nitrogen Nutrition Index (NNI)

Plant biomass samples were taken at GS30, GS37, and GS65 in all conventional treatments DS + 0N, and SM + 0N and in two control treatments (0N and 280N). Plant biomass sampling was done according to Aranguren et al. [11]. The biomass was measured and N concentration was determined following Kjeldahl's method [33] to calculate the Nitrogen nutrition index (NNI) [1]:

$$\text{NNI} = \frac{Na}{Nc} \quad (1)$$

where *Na* represents the present wheat N uptake, and *Nc* represents the critical N uptake that corresponds to the present shoot wheat biomass *W* (t ha^{-1}) [35]:

$$Nc = 5.35 \times W^{-0.442} \quad (2)$$

when the NNI values are close to one, wheat has an optimum N status; values lower than 0.8 indicate N deficiency, and values higher than one indicate non-limiting N.

2.4. Crop Sensors for Following Crop N Status

Crop sensor readings (CSR) were taken with a Yara N-Tester™ (Yara International ASA, Oslo, Norway) and RapidScan CS-45 (Holland Scientific, Lincoln, NE, USA) at GS30, GS37, and GS65 [26] in four pseudo-replications of all treatments (Table 1). The Yara N-Tester™ is a chlorophyll meter that measures and processes the ratio of the light transmitted at 650 and 940 nm wavelengths, in addition to the ratio determined with no sample, to produce a digital reading. It is a clip-on hand-held tool whose measurement point is placed in the middle of the blade of the youngest fully developed leaf. To acquire a representative value for each measured treatment, thirty random measurements were recorded. The RapidScan CS-45 is a ground-based active crop canopy sensor that measures crop reflectance at 670, 730, and 780 nm and provides the NDVI and NDRE (Equations (3) and (4)). Plot measurements were taken when the sensor was passed over the crop at approximately 1 m at a constant walking speed. Two rows per plot were scanned, and the NDVI and NDRE values were averaged to generate a value for the plot.

We refer to the Yara N-Tester™ measurements as abs_N-Tester. We refer to the RapidScan CS-45 measurements as abs_NDVI and abs_NDRE. Measurements with both tools were taken as described by Aranguren et al. [11]:

$$\text{NDVI} = \frac{R_{NIR780} - R_{RED670}}{R_{NIR780} + R_{RED670}} \tag{3}$$

$$\text{NDRE} = \frac{R_{NIR780} - R_{RED-EDGE730}}{R_{NIR780} + R_{RED-EDGE730}} \tag{4}$$

The normalized values for crop sensor readings (nor_CSR: nor_N-Tester, nor_NDVI, and nor_NDRE) were calculated according to Aranguren et al. [11]:

$$\text{nor}_{CSR} = \frac{CSR}{CSRoverfertilized} \, 100 \tag{5}$$

Thus, each absolute crop sensor reading (CSR; abs_N-Tester, abs_NDVI, and abs_NDRE) was divided by the CSR values of the overfertilized plot (280N; CSR_overfertilized) at the same growing stage and in the same growing season [36].

2.5. Models to be Fitted

Based on a literature review, the models for predicting NNI [14,20,22], yield [19,24], and GPC [3,25] in cereals from crop sensors readings were selected.

2.5.1. The Linear Model

This model assumes that the NNI, yield, and GPC increase steadily with the CSR (or nor_CSR). The NNI is defined as follows:

$$Y = a + b \times CSR \tag{6}$$

where Y is the NNI (or yield or GPC), CSR (or nor_CSR) is the measured value with the crop sensor, and a and b are the parameters of the linear trend that are estimated when the model is fitted to the experimental data.

2.5.2. The Exponential Model

This model does not feature a constant increase in *NNI*, yield, or GPC with CSR (or nor_CSR) (unlike the linear model) and is defined as follows:

$$Y = a + e^{b \times CSR} \tag{7}$$

where *Y* is *NNI* (or yield or GPC), *CSR* (or nor_CSR) is the measured value with the crop sensor, and a and b are the parameters of the exponential trend that is estimated when the model is fitted to the experimental data.

The three-year dataset was divided into two subsets: 75% for fitting the coefficients of determination and 25% for the validation dataset. The 25% dataset was always taken from the same block in the field experiment. The coefficients of determination (R^2) were calculated using the R 3.2.5 software [37]. R^2 was calculated for the relationship between the CSR (abs_N-Tester, abs_NDVI, and abs_NDRE) and NNI and between nor_CSR (nor_N-Tester, nor_NDVI, and nor_NDRE) and NNI in each growing stage (GS30, GS37, and GS65). R^2 was calculated for the relationship between the CSR and NNI and between nor_CSR and the NNI for all growing stage readings together (general model). R^2 was calculated for the relationship between the CSR and the yield and between nor_CSR and the yield for each growing stage (GS30, GS37, and GS65). R^2 was calculated for the relationship between the CSR and GPC and between nor_CSR and GPC for each growing stage (GS30, GS37, and GS65).

Only when the above-mentioned relationships were statistically significant were the relationships plotted. Moreover, when these relationships were significant, the NNI values predicted from the different indexes and models were plotted against the NNI values measured from the remaining samples (25%) using the R 3.2.5 software [37].

The output of the models was assessed by comparing the R^2, RMSE (root mean square error), and AIC (Akaike Information Criterion [38]). The RMSE defines the best-fit function that captures the relationship between NNI, yield, or GPC and CSR (or nor_CSR), which is defined as follows:

$$\text{RMSE} = \sqrt{\frac{1}{N}\sum_{}^{N}(Y - Y')^2} \tag{8}$$

where Y is the measured NNI, and Y' is the estimated NNI.

The AIC describes to what degree the model is explained by the data. The models were compared, and that with the least amount of information loss was used [39]. The models close to reality had lower AIC values [14].

The highest precision and accuracy of the model for predicting crop N status (NNI), yield, or GPC was chosen based on (i) the highest R^2, and (ii) the lowest RMSE and AIC. In the results of the models, the highest value was the R^2 and the lowest was the RMSE and AIC in all cases. Therefore, only the R^2 will be mentioned in the results.

3. Results

3.1. Relationship between the NNI and Crop Sensor Readings

The correlations were fitted between the absolute and normalized CSR and NNI for each different growing stage (GS30, GS37, and GS65; Figures 2 and 3), as well as a general correlation across growing stages (Figure 4). For growing stage-specific models, the RapidScan CS-45 indexes predicted a better NNI (Figures 2 and 3) than the Yara N-Tester™ for both models (linear and exponential). The Yara N-Tester™ did not present a significant relationship to NNI, thus, relationships were not plotted. With RapidScan, the exponential models predicted the NNI more successfully than the linear models in all cases (a higher R² and lower AIC and RMSE). For absolute values, the abs_NDVI values better explained the NNI variability (Figure 2a–c) than the abs_NDRE values (Figure 3a,b) in every growing stage (R² values higher than 52 in all cases). The relationship between abs_NDRE at GS65 and the NNI was not significant, thus, it was no plotted. For the normalized values, the nor_NDRE (Figure 3d,e) values fit the NNI slightly better at GS37 and GS65 ($R^2 = 0.7$) than the nor_NDVI values (Figure 2e,f). However, at GS30, nor_NDVI predicted the NNI better than nor_NDRE (Figures 2d and 3c). The general model (Figure 4), especially the abs_NDVI values (Figure 4a), had good accuracy when estimating the NNI ($R^2 = 0.7$, similar to the models for different growing stages). In the general model, there were no significant relationships between the Yara N-Tester™ and NNI, like that found in the growing stage-specific models (data not shown).

Figure 2. Relationship between the NNI (Nitrogen Nutritional Index) and abs_NDVI values at GS30 (**a**), GS37 (**b**), and GS65 (**c**) and between the NNI and nor_NDVI values at GS30 (**d**), GS37 (**e**), and GS65 (**f**). Two models were fitted: linear, solid line; exponential, dashed line. **, *** Significant at 0.01 and 0.001 probability levels, respectively. +, 0N; ×, 40N + 0N; ×, 40N + 40N; ×, 40N + 80N; ×, 40N + 120N; ×, 40N + 160N; o, DS + 0N; o, SM + 0N. abs, absolute values; nor, normalized values; NDVI, Normalized Difference Vegetation Index.

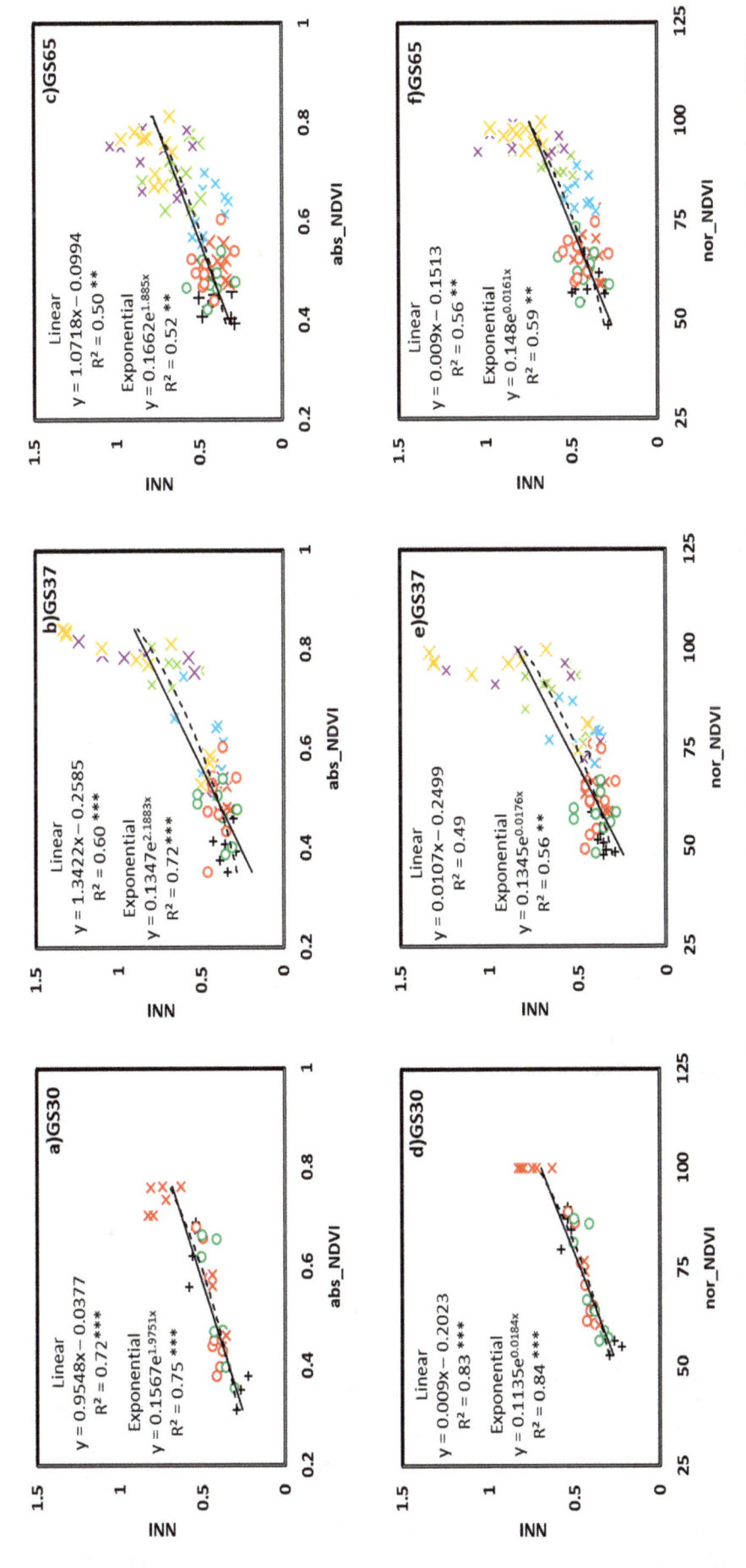

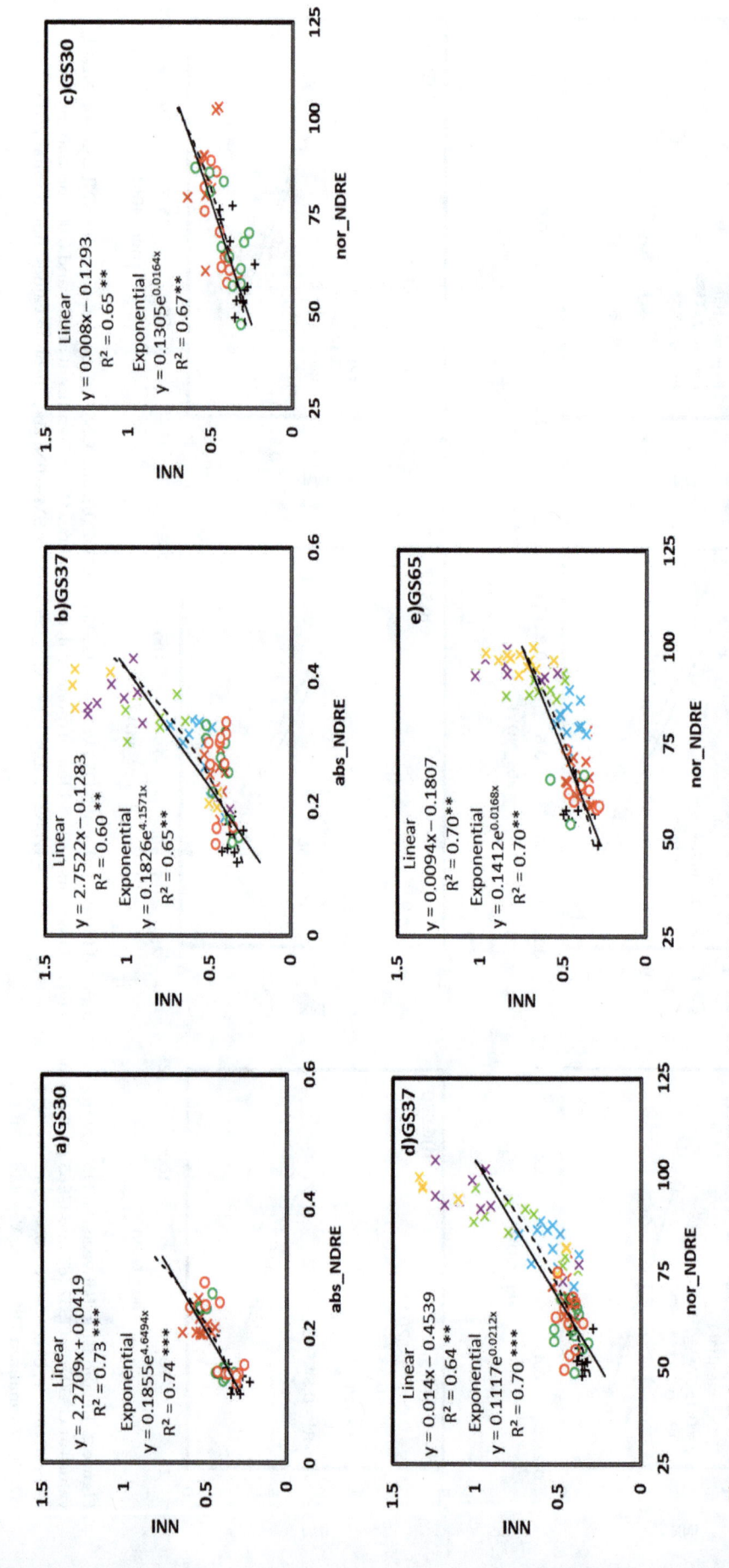

Figure 3. Relationship between the NNI (Nitrogen Nutritional Index) and abs_NDRE values at GS30 (**a**) and GS37 (**b**) and between the NNI and nor_NDRE values at GS30 (**c**), GS37 (**d**), and GS65 (**e**). Two models were fitted: linear, solid line; exponential, dashed line. **, *** Significant at 0.01 and 0.001 probability levels, respectively. +, 0N; ×, 40N + 0N; ×, 40N + 40N; ×, 40N + 80N; ×, 40N + 120N; ×, 40N + 160N; o, DS + 0N; o, SM + 0N. abs, absolute values; nor, normalized values; NDRE, Normalized Difference Red Edge Index.

Figure 4. Relationship between the abs_NDVI (**a**) and abs_NDRE (**b**) values for all growing stages measurements together and the NNI (Nitrogen Nutritional Index) and the nor_NDVI (**c**) and nor_NDRE (**d**) values for all growing stage measurements together and the NNI. Two models were fitted: linear, solid line; exponential, dashed line. *, **, *** Significant at 0.05, 0.01 and 0.001 probability levels, respectively: +, 0N; ×, 40N + 0N; ×, 40N + 40N; ×, 40N + 80N; ×, 40N + 120N; ×, 40N + 160N; o, DS + 0N; o, SM + 0N. abs, absolute values; nor, normalized values; NDVI, Normalized Difference Vegetation Index; NDRE, Normalized Difference Red Edge Index.

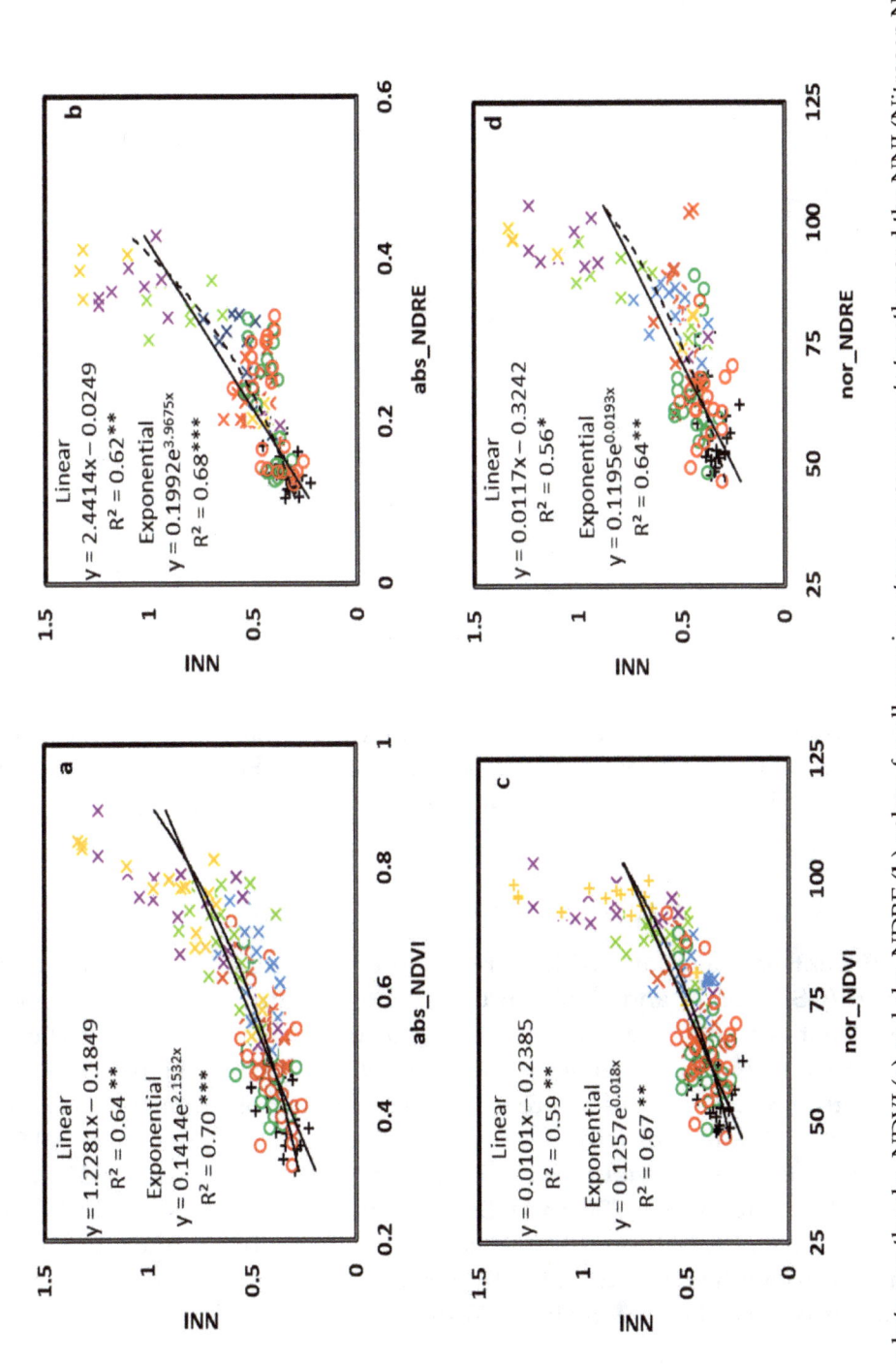

Saturation effects were detected for both the NDVI and NDRE indexes when NNI > 0.8 (Figures 2–4). The RapidScan CS-45 indexes did not increase more than 0.8 for the NDVI (Figures 2 and 4) or more than 0.4 for the NDRE (Figures 3 and 4), even though the NNI values continued increasing above 0.8. Although NDVI and NDRE reached saturation at a similar point, the NDRE value range was slightly wider for NDVI values around 0.8 (0.35–0.40; Figure 5). The values located from NNI = 0.8 to NNI = 1.4 were not quantified by RapidScan CS-45 (Figures 2–4).

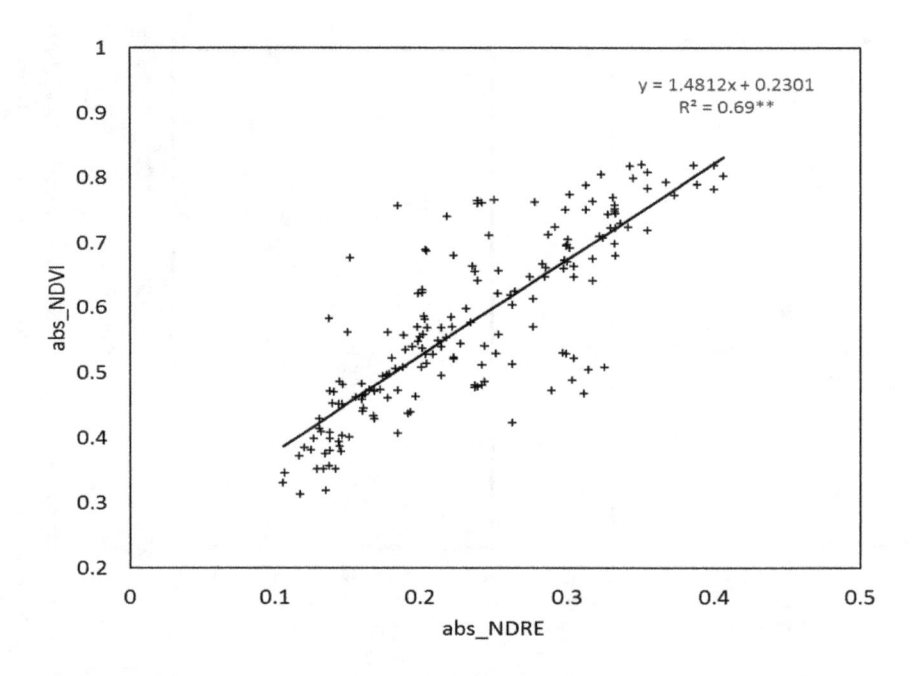

Figure 5. Relationship between the abs_NDVI and abs_NDRE values from the three growing seasons. abs, absolute values; nor, normalized values; NDVI, Normalized Difference Vegetation Index; NDRE, Normalized Difference Red Edge Index. **, significant at 0.01 probability level.

3.2. Relationship between the Yield and GPC and Crop Sensor Readings

Correlation coefficients of the relationship between the absolute and normalized CSR at each different growing stage (GS30, GS37, and GS65) and grain yield (Figure 6) and GPC were fitted. The lineal models and exponential models predicted similar yields from CSR. The yield prediction capacity was high with abs_NDVI at GS65 (Figure 6a; $R^2 = 0.72$), nor_NDVI at GS37 (Figure 6b; $R^2 = 0.76$), and nor_NDRE at GS37 (Figure 6c; $R^2 = 0.70$), whereas the yield prediction capacity with abs_N-Tester at GS37 (Figure 6d; $R^2 = 0.53$) and that with nor_ N-Tester at GS65 (Figure 6e; $R^2 = 0.57$) was low. The remaining relationships between yield and CSR (abs_NDVI at GS30 and GS37; abs_NDRE at GS30, GS37 and GS65; abs_N-Tester at GS30 and GS65) and nor_CSR (nor_NDVI at GS30 and GS65; nor_NDRE at GS30 and GS65; nor_N-Tester at GS30 and GS37) were not significant (data not shown). The GPC prediction capacity with CSR and that with nor_CSR was not significant in any of the cases (the best relationship was observed between abs_N-Tester at GS65 ($R^2 = 0.35$)).

Figure 6. Relationship between the yield (kg ha^{-1}) and abs_NDVI values at GS65 (**a**), nor_NDVI at GS37 (**b**), nor_NDRE at GS37 (**c**), abs_N-Tester at GS37 (**d**), and nor_N-Tester at GS65 (**e**). Two models were fitted: linear, solid line; exponential, dashed line. *, *** Significant at 0.05 and 0.001 probability levels, respectively. Initial fertilizers: ×, conventional; o, dairy slurry; ◊, sheep manure. N rate at GS30: red, +0N; blue, +40N; green, +80N; purple, +120N; orange, +160N. abs, absolute values; nor, normalized values; NDVI, Normalized Difference Vegetation Index; NDRE, Normalized Difference Red Edge Index.

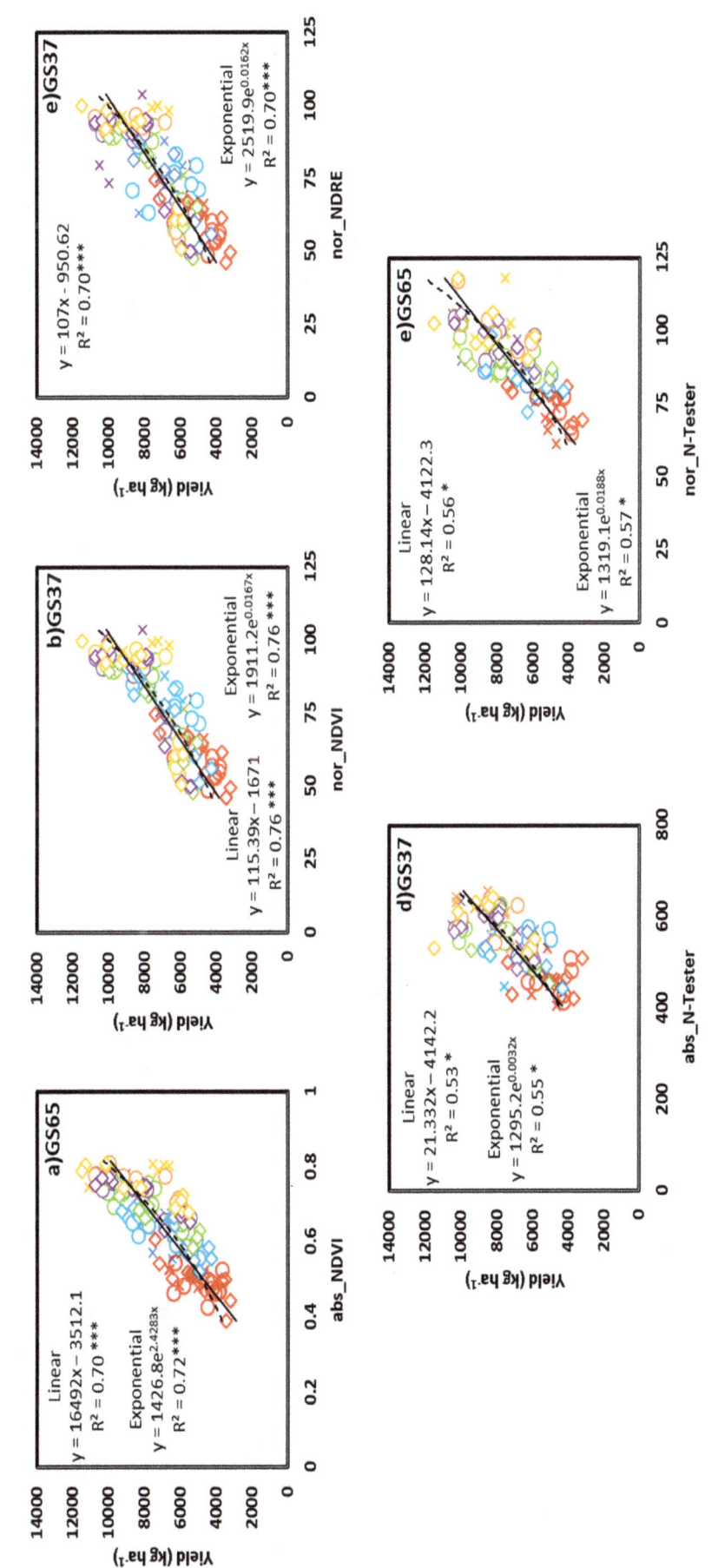

3.3. NNI Estimation Perfomances from Proximal Sensing Tools

The NNI values predicted from the different indexes and models were plotted against the measured NNI values from the remaining samples (25%) when the correlations were significant (Figures 7–9). In the models specific to the growing stage, the order of the accuracy of the correlations for validation was similar to the accuracy of the prediction correlations (Figures 7 and 8). There was a significant agreement between the estimated NNI and the measured NNI for the vast majority of the correlations, with the exception of nor_NDRE at GS30 (Figure 8b). NDVI had a greater potential for predicting NNI than NDRE, as represented by its higher R^2 and lower RMSE and AIC. In the general model (Figure 9), there was a significant agreement between the estimated NNI and the measured NNI with abs_NDVI and abs_NDRE (Figure 9a,c), especially with abs_NDVI in the exponential model (Figure 9a). For nor_NDRE and nor_NDVI, the predicted NNI and the measured NNI did not agree (Figure 9b,d). NDVI had a high potential for predicting NNI, especially with the exponential model ($R^2 = 0.74$; RMSE = 0.12; AIC = −89), even in the general model.

3.4. Grain Yield and GPC Estimation Perfomance from Proximal Sensing Tools

The predicted yield values from the different indexes and models were plotted against the measured yield values from the remaining samples (25%) when the correlations were significant (Figure 10). Figure 10 shows that there was a significant agreement between the estimated yields and the measured yields in all cases. However, for abs_N-Tester at GS37 and nor_N-Tester at GS65, the agreement was lower than that for nor_NDVI and nor_NDRE at GS37 and abs_NDVI at GS65, as shown by the higher R^2 values and lower RMSE and AIC values. Since no index or model could predict the GPC values, GPC estimation performance was studied.

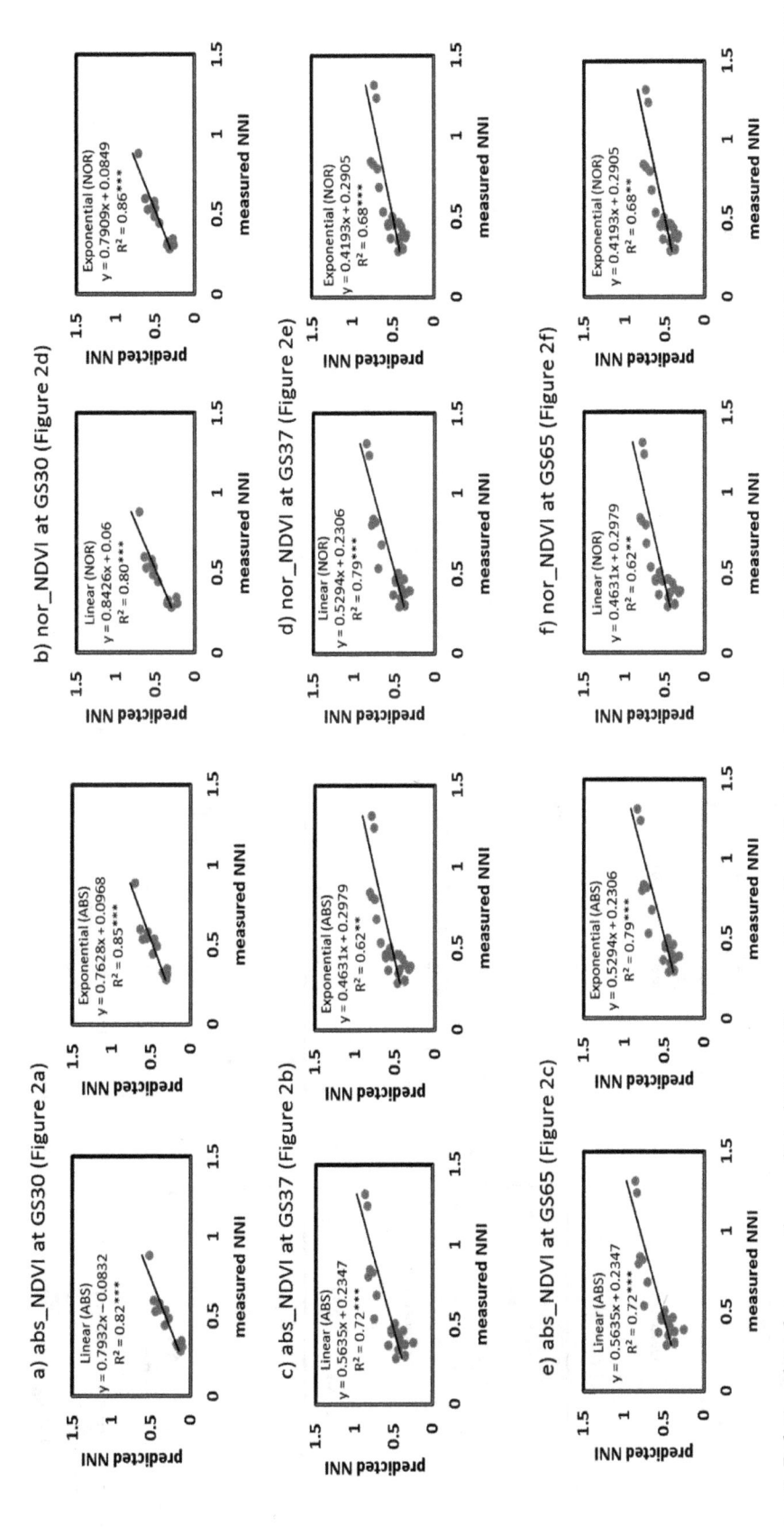

Figure 7. Relationships between the predicted NNI (Nitrogen Nutritional Index) values (from abs_NDVI at GS30 (**a**), nor_NDVI at GS30 (**b**), abs_NDVI at GS37(**c**), nor_NDVI at GS37 (**d**), abs_NDVI at GS65 (**e**) and nor_NDVI at GS65 (**f**) values in the growing stage-specific models) and the measured NNI values from 25% of the samples. **, *** Significant at 0.01 and 0.001 probability levels, respectively. abs, absolute values; nor, normalized values; NDVI, Normalized Difference Vegetation Index.

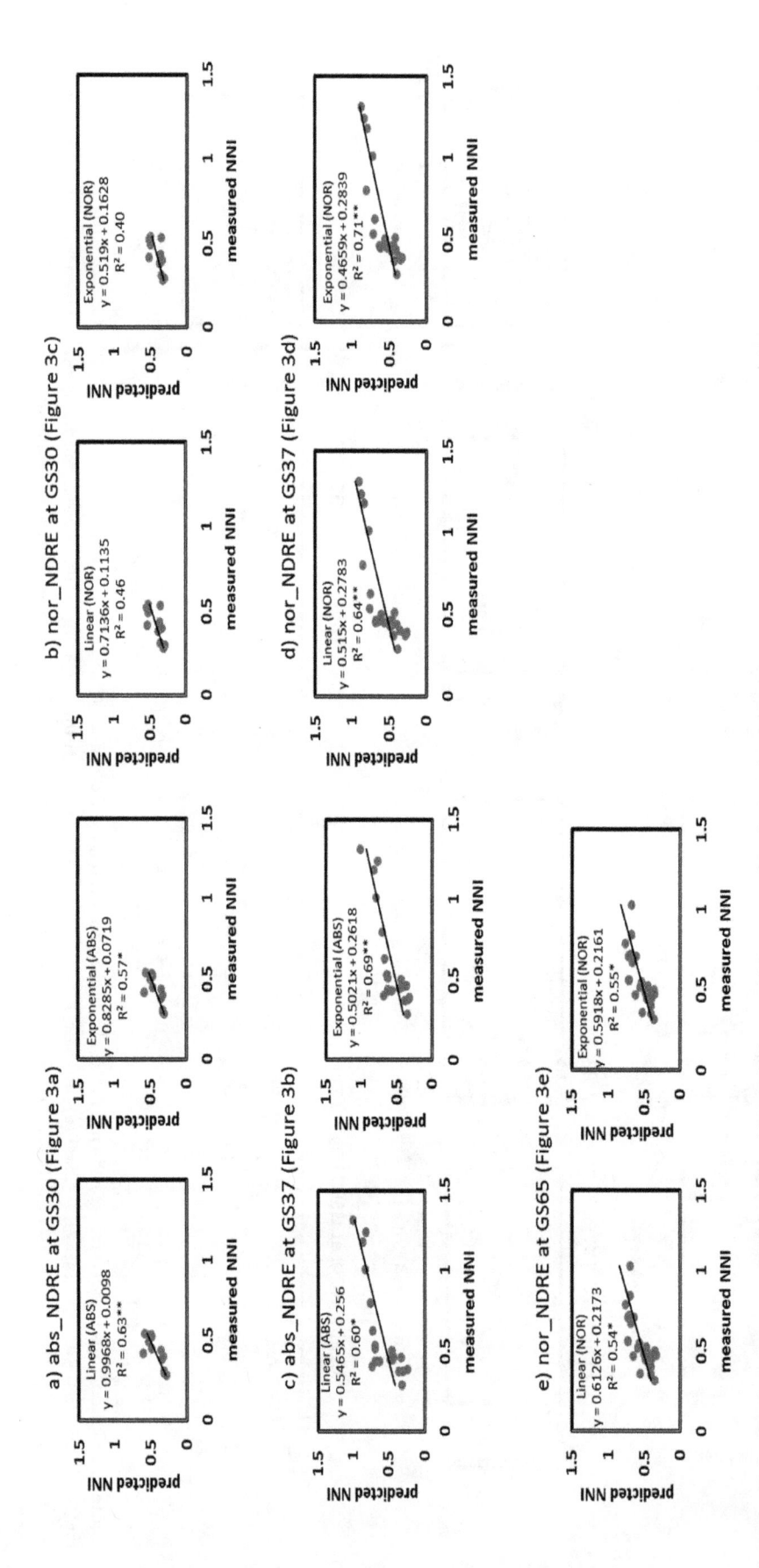

Figure 8. Relationships between the predicted NNI (Nitrogen Nutritional Index) values (from abs_NDRE at GS30 (**a**), nor_NDRE at GS30 (**b**), abs_NDRE at GS37(**c**), nor_NDRE at GS37 (**d**), and nor_NDRE at GS65 (**e**) values in the growing stage-specific models) and the measured NNI values from 25% of the samples. *, ** Significant at 0.05 and 0.01 probability levels, respectively. abs, absolute values; nor, normalized values; NDRE, Normalized Difference Red Edge Index.

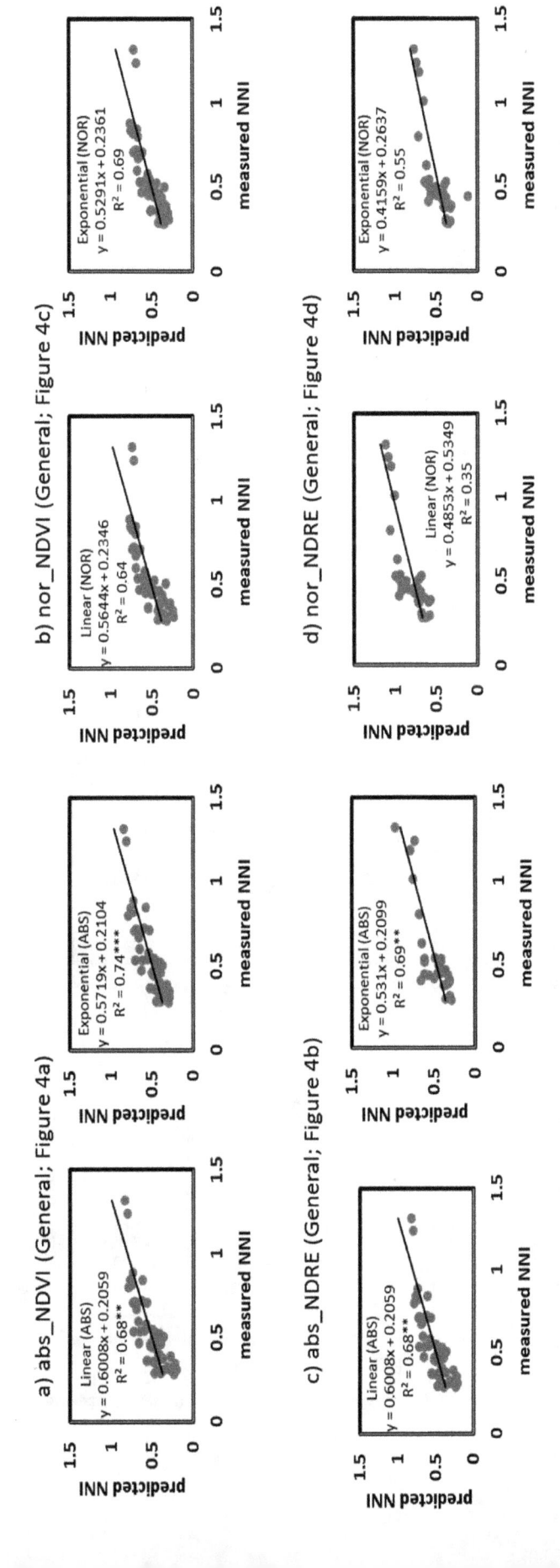

Figure 9. Relationships between the predicted NNI (Nitrogen Nutritional Index) values (from abs_NDVI (**a**) and nor_NDVI (**b**) and abs_NDRE (**c**) and nor_NDRE (**d**) values in the general models) and the measured NNI values from 25% of the samples. **, *** Significant at 0.01 and 0.001 probability levels, respectively. abs, absolute values; nor, normalized values; NDVI, Normalized Difference Vegetation Index; NDRE, Normalized Difference Red Edge Index.

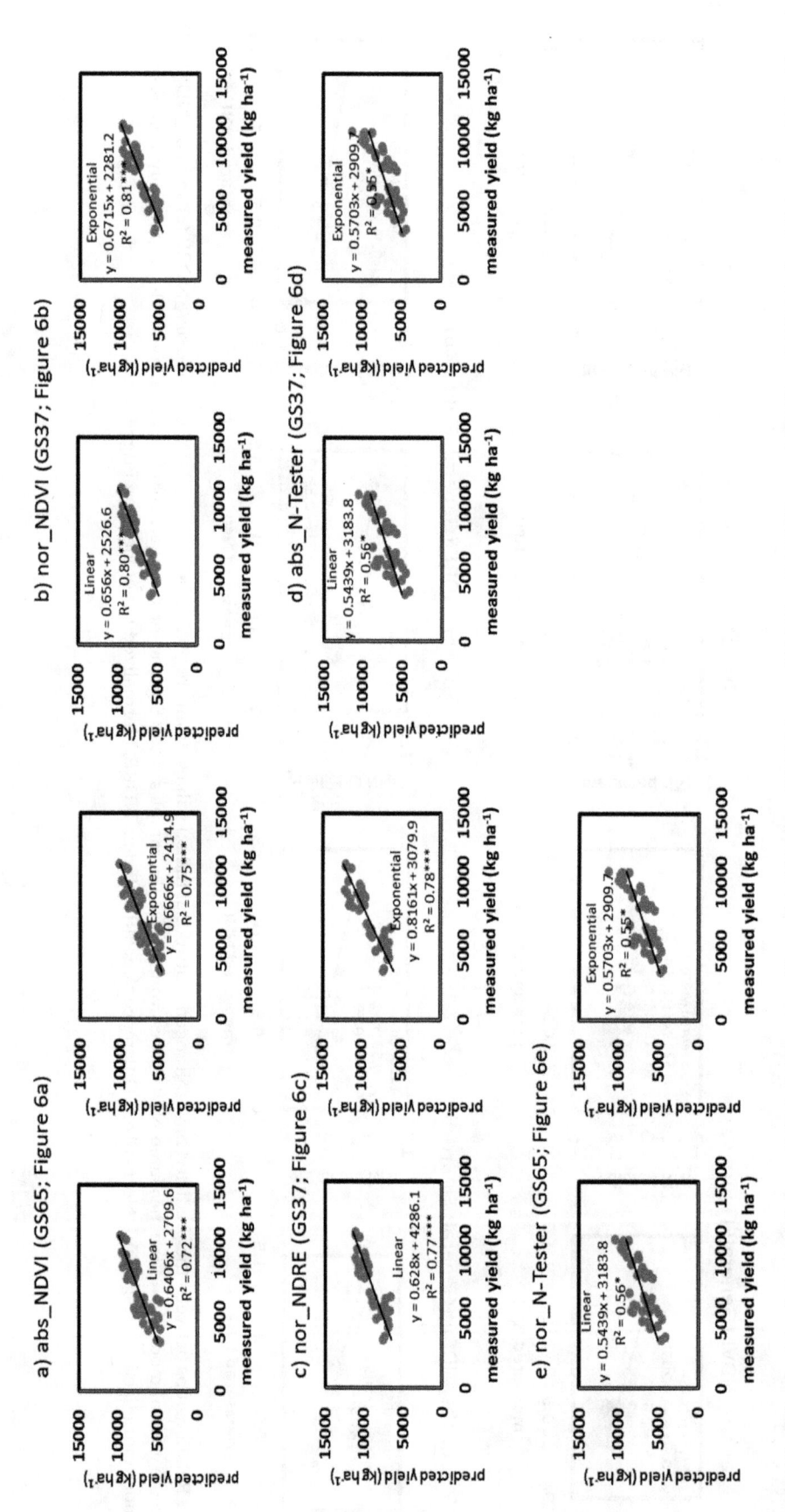

Figure 10. Relationships between the predicted yield values (from abs_NDVI at GS65 (**a**), nor_NDVI at GS37 (**b**), nor_NDRE at GS37 (**c**) abs_N-Tester at GS37 (**d**) and nor_N-Tester (**e**) and the measured yield values from 25% of the samples. *, *** Significant at 0.05 and 0.001 probability levels, respectively. abs, absolute values; nor, normalized values; NDVI, Normalized Difference Vegetation Index; NDRE, Normalized Difference Red Edge Index.

4. Discussion

The NDVI has been the most commonly used vegetation index in agriculture over the last four decades [40] and is a common measure for determining crop N status. The vast majority of the models for predicting the in-season N rate use NDVI [41]. Similar to the results of the present study, Xue et al. [42] described a good relationship between N status and the vegetative period of rice (R^2 = 0.70–0.90), and Cao et al. [41] found that the NDVI explained 47% of NNI changeability across growing stages and growing seasons in wheat.

Usually, in scientific studies, ground-based values are normalized with an overfertilized reference strip where non-limiting N has been applied. However, this approach has limitations because it is not easy for a control fringe to be representative of the field. Moreover, using normalized values would make the use of these tools more difficult for the farmer. Furthermore, Ravier et al. [14] showed that it is not easy to ensure that an overfertilized fringe is not N deficient, thereby problematizing the use of normalized data. In our case, the absolute values were normalized using a strip fertilized with 280 kg N ha^{-1}. However, these overfertilized treatments did not obtain NNI ≥ 1 in some situations. Moreover, when extending the method to large scale tools as the satellite, the utilization of absolute values is convenient, as having an overfertilized strip at each field for data normalization would be a challenging issue. However, if the measurements are not normalized it would be necessary to adjust the crop sensor measurements to different conditions, locations, and varieties [43].

Many authors have noted that the correlations between the NNI and proximal sensors fit better with growing stage-specific models. Sembiring et al. [44] and Mistele and Schmidhaltel [18] (using spectral reflectance) and Cao et al. [45], Ravier et al. [14] (using chlorophyll meters), and Lu et al. [20] (using RapidScan CS-45 sensors) have showed that the highest diagnostic accuracy obtained for cereals differs depending on the growing stage. In our study, the general model, especially abs_NDVI values, showed good accuracy in its NNI estimations, similar to the models for different growing stages (Figure 4). The accuracy of the correlations for validation was also high for abs_NDVI (Figures 7 and 9a), especially in the exponential model. Remarkably, it would thus be easier to use a unique model for on-farm implementations than growing stage-specific models. As in the present study, other authors have also supported using general models with active canopy sensors for winter wheat [46] and with chlorophyll meters for durum wheat [13].

Saturation effects were detected for both the NDVI (around 0.8) and NDRE (around 0.4) indexes when NNI > 0.8. Although NDVI and NDRE reached saturation at a similar point, the NDRE value range is slightly wider for NDVI values around 0.8 (0.35–0.40; Figure 5). However, the NNI threshold values needed to achieve the maximum yields in wheat proposed by Ravier et al. [1] were always NNI < 0.8 or lower. In that case, the saturation effect of the RapidScan CS-45 indexes would not be a problem, as the values related to NNI = 0.8 were close to NDVI = 0.8 and NDRE = 0.37 (Figures 2–5). Therefore, the NDVI values obtained in our conditions would be useful for us making a good nitrogen nutritional diagnosis. The saturation effects on NDVI have been also shown by other authors for winter wheat [41,47] and for rice [48]. It has been reported that when crops achieve a critical canopy or critical chlorophyll content, the NDVI saturation effect is relevant [41,49,50]. This saturation effect is relevant when the canopy is very close and the NIR and visible light break into the crop canopy differently, thereby diminishing the normalization effect of the calculations [16]. When the canopy is close, the NIR reflectance increases while the red reflectance hardly changes [16]. The transmittance of the visible light through the canopy is low, so it is dominated by the top leaves of the plants [16]. However, the NIR detects the biomass bellow as it has a higher transmittance through the crop canopy [16]. The saturation effect can be reduced by using red-edge-based vegetation indices, such as NDRE [46]. Indices including red-edge channels could have higher sensitivity to chlorophyll content in crops, as Cao et al. [46] and Zhang et al. [24] detected that the red-edge bands were more suitable for determining crop N status than the NDVI, and Sharma et al. [50] concluded that NDRE would be better for developing late-season N application algorithms. Conversely, Bonfil [21] showed that the NDVI reaches the same accuracy as the NDRE with RapidScan CS-45 in wheat, similar to the present study.

The yield potential may vary between growing seasons because of the temporal variability in rainfall, temperature, or relative humidity [1]. In this study, the rainfall patterns were very different between years (data not shown), giving more variability to the data. Royo et al. [51] and Martí et al. [19] concluded that the prediction of wheat yield is better when the absolute NDVI readings from later in the season (flowering; GS60–GS65) are used (when yield estimates stabilize). In late growing stages, crop development is advanced in its phenology, and fewer abiotic effects can affect the grain yield. However, similar to our results, Magney et al. [23] showed that the strongest predictions of wheat yield are made prior to heading (before GS50). Otherwise, the yield predictive capacity of the NDVI decreases during grain filling [23].

The GPC prediction capacity with CSR and nor_CSR was not possible in the present study. The best relationship was observed between abs_N-Tester at GS65, where the GPC values could be partially explained ($R^2 = 0.36$). GPC is a product of the N assimilated by the crop prior to grain filling and of the environmental conditions that the crop undergoes in that period [52]. Contrary to our results, the literature showed the potential of using chlorophyll meters to estimate the GPC in wheat [7,9]. As also verified by other authors [3], in this study, the NDRE and NDVI could not explain the GPC variability in any of the growing stages. Magney et al. [23] showed that the NDVI's predictive capacity for GPC never exceeds 0.2. As much as 75%–90% of the total N in plants at harvest may come from the preanthesis (growing stage prior to GS60) N uptake in cereals [53], significantly influencing grain quality [54]. However, it is not possible for crop sensors to estimate N translocation efficiency from the vegetative portion to the grain.

As the results showed, both tools have different behaviours in the prediction of NNI, yield, and GPC. These differences can be explained by the intrinsic differences between the proximal sensing tools, which sense different physical variables and different targets. The RapidScan CS-45 measurements are related to the photosynthetically active crop canopy biomass, while the Yara N-Tester measurements are related to leaf chlorophyll content. On the other hand, RapidScan CS-45 measures the crop reflectance, and Yara N-Tester measures the transmittance of the light on a particular leaf. The operations for the measurements are also different; the measurements using RapidScan CS-45 are simpler, faster, and cover a much larger area as they measure the whole canopy of the crop and thus better represent spatial variability. However, the Yara N-Tester only measures the central zone of the last fully developed leaf, even if this leaf has a relevant role in plant N nutrition.

N fertilization recommendations must be made based on remote sensing indexes that were studied and developed for NNI [6]. NNI measurements require great time and labour and are not instantaneous as subsequent laboratory analyses are needed. Active crop sensor measurements, on the other hand, are not invasive, their information obtained for crop N status is instantaneous and well correlated with the NNI, and measurements can be taken at many time-points during the growing season. Following crop N status is not only important in conventional fertilization but also when organics are applied as initial fertilizers. Organic manures differ in their physical and chemical characteristics; they possess different N mineralization patterns and, therefore, leave uncertain quantities of N available for the crop [10]. Moreover, the same organic fertilizer may have different N release dynamics depending on the year, especially considering the long period between the application and beginning of N uptake by the plant (3–4 months).

Some authors have noted that having only two or three wavebands are a limitation to developing optimum vegetation indices [55]. These authors have suggested the use of tools with more wavebands to calculate more complex indexes or indexes that also include green. Satellite remote sensing, where the waveband spectrum is more complete, may have the potential to improve ground-based sensor performance. However, for the accurate interpretation of satellite-based data necessary for making large-scale N fertilization recommendations, we must first understand the information obtained in ground-based areas through field-trials, where different fertilization strategies are tested and indexes are measured at various growing stages, as done in this study.

5. Conclusions

This study demonstrated that RapidScan CS-45 indexes are able to follow the NNI throughout the entire wheat growing season. A single model (for all growing stages) with absolute NDVI data can, therefore, be used for NNI prediction. The RapidScan CS-45 indexes were able to predict yield with normalized values at GS37 better than at GS65. RapidScan CS-45 and Yara N-Tester^{TM} were not able to predict GPC. Data normalization improved the model for yield but not for NNI prediction. Therefore, following crop N status throughout the growing season using proximal sensing may allow for better adjustment of the N fertilizer to the crop requirements.

Author Contributions: M.A., A.C., and A.A. contributed to the design of this experiment. Field work was accomplished by M.A. and A.C.; M.A., A.C., and A.A. contributed to the interpretation of the data; M.A. and A.A. wrote the paper. M.A., A.C., and A.A. revised the paper critically and approved the final version. All authors have read and agreed to the published version of the manuscript.

References

1. Ravier, C.; Meynard, J.M.; Cohan, J.P.; Gate, P.; Jeuffroy, M.H. Early nitrogen deficiencies favor high yield, grain protein content and N use efficiency in wheat. *Eur. J. Agron.* **2017**, *89*, 16–24. [CrossRef]
2. Raun, W.R.; Johnson, G.V.; Stone, M.L.; Solie, J.B.; Lukina, E.V.; Thomason, W.E.; . Schepers, J.S. In-season prediction of potential grain yield in winter wheat using canopy reflectance. *Agron. J.* **2001**, *93*, 131–138. [CrossRef]
3. Hansen, P.M.; Jørgensen, J.R.; Thomsen, A. Predicting grain yield and protein content in winter wheat and spring barley using repeated canopy reflectance measurements and partial least squares regression. *J. Agric. Sci.* **2003**, *139*, 307–318. [CrossRef]
4. Zhao, B.; Duan, A.; Ata-Ul-Karim, S.T.; Liu, Z.; Chen, Z.; Gong, Z.; Zhang, J.; Xiao, J.; Liu, Z.; Anzhen Qin, A.; et al. Exploring new spectral bands and vegetation indices for estimating nitrogen nutrition index of summer maize. *Eur. J. Agron.* **2018**, *93*, 113–125. [CrossRef]
5. Lemaire, G.; Gastal, F. Quantifying Crop Responses to Nitrogen Deficiency and Avenues to Improve Nitrogen Use Efficiency. In *Crop. Physiology: Applications for Genetic Improvement and Agronomy*; Sadras, V.O., Calderini, D.F., Eds.; Academic Press: San Diego, CA, USA, 2009; pp. 171–211.
6. Ravier, C.; Jeuffroy, M.H.; Gate, P.; Cohan, J.P.; Meynard, J.M. Combining user involvement with innovative design to develop a radical new method for managing N fertilization. *Nutr. Cycl. Agroecosyst.* **2018**, *110*, 117. [CrossRef]
7. Denuit, J.P.; Olivier, M.; Goffaux, M.J.; Herman, J.L.; Goffart, J.P.; Destain, J.P.; Frankinet, M. Management of nitrogen fertilization of winter wheat and potato crops using the chlorophyll meter for crop nitrogen status assessment. *Agronomie* **2002**, *22*, 847–854. [CrossRef]
8. López-Bellido, R.J.; Shepherd, C.E.; Barraclough, P.B. Pedicting post-anthesis N requeriments of bread wheat with Minolta SPAD meter. *Eur. J. Agron.* **2004**, *20*, 313–320. [CrossRef]
9. Arregui, L.M.; Lasa, B.; Lafarga, A.; Irañeta, I.; Baroja, E.; Quemada, M. Evaluation of chlorophyll meters as tools for N fertilization in winter wheat under humid Mediterranean conditions. *Eur. J. Agron.* **2006**, *24*, 140–148. [CrossRef]
10. Aranguren, M.; Castellon, A.; Aizpurua, A. Topdressing nitrogen recommendation in wheat after applying organic manures: The use of field diagnostic tools. *Nutr. Cycl. Agroecosys.* **2018**, *110*, 89–103. [CrossRef]
11. Aranguren, M.; Castellón, A.; Aizpurua, A. Crop Sensor-Based In-Season Nitrogen Management of Wheat with Manure Application. *Remote Sens.* **2019**, *11*, 1094. [CrossRef]

12. Ortuzar-Iragorri, M.A.; Aizpurua, A.; Castellón, A.; Alonso, A.; Estavillo, J.M.; Besga, G. Use of an N-tester chlorophyll meter to tune a late third nitrogen application to wheat under humid Mediterranean conditions. *J. Plant. Nutr.* **2017**, *41*, 627–635. [CrossRef]
13. Debaeke, P.; Rouet, P.; Justes, E. Relationship between normalized SPAD index and the nitrogen nutrition index: Application to durum wheat. *J. Plant. Nutr.* **2006**, *29*, 275–286. [CrossRef]
14. Ravier, C.; Quemada, M.; Jeuffroy, M.H. Use of a chlorophyll meter to assess nitrogen nutrition index during the growth cycle in winter wheat. *Fields Crops Res.* **2017**, *214*, 73–82. [CrossRef]
15. Ortuzar-Iragorri, M.A.; Alonso, A.; Castellón, A.; Besga, G.; Estavillo, J.M.; Aizpurua, A. N-tester use in soft winter wheat: Evaluation of nitrogen status and grain yield prediction. *Agron. J.* **2005**, *97*, 1380–1389. [CrossRef]
16. Knipling, E.B. Physical and physiological basis for the reflectance of visible and near-infrared radiation from vegetation. *Remote Sens. of Environ.* **1970**, *1*, 155–159. [CrossRef]
17. Padilla, F.M.; de Souza, R.; Peña-Fleitas, M.T.; Grasso, R.; Gallardo, M.; Thompson, R.B. Influence of time of day on measurement with chlorophyll meters and canopy reflectance sensors of different crop N status. *Precision Agric.* **2019**, *20*, 1087–1106. [CrossRef]
18. Mistele, B.; Schmidhalter, U. Estimating the nitrogen nutrition index using spectral canopy reflectance measurements. *Eur. J. Agron.* **2008**, *29*, 184–190. [CrossRef]
19. Marti, J.; Bort, J.; Slafer, G.A.; Araus, J.L. Can wheat yield be assessed by early measurements of Normalized Difference Vegetation Index? *Ann. Appl. Biol.* **2007**, *150*, 253–257. [CrossRef]
20. Lu, J.; Miao, Y.; Shi, W.; Li, J.; Yuan, F. Evaluating different approaches to non-destructive nitrogen status diagnosis of rice using portable RapidSCAN active canopy sensor. *Sci. Rep.* **2017**, *7*, 14073. [CrossRef]
21. Bonfil, D.J. Monitoring Wheat Fields by RapidScan: Accuracy and Limitations. In Proceedings of the Conference on Precision Agriculture 2017, John McIntyre Centre, Edinburgh, UK, 16–20 July 2017.
22. Cao, Q.; Miao, Y.; Wang, H.; Huang, S.; Cheng, S.; Khosla, R.; Jiang, R. Non-destructive estimation of rice plant nitrogen status with CropCircle multispectral active canopy sensor. *Fields Crops Res* **2013**, *154*, 133–144. [CrossRef]
23. Magney, T.S.; Vierling, L.A.; Eitel, J.U.H.; Huggins, D.R.; Garrity, S.R. Response of high frequency Photochemical Reflectance Index (PRI) measurements to environmental conditions in wheat. *Remote Sens. Environ.* **2016**, *173*, 84–97. [CrossRef]
24. Zhang, K.; Ge, X.; Shen, P.; Li, W.; Liu, X.; Cao, Q.; Zhu, Y.; Cao, W.; Tian, Y. Predicting rice grain yield based on dynamic changes in vegetation indexes during early to mid-growth stages. *Remote Sens.* **2019**, *11*, 387. [CrossRef]
25. Bijay-Singh, S.R.K.; Jaspreet-Kaur, J.M.L.; Martin, K.L.; Yadvinder-Singh, V.S.; Chandna, P.; Choudhary, O.P.; Gupta, R.K.; Thind, H.S.; Jagmohan-Singh, U.H.S.; Khurana, H.S.; et al. Assessment of the nitrogen management strategy using an optical sensor for irrigated wheat. *Agron. Sustain. Dev.* **2011**, *31*, 589–603. [CrossRef]
26. Zadoks, J.C.; Chang, T.T.; Konzak, C.F. A decimal code for growth stages of cereals. *Weed Res.* **1974**, *4*, 415–421. [CrossRef]
27. *Climates of the World and Their Agricultural Potentialities*; J. Papadakis: Buenos Aires, Argentina, 1966.
28. Gee, G.W.; Bauder, J.W. Particle-Size Analysis. In *Methods of Soil Analysis: Part 1. Physical and Mineralogical Methods*; Klute, A., Ed.; SSSA: Madison, AL, USA, 1986; pp. 383–411.
29. Soil Survey Staff. *Soil Taxonomy: A Basic System of Soil Classification for Making and Interpreting Soil Surveys*, 2nd ed.; Natural Resources Conservation Service; U.S. Department of Agriculture Handbook 436: Washington, DC, USA, 1999.
30. MAPA. *Métodos Oficiales de Análisis. Tomo III*; Ministerio de Agricultura; Pesca y Alimentación: Madrid, Spain, 1994.
31. Walkey, A.; Black, I.A. An examination of Degtjareff method for determining soil organic matter and a proposed modification of the chromic and titration method. *Soil Sci.* **1934**, *37*, 29–37. [CrossRef]
32. Soil Survey Staff. *Illustrated Guide to Soil Taxonomy*; U.S. Department of Agriculture, Natural Resources Conservation Service, National Soil Survey Center: Lincoln, Nebraska, 2015.

33. AOAC, Association of Official Analytical Chemists International. Plants, 24, 127. In *Official Methods of AOAC International*, 16th ed.; Patricia, C., Ed.; AOAC International: Gaithersburg, MD, USA, 1999.
34. Teller, G.L. Non-protein nitrogen compounds in cereals and their relation to the nitrogen factor for protein in cereals and bread. *Cereal Chem.* **1932**, *9*, 261–274.
35. Justes, E.; Mary, B.; Meynard, J.M.; Machet, J.M.; Thelier-Huché, L. Determination of a critical nitrogen dilution curve for winter wheat crops. *Ann. Bot.* **1994**, *74*, 397–407. [CrossRef]
36. Wang, G.; Bronson, K.F.; Thorp, K.R.; Mon, J.; Badaruddin, M. Multiple leaf measurements improve effectiveness of chlorophyll meter for durum wheat nitrogen management. *Crop. Sci.* **2014**, *54*, 817. [CrossRef]
37. R Core Team. *R: A Language and Environment for Statistical Computing*; R Foundation for Statistical Computing: Vienna, Austria, 2013.
38. Akaike, H. Information Theory and an Extension of the Maximum Likelihood Principle. In *Proceedings of the Second International Symposium on Information Theory, Tsahkadsor, Armenia, September 2–8, 1971*; Petrov, B.N., Csaki, F., Eds.; Akademiai Kiado: Budapest, Hungary, 1973; pp. 267–281.
39. Burnham, K.P.; Anderson, D.R. *Model Selection and Inference: A Practical Information-Theoretic Approach*, 2nd ed.; Springer: New York, NY, USA, 2002. [CrossRef]
40. Tucker, C.J. Red and photographic infrared linear combinations monitoring vegetation. *J. Remote Sens. Environ.* **1979**, *8*, 127–150. [CrossRef]
41. Cao, Q.; Miao, Y.; Feng, G.; Gao, X.; Li, F.; Liu, B.; Yue, S.; Cheng, S.; Ustin, L.U.; Khosla, R. Active canopy sensing of winter wheat nitrogen status: An evaluation of two sensor systems. *Comput. Electron. Agric.* **2015**, *112*, 54–67. [CrossRef]
42. Xue, L.; Li, G.; Qin, X.; Yang, L.; Zhang, L. Topdressing nitrogen recommendation for early rice with an active sensor in south China. *Precision Agric.* **2014**, *15*, 95–110. [CrossRef]
43. Craigie, R.; Yule, I.; McVeagh, P. *Crop sensing for nitrogen management*; Foundation for Arable Research: Christchurch, New Zealand, 2013.
44. Sembiring, H.; Lees, H.L.; Raun, W.R.; Johnson, G.V.; Solie, J.B.; Stone, M.L.; DeLeon, M.J.; Lukina, E.V.; Cossey, D.A.; LaRuffa, J.M.; et al. Effect of growth stage and variety on spectral radiance in winter wheat. *J. Plant. Nutr.* **2000**, *23*, 141–149. [CrossRef]
45. Cao, Q.; Miao, Y.; Gao, X.; Liu, B.; Feng, G.; Yue, S. Estimating the Nitrogen Nutrition Index of Winter Wheat Using an Active Canopy Sensor in the NorthChina Plain. In Proceedings of the 1st International Conference Agro-Geoinformatics, Agro-Geoinformatics 2012, Shanghai, China, 2–4 August 2012; IEEE Computer Society: Shanghai, China, 2012; pp. 178–182.
46. Cao, Q.; Miao, Y.; Shen, J.; Yuan, F.; Cheng, S.; Cui, Z. Evaluating Two Crop Circle Active Canopy Sensors for In-Season Diagnosis of Winter Wheat Nitrogen Status. *Agronomy* **2018**, *8*, 201. [CrossRef]
47. Erdle, K.; Mistele, B.; Schmidhalter, U. Comparison of active and passive spectral sensors in discriminating biomass parameters and nitrogen status in wheat cultivars. *Field Crops Res.* **2011**, *124*, 74–84. [CrossRef]
48. Gnyp, M.L.; Miao, Y.; Yuan, F.; Ustin, S.L.; Yu, K.; Yao, Y.; Huang, S.; Bareth, G. Hyperspectral canopy sensing of paddy rice aboveground biomass at different growth stages. *Field Crops Res.* **2014**, *155*, 42–55. [CrossRef]
49. Moriondo, M.; Maselli, F.; Bindi, M. A simple model of regional wheat yield based on NDVI data. *Eur. J. Agron.* **2007**, *26*, 266–274. [CrossRef]
50. Sharma, L.K.; Bu, H.; Denton, A.; Franzen, W.F. Active-Optical Sensors Using Red NDVI Compared to Red Edge NDVI for Prediction of Corn Grain Yield in North Dakota, U.S.A. *Sensors* **2015**, *15*, 27832–27853. [CrossRef]
51. Royo, C.; Aparicio, N.; Villegas, D.; Casadesus, J.; Monneveux, P.; Araus, J.L. Usefulness of spectral reflectance indices as durum wheat yield predictors under contrasting Mediterranean conditions. *Int. J. Remote Sens.* **2003**, *24*, 4403–4419. [CrossRef]
52. Masclaux-Daubresse, C.; Daniel-Vedele, F.; Dechorgnat, J.; Chardon, F.; Gaufichon, L.; Suzuki, A. Nitrogen uptake, assimilation and remobilization in plants: Challenges for sustainable and productive agriculture. *Ann. Bot.* **2010**, *105*, 1141–1157. [CrossRef]

53. Dupont, F.M.; Altenbach, S.B. Molecular and biochemical impacts of environmental factors on wheat grain development and protein synthesis. *J. Cereal Sci.* **2003**, *38*, 133–146. [CrossRef]
54. Montemurro, M.; Convertini, G.; Ferri, D. Nitrogen Application in Winter Wheat Grown in Mediterranean Conditions: Effects on Nitrogen Uptake, Utilization Efficiency, and Soil Nitrogen Deficit. *J. Plant. Nutr.* **2007**, *30*, 1681–1703. [CrossRef]
55. Colaço, A.F.; Bramley, R.G.V. Do crop sensors promote improved nitrogen management in grain crops? *Field Crops Res.* **2018**, *218*, 126–140.

Development of an Autonomous Electric Robot Implement for Intra-Row Weeding in Vineyards

David Reiser [1,*], **El-Sayed Sehsah** [2], **Oliver Bumann** [1], **Jörg Morhard** [1] **and Hans W. Griepentrog** [1]

[1] Institute of Agricultural Engineering, Hohenheim University, Stuttgart 70599, Germany; oliver.bumann@uni-hohenheim.de (O.B.); morhardj@uni-hohenheim.de (J.M.); hwgriep@uni-hohenheim.de (H.W.G.)

[2] Department of Agricultural Engeenering, Faculty of Agriculture, Kafrelsheikh University, Kafrelsheikh 33516, Egypt; sehsah_2000@yahoo.de

* Correspondence: dreiser@uni-hohenheim.de

Abstract: Intra-row weeding is a time consuming and challenging task. Therefore, a rotary weeder implement for an autonomous electrical robot was developed. It can be used to remove the weeds of the intra-row area of orchards and vineyards. The hydraulic motor of the conventional tool was replaced by an electric motor and some mechanical parts were refabricated to reduce the overall weight. The side shift, the height and the tilt adjustment were performed by linear electric motors. For detecting the trunk positions, two different methods were evaluated: A conventional electromechanical sensor (feeler) and a sonar sensor. The robot performed autonomous row following based on two dimensional laser scanner data. The robot prototype was evaluated at a forward speed of 0.16 ms^{-1} and a working depth of 40 mm. The overall performance of the two different trunk detection methods was tested and evaluated for quality and power consumption. The results indicated that an automated intra-row weeding robot could be an alternative solution to actual machinery. The overall performance of the sonar was better than the adjusted feeler in the performed tests. The combination of autonomous navigation and weeding could increase the weeding quality and decrease power consumption in future.

Keywords: autonomous robot; agriculture; viticulture; electric weeder; sonar; intra-row; under-vine; row following

1. Introduction

The impact of chemical pesticides became an important global issue for the sustainability of the food production system. More and more organic food is produced and has become popular all over the world. One reason for the use of chemical pesticides is to reduce weeds, as they can be responsible for high yield losses [1]. However, to leave them untreated is not an option, as weeds could be responsible for a decrease of up to 40% of yield [2,3]. The importance of mulching and weeding in vineyards was already conducted in different studies. It was proclaimed that organic mulching was the most sustainable practice for yearly vine production [4]. In order to reduce chemical weed control, mechanical weeding approaches are promising alternatives. Mechanical weed control can be conducted between the tree/crop rows (inter-row) and within the tree/crop rows (intra-row). The main challenge of mechanical weed control is the realization at the intra-row area [5]. Weeding close to the crop/trunk and the use of intra-row weeding tools needs a very accurate steering for not damaging the crop/trunk [6]. Therefore, accurate guidance systems are needed. During the last decades, navigation has been improved by the use of new automatic row guidance systems using feelers, Global Navigation Satellite Systems (GNSS), distance sensors and cameras [7–9]. The development of these technologies has created wide opportunities for weed management and may become a key element of modern weed control [10]. Anyhow, in order to increase the selectivity of the weed control and reduce the crop plants damage, intelligent autonomous weeding systems are needed. Identifying individual weeds and crops is possible by the use of mechanical feelers, GNSS coordinates, sonar sensors, laser scanners, or cameras [11–14]. Automation offers the possibility to determine and differentiate crops from weeds and at the same time, to remove the weeds with a precisely controlled device. Automated weeding implements developed [15–18], provide examples of how the control of mechanical weeding can be performed. Real-time weeding robots have the potential to control intra-row weeds plus reduce the reliance on herbicides and hand weeding [19,20].

For reducing labor costs of mechanical weed control the whole task has to be automated. This approach could result in self-guided, self-propelled and autonomous machines that could cultivate crops with minimal operator intervention. Weeding robots can improve labor productivity, solve labor shortage, improve the environment of agricultural production, improve work quality, reduce energy input, improve resource management, and help farmers to change their traditional working methods and conditions [21]. Introducing self-governing mobile robots in agriculture, forestry and landscape conservation is dependent on the natural environment and the structure of the facilities. The solution proposed is the use of little and light machines, like unmanned vehicles that are self-propelled, autonomous and low powered [22]. Mobile robots must always be aware of its surroundings such as unpredicted obstacles [23]. To this scope, real-time environment detection could be performed using different sensor systems.

There were several autonomous machines and robots developed in the past [24]. Zhang et al. investigated the performance of a mechanical weeding robot [25]. They used machine vision to locate crop plants and a side shift mechanism to move along the crop rows. The intra-row weeds were controlled by moving hoes that could open and close to prevent damage to crop plants. Another study by Nørremark et al. developed an automatic GNSS-based intra-row weeding machine using a cycloid hoe [13]. The weeder used eight rotating tines, which could be released to tolerate crop plants. This automated weeding system mainly used Real Time Kinematic (RTK)-GNSS to navigate and control the side-shifting frame plus a second RTK-GNSS to control the rotating tines. Astrand and Baerveldt investigated the performance of a vision-based intra-row robotic weed control system [16]. The robotic system implemented two vision systems to guide the robot along the crop row and to discriminate between crops and weed plants. The weeding tool consisted of a rotating wheel connected to a pneumatic actuator that lifted and lowered the weeding tool to tolerate crop plants. Robots for weeding in vineyards were also developed, mainly focusing on trunk detection [26,27]. Weeds in vineyards can cause significant reductions in vine growth and grape yields [11]. Conventional control methods rely on herbicide applications in the vine rows and the area between them (middles), or a combination of herbicide strip application in the vine row and mowing or disking of the middles [28]. For a mobile robot to navigate in between the vine or tree rows, it must detect the position of the rows first, by means of detecting the trunks. The complete integration of an under-vine rotary weeder in an autonomous robot system for active intra-row weeding in vineyards and orchards was not done until now.

The overall goal of the current research was to develop and test the performance of a rotating electrical tiller weeder mechanism, built for automated intra-row weeding for vineyards. The system should be integrated in an autonomous robot platform and should be evaluated under controlled soil bin and outdoor conditions. The power requirement of the rotary blades mechanism and the total power for the autonomous robotic machine were of interest. The second interest was to test and compare two control strategies for detecting trunks, using a sonar sensor and a feeler.

2. Materials and Methods

2.1. Mechanics and Electronics

An electric tiller head rotary weeder cultivator was manufactured and designed in the laboratory of the Institute of Agricultural Engineering, Hohenheim University and was mounted to an autonomous caterpillar robot called "phoenix" [14] (see Figure 1). The prototype electric rotary weeder was built up using a rotary weeder from Humus Co. called "Humus Planet" (Humus, Bermatingen, Germany). This tool originally operated with a hydraulic motor and was redesigned to get driven by an electric DC motor. A 1 kW 48DC brushed motor model ZY1020 (Ningbo Jirun Electric Machine Co.,Ltd., Nigbo, China) was used. The motor had a nominal torque of 2 Nm and a rated speed of 3200 rpm. The motor was mounted using a gearbox reducing with 30:1 ratio to produce a nominal torque of 60 Nm and a motor speed of 106 rpm at the rotary weeder. The gearbox was directly connected with a drive shaft to the rotary weeder. The motor and the gearbox were connected using a v-belt. The developed rotary weeder was fixed on a metal bar together with the drive unit.

Figure 1. Autonomous robot called "phoenix" with attached rotary weeder implement and sensors.

The whole implement was attached to the robot using a coupling triangle. The implement could be moved up, down, left, right and could be tilted with the help of three Linak LA36 actuators (Linak, Nordborg, Denmark). They were capable of creating a pull/push force between 1700 N and 4500 N. This was sufficient to move the implement under all circumstances. The robot system was driven by two brushless motors HBL5000 (Golden Motor Technology Co., Ltd., Jiangsu, China), creating a maximum driving power of 10 kW, sufficient to move the 500 kg weight of the robot plus implement even in harsh terrain. The two times 80 × 15 cm belt system of the phoenix minimized soil compaction and provided enough grip and pull force for mechanical weeding. The width of the robot was 1.70 m. The implement could move 30 cm side wards, to move the tool inside the crop row. The power of the whole robot system was provided by batteries, with a nominal voltage of 48 V and 304 Ah. The implement motor was powered by a motor controller Model LB57 (Yongkang YIYUN Elektronic Co., Ltd., Zhejiang, China). The drive motors were controlled by two ACS48S motor controllers (Inmotion Technologies AB, Stockholm, Sweden). The whole robot was controlled and programmed with an embedded computer with i5 processor, 4 GB Ram and 320 GB disk space.

2.2. Sensors

For following the tree and vine rows, a LMS 111 (Sick, Waldkirch, Germany) 2D laser scanner was used [29]. The mounting of the scanner was in the front of the vehicle to enable a security stop if an obstacle appeared in front of the robot. Additionally, there were four security switches attached to the robot, disabling the drive and implement motors in case of an emergency. For gaining the heading and traveled distance, an Inertial Measure Unit VN100 (VectorNav, Dallas, TX, USA) was fused with the hall sensor information of the drive motors [30]. The traveling distance was defined by the hall sensors, and the heading was estimated with the IMU. For guiding the implement, two different methods and sensors were tested. The first method used a standard feeler, a modified version of the original feeler of the rotary weeder. As the second method, a sonar sensor pico + 100/U (microsonic, Dortmund, Germany) was used to detect the trunks [31]. This sensor was used to detect the trunks, because it could be placed at any position at the robot. The signal could directly replace the input signal of the feeler without any additional software changes.

The specifications of the different sensor systems can be found in Table 1. The data of the feeler and the sonar sensor were digitalized using an analog digital converter from RedLab 1208LS (Meilhaus Electronic, Alling, Germany). This information could be used to send control signals to the implement and the linear motors.

Table 1. Sensor specifications of the used sensors of the robot.

Sensor	Specification	Value
LMS 111 [29]:	Operating range:	0.5 m to 20 m
	Field of view/scanning angle:	270°
	Data rate:	25 Hz
	Angular resolution:	0.5°
	Systematic error:	± 30 mm
	Statistical error (1σ):	12 mm (0.5–10 m)
Pico + 100/U [31]:	Operating range:	0.12 to 1.0 m
	Field of view/scanning angle:	22°
	Data rate:	10 Hz
	Distance measurement accuracy:	± 1%
VN100 [30]:	Data rate:	40 Hz
	Angular resolution:	0.05°
	Accuracy heading:	2.0°
	Accuracy pitch/roll:	1.0°

2.3. Software

The control and data logging of the whole robot system and implement was programmed using ROS Indigo-middleware [32]. The embedded computer ran under Ubuntu 14.04. All parts of the guidance software were programmed using C/C++. All data created were pushed to a so called "topic" which could be time stamped and recorded with the ROS system. This data could be analyzed for evaluating the power consumption, the movement of the robot and the implement behavior. The movement and the transformations between the robot base and the implement was performed using the Transform Library of ROS [33].

The guidance of the robot was performed with the use of 2D laser scanner data. First the data were converted to Cartesian coordinates and were filtered with a range filter of 5 m. The points were separated into two clusters, one for the left and one for the right side of the row. This was done by separating all points exactly at the laser scanner position into two point clouds (left and right). Afterwards the points were filtered using a Euclidean Clustering method using a distance threshold of 1.5 m. The resulting cluster with the highest point cloud number was estimated as the row point cluster. This clustering helped to get the row filtering more robust. The resulting points of each side were afterwards used to estimate a best fit line with the use of a RANSAC algorithm [34]. This resulted in two estimated lines, one for each row side. The median of the two resulting lines was used to define the row direction and the next goal point. When one line was missing, or a fixed distance to the row was needed, like in this weeding application, a fixed offset to one side could be set.

In this application a side offset of 0.15 m to the right side detected row was used. Therefore, the implement could move 0.15 m inside the crop row, as the maximum side shift was limited to 0.3 m. The forward speed was controlled by the software and just an active angle adjustment was done based on the line following algorithm of the laser scanner. The robot drive motor controller and the implement motors were controlled with CAN bus signals. The update rate of the motor control was fixed to 10 Hz.

The control of the implement was realized in a separate ROS node, analyzing the values of the switch and the sonar sensor. As soon as the feeler or the sonar detected an obstacle like a trunk, the linear actuator was shifted "inside". When there was no obstacle detected, the actor moved "outside" until the maximum outer position was reached. The robot program could be started and stopped using a basic joystick. There were two different programs available, one using just the sonar value and one the feeler signals. The feeler signal just included a digital signal. The output of the sonar sensor was an analog signal between 0 and 10 V. This sensor output correlated with the object distance, detected by the sensor. As soon as the limit distance of 5V was read in the ADC Redlab, a digital input signal was triggered, shifting the actuator "inside".

The height of the implement could be fixed to one depth, or could use the actual motor current of the implement to control the depth dependent on the motor torque. This could help to minimize the necessary force and prevent the rotary tiller weeder to get stuck. In the indoor soil bin, the height was fixed, as the surface was even and without any huge disturbances. In the outdoor tests the adaptive height adjustment variance was used to adjust the implement to the soil surface.

The following Figure 2 shows the ROS-visualization tool "rviz" used to visualize the live data, acquired at the robot system. The robot pose was visualized in real time together with the implement pose, the laser scanner data, the sonar data and the next waypoint the robot wanted to follow. This helped to debug the system while programming and provided feedback to the user.

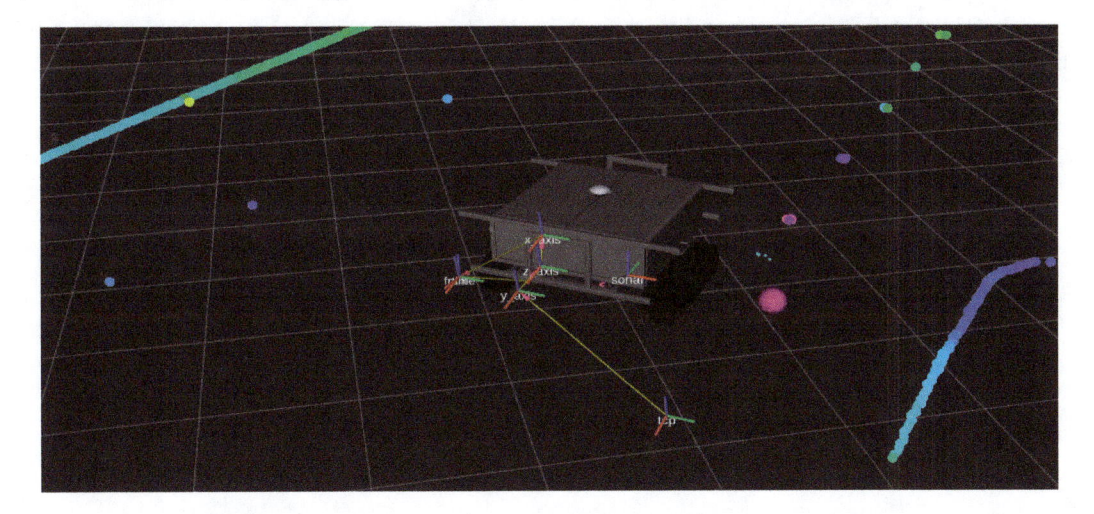

Figure 2. ROS-Software environment visualization "rviz" of robot position, implement position and the laser scanner points (small dots) and sonar data (purple ball).

The tilled area in the vineyard was analyzed by using the open source software ImageJ version 5.22 [35]. For further processing of the gained data, the software Microsoft Excel (Microsoft Corporation, Redmond, WA, USA) was used.

2.4. Test Environments

The indoor experiment was performed at the Hohenheim University soil bin laboratory. The indoor soil bin had 46 m length, 5 m width and 1.2 m depth. The mixture of the soil content was 73% sand, 16% silt and 11% clay. The soil was prepared by a rotary harrow, first loosening the soil and afterwards re-compacting the upper soil layer using a flat roller. 22 different plots were prepared and the soil was marked and oriented by using poles of wood with the size of 3 × 3 × 160 cm. The poles were separated with 1.5 m distance and simulated the trees/trunks for the indoor test. The total area evaluated for the treatment quality between every two poles was 0.45 m^2 (1.5 m length, 0.3 m row width). This area was divided into 18 rectangles marked on the soil by using chalk. After the experiment, the rectangles with no interaction with the tiller were defined as non-tilled area. Three main soil parameters were recorded for each individual plot. The penetration resistance, soil moisture content and soil shear strength conditions were measured with an H-60 Hand-Held Vane Tester (GEONOR, Inc., Augusta, GA, USA). The specific soil moisture was logged using a TRIME-PICo 64 Time-Domain-Reflectometry sensor (IMKO Micromodultechnik GmbH, Ettlingen, Germany). The penetration force was measured with an Eijkelkamp penetrometer logger (Eijkelkamp, Giesbeek, The Netherlands).

It was decided to use the absolute rotational speed for the implement and a forward speed for the robot of 0.16 m s^{-1}. 11 plots were treated using a feeler and the remaining 11 plots were treated while detecting the poles with the sonar sensor. The position of the sonar and the feeler was placed 0.5 m before the weeder, so that the implement could move out of the way even without stopping the robots forward speed. The goal distance between the robot and the row was fixed to 0.15 m in the row following software. When a pole was detected, the electric actuator shifted the rotary out of the row, to prevent damage to the poles. As soon as the pole was passed, the actuator got shifted back inside the row. The linear motor used for the side shift had a maximum speed of 68 mm/s with no load and 52 mm/s at maximum load. On average this resulted in a travel speed of 60 mm/s of the linear LA36 actuator. As the implement was set up to move 0.15 m to the intra row area, at least a time difference between pole detection and arrival of the tool of 2.5 s were needed. As the traveling speed of the robot was fixed to 0.16 m/s, it was necessary to detect the tree at least 0.4 m before the implement. For gaining some buffer, the pole detection was set up at a distance of 0.5 m. The time delay for the shifting mechanism was fixed to 2.5 s in the software to get as close as possible to the poles and therefore to potential trunks. The following Figure 3 describes the mechanical and electrical components of the two control methods using the sonar sensor and the feeler.

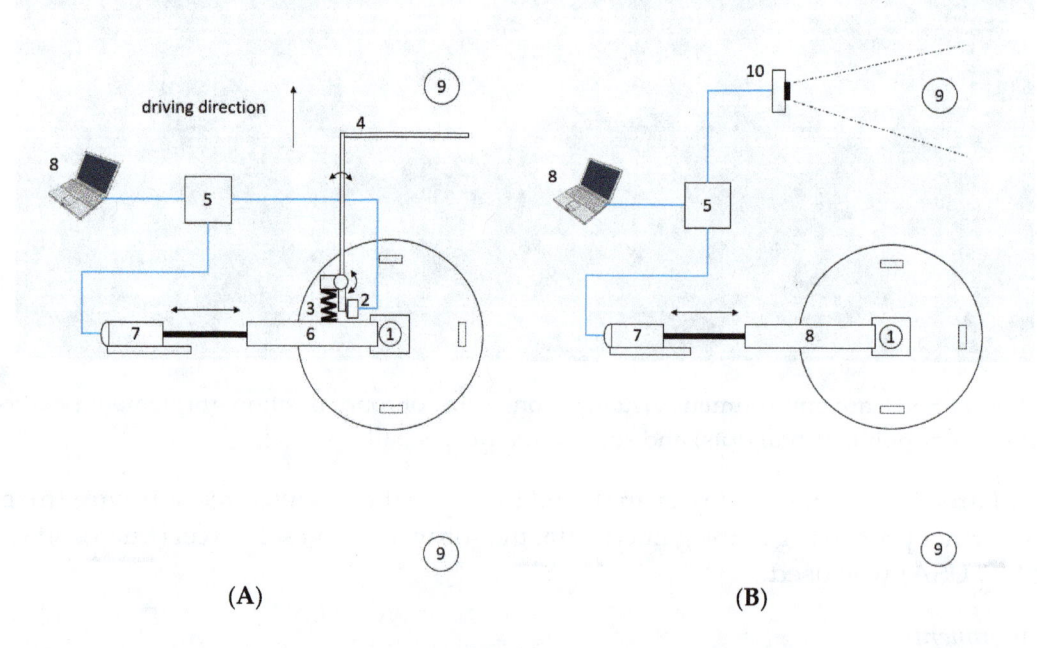

Figure 3. The two different control methods: (**A**) Feeler controlled unit and (**B**) Sonar sensor controlled actuator for the developed autonomous electric weeder. 1–Electric weeder; 2–Switch; 3–Spring; 4–Feeler; 5–Embedded controller; 6–Bar for implement; 7–Linear actuator; 8–PC; 9–Trunks (poles); 10–Sonar sensor.

For each treatment, the power consumption and the accuracy of the processing procedure were evaluated estimating the treated and the non-treated area. The better control method tested in the indoor laboratory was evaluated at the vineyard of Hohenheim University (48.710115 N, 9.212913 E) on 9 October 2018 (see Figure 4). The following figure shows the two different test areas for the autonomous weeding robot.

Figure 4. The two test environments– (**A**) indoor laboratory and (**B**) outdoor vineyard.

3. Results and Discussion

3.1. Comparison of Feeler and Sonar

Over the plots of the indoor trial, the soil moisture varied between 13.1% and 7.4% by volume. The shear stress varied between 59 and 82 MPa. The penetration resistance varied between 2 and 4 MPa for the depths of the first 10 cm. The detailed results of the soil physical properties for the laboratory test can be found in Table 2.

Table 2. Soil physical properties for the evaluation of the sonar and the feeler method in the soil bin and the resulting tilled area (indoor laboratory test).

Plot	Sensor	Soil Penetration, MPa (0–10 cm Depth)	Moisture Vol. %	Soil Temp., °C	Shear Stress, MPa	Tilled Area, %
t1	Sonar	2	13.8	20.1	61.5	94
t2	Sonar	2	12.7	20.4	61	83
t3	Sonar	3.8	9.09	20.8	61	89
t4	Sonar	4	9.7	20.9	59	100
t5	Sonar	3	8.8	21	60.5	89
t6	Sonar	2	7.4	21.3	56	94
t7	Sonar	2	13.2	21.9	60.5	72
t8	Sonar	2.5	9.8	22	69.5	89
t9	Sonar	2.5	11.5	22.2	66.5	61
t10	Sonar	2.2	8.8	22.4	67.5	83
t11	Sonar	2.4	9.5	22.5	68	50
t1	Feeler	3.8	10.1	22.6	62	50
t2	Feeler	3	11.9	22.7	66	61
t3	Feeler	3.5	12.07	22.9	79.5	61
t4	Feeler	2.5	13.1	23	82	94
t5	Feeler	4	9.8	23.6	70	78
t6	Feeler	4	8.8	23.7	81	78
t7	Feeler	4	9.5	24	70	56
t8	Feeler	4	9.7	24	73	44
t9	Feeler	3.5	10.07	24	75	78
t10	Feeler	4	9.7	24.8	79	56
t11	Feeler	3.8	9.7	24.8	79	61

3.2. Efficiency of the Control Algorithm

The percentages of the tilled area between the poles (trunks) were used as parameters to indicate the efficiency of weed control in the current study. Table 2 indicates that the sonar treatment performed better than the feeler. Only at the plots t9 and t11 the feeler performed better than the sonar. In all other plots sonar performed equal or better compared to the feeler solution. The overall performance of both methods was at least over 50% of the intra-row area, at some plots reaching a tilled area of almost 100%. The average of the tilled area for the feeler was 65% and for the sonar it was 82%.

The results of the tilled area indicated a high success rate of both methods. However, the sonar control method outperformed the feeler control. One reason for this was that the sonar signals of the poles could directly be converted to a control message. The feeler sometimes got stuck at the poles for some seconds, causing wrong signals to the actuator and made it hard for the software to define the exact pole position. Additionally, the long lever of the feeler caused vibrations, sometimes triggering the switch even when there was no pole. With no other disturbances e.g., of weeds and branches, the sonar worked reliably and without any failure. Therefore, it was suggested, to test this control method under outdoor conditions later on.

The signals from the sonar sensor over the travelled distance in the soil bin can be seen in the following Figure 5. The gained signals showed a peak at the pole positions. At the beginning the sonar values hit the edges of the poles, causing wrong distance signals. This effect could be seen at both sides of the poles. However, the minimum value of the sticks was quite accurate in distance and position. The signals always formed small peaks with similar values. This shape could be used for further filtering of the exact trunk positions in the future and separate high weed spots from trunks. All poles triggered at least one signal of less than 0.5 m distance, which triggered the side shift of the implement. Therefore, the sonar could detect all poles at the indoor test.

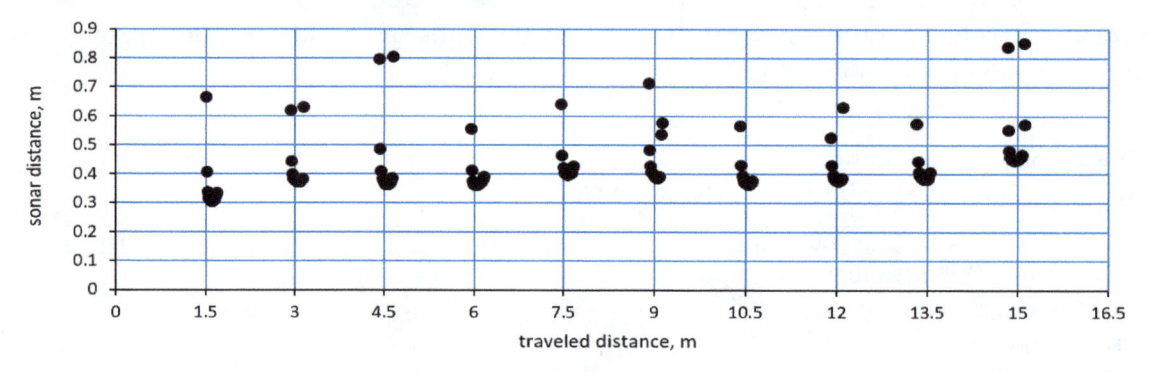

Figure 5. Sonar values detected at the indoor test at the laboratory (soil bin).

In the following outdoor test just the sonar detection was evaluated. Figure 6 shows the sonar detections based on the test in the vineyard of Hohenheim University. In this case the trunks were not spaced out evenly like in the indoor tests. Sometimes individual trunks were missing, or were closer together (less than 0.5 m), causing that in this case the intra-row area could not be entered by the tilling device. Additionally, there were shoots and weeds causing disturbances. However, with small height adjustments of the sonar sensor, the system worked well in the outdoor vineyard. The linearity of the row was not given, because of curved trunks, as they were not in a straight line as the wooden poles.

However, all vines were successfully detected, causing no issues with hitting the vines basal area with the implement or similar. But the within-row spacing of the vines was not always suitable for the 0.3 m diameter electric weeder head, as there was not enough space to enter the intra row area between some trunks. Some of them were close to 0.5 m distance, causing the system to stay outside of the crop row. Using different sizes of weeder heads would solve this issue. At the spots with bigger gaps, the system performed well and did not cause any problems at all. When analyzing the tilled area of the outdoor test using the ImageJ software, the min value reached 57%, the max values 98% of the tilled area. Areas where the tiller head could not enter were excluded from the analysis.

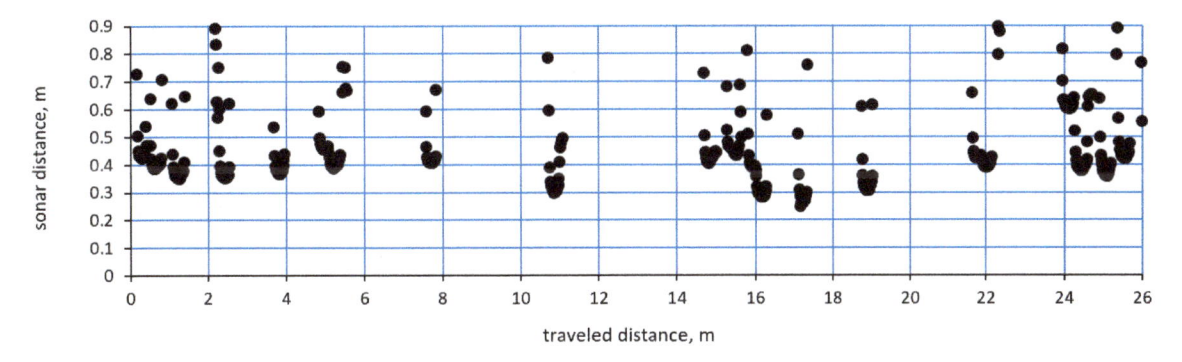

Figure 6. Sonar values detected at the outdoor test (vineyard).

It could be shown, that the outdoor test did perform well under the given circumstances. However, the environment was not sloped, which could bring more issues to the trunk detection. At high angles, shoots and leaves could hang more into the area of the sonar sensor. The automatic trunk detection would work in ordered and well cultivated vineyards. More complex examples would need an additional tree detection system like a laser scanner or camera, which could filter out the sonar values depending on the environment and context. Additionally, the trajectory of the implement could be planned dependent on the robot forward speed. This could make a more flexible path planning possible. However, the tilled area and the detection rate of the system were sufficient for the evaluated test area.

3.3. Power Requirement for the Autonomous Robot

The maximum and average power requirement for the implement guidance, actuator motor and belt drive of the autonomous robot is shown in the following Figure 7. In the indoor test, the power consumption increased with higher soil penetration resistance and shear stress. The overall power consumption for the implement was smaller than in the outdoor test. Here, an additional factor was the vegetation layer of weeds, which caused higher power consumption. Because of the implemented algorithm for limiting the working depth depending on the implement motor current, the power consumption peaks in the outdoor test were smaller than in the indoor test. When the soil in the indoor experiment would be compacted more, this difference between indoor and outdoor power consumption for the implement could be equalized.

In all tests the right drive motor of the robot needed more power than the left one. This was quite logical, as the implement was attached to the right side of the robot. The inertial forces of the tiller caused the robot to move to the right. To overcome this momentum, the row follower sent correction signals to the drive motors, forcing the right motor to consume more current. This effect could be seen at all performed tests. In the outdoor experiment, the slope of the tested area was not completely horizontal, causing higher power consumption for the drive motors. The overall power consumption of the machine was 1.8 KW in the soil bin and 2.2 kW in the outdoor field experiment. This would allow this machine with the given battery power to work around 8.1 h and 6.6 h depending on the slope and weed density. With a fixed speed evaluated in the experiments, the system could perform an intra-row weeding of 500 m/h.

When comparing the power consumption of the whole machine with the official power requirement provided by the rotary tiller manufacturer, the energy reduction is immense. The official power requirement for the tiller was set up with 8.1 kW by the company. Comparing the power consumption of the whole robot, the savings are more than 5.9 kW, meaning a 73% energy reduction. When we just compare the attachment, a 1 kW motor was enough to drive the implement in the performed experiments. When driving the robot system in a sloped environment, the power consumption of the drive motors would increase. However, it is expected, that the power consumption of the actuator would not differ in sloped environments.

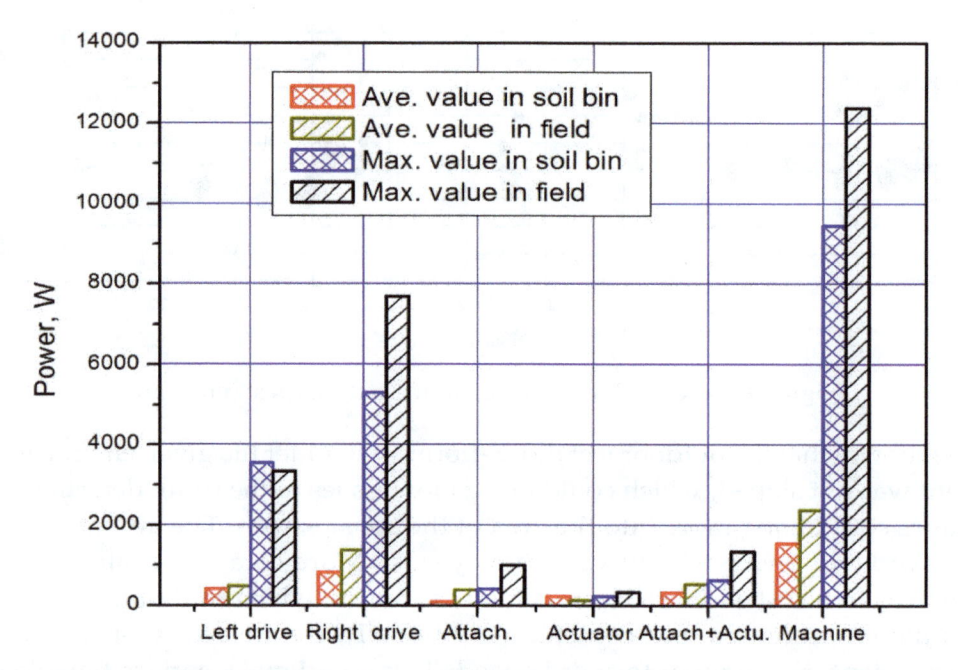

Figure 7. Overall power consumption of the robot for the weeding task.

3.4. Outlook

It could be shown that with an electrical implement a high power reduction could be possible, even when using commercial weeding tools. Combined with autonomous robots, this system could be highly energy efficient and sustainable. However, the system could be improved in the future. When using a higher speed for the linear side shift control, the overall speed of the machine could be increased, depending on the rotary speed of the rotary tiller and the side shift performance. Additionally, the tillage quality of the system could be increased, when the robot would stop at each trunk, to move the rotary weeder out of the row. Afterwards the robot could move a little bit forward and move the tool back into the row as soon as the trunk has passed. This would result in a rectangular trajectory, just excluding the trunk. Between the trunks, the speed could be increased until the next trunk is detected.

A different solution for detecting the trunks could be to use the information of the navigation laser scanner in front of the robot. Together with the odometry value, the software could increase the accuracy and plan the best trajectory for the implement based on speed, trunk spacing and weed density. This could help optimize path and work quality for weeding tasks in vineyards and orchards. In combination with a feeler or a sonar sensor this information could be filtered and optimized. This could help to increase the robustness and to not be disturbed from shoots and higher weeds.

4. Conclusions

An autonomous tiller weeder for intra-row weeding in vineyards was developed and tested at the University of Hohenheim. The system was built up using a commercial rotary tiller weeder. It was mounted on a mobile robot platform. For the side shift of the system, linear electrical motors were used. The trunk detection was performed with two different methods, using a feeler and a sonar. Both methods performed well and did not harm any trunks. The overall tilled area between the trunks with the sonar was higher than with the feeler, caused by vibrations and the design of the feeler. The autonomous row following based on the laser scanner enabled the machine to follow the rows accurately. The energy consumption of the whole system was evaluated. It could be shown that electrical driven machines could perform autonomous weeding in vineyards, which is energy efficient. Therefore, autonomous intra-row weeding machines could save energy and time for workers.

Author Contributions: The experiment planning was done by the first and second author. The programming of the software was done by the first author. The writing was performed by the first, the second and fourth author. The experiments were conducted by the first, second and third author. The first and second author analyzed and processed the data. The mechanical development of the frame and design was performed by the first and the third author. The whole experiment was supervised by the fourth and fifth author.

Acknowledgments: The project was conducted at the Department for Technology in Crop Production at the University of Hohenheim (Stuttgart, Germany). The companies Mädler and Linak supported the development of the robot with materials and actuators. The authors want to thank Nikolaus Merkt for his support in conducting the outdoor experiments in the university vineyard of Hohenheim

References

1. Peteinatos, G.G.; Weis, M.; Andújar, D.; Rueda Ayala, V.; Gerhards, R. Potential use of ground-based sensor technologies for weed detection. *Pest Manag. Sci.* **2014**, *70*, 190–199. [CrossRef] [PubMed]
2. Andújar, D.; Dorado, J.; Fernández-Quintanilla, C.; Ribeiro, A. An approach to the use of depth cameras for weed volume estimation. *Sensors* **2016**, *16*, 972. [CrossRef] [PubMed]
3. Oerke, E.-C. Crop losses to pests. *J. Agric. Sci.* **2006**, *144*, 31. [CrossRef]
4. Susaj, L.; Susaj, E.; Belegu, M.; Mustafa, S.; Dervishi, B.; Ferraj, B. Effects of different weed management practices on production and quality of wine grape cultivar Kallmet in North-Western Albania. *J. Food Agric. Environ.* **2013**, *11*, 379–382.
5. Bowman, G. *Steel in the Field—A Farmer's Guide to Weed Management Tools*; SARE Outreach: College Park, MD, USA, 1997; ISBN 188862602X.
6. Van Der Weide, R.Y.; Bleeker, P.O.; Achten, V.T.J.M.; Lotz, L.A.P.; Fogelberg, F.; Melander, B. Innovation in mechanical weed control in crop rows. *Weed Res.* **2008**, *48*, 215–224. [CrossRef]
7. Barawid, O.C.; Mizushima, A.; Ishii, K.; Noguchi, N. Development of an autonomous navigation system using a two-dimensional laser scanner in an orchard application. *Biosyst. Eng.* **2007**, *96*, 139–149. [CrossRef]
8. Reiser, D.; Paraforos, D.S.; Khan, M.T.; Griepentrog, H.W.; Vázquez-Arellano, M. Autonomous field navigation, data acquisition and node location in wireless sensor networks. *Precis. Agric.* **2017**, *18*, 279–292. [CrossRef]
9. Andújar, D.; Escolà, A.; Rosell-Polo, J.R.; Fernández-Quintanilla, C.; Dorado, J. Potential of a terrestrial LiDAR-based system to characterise weed vegetation in maize crops. *Comput. Electron. Agric.* **2013**, *92*, 11–15. [CrossRef]
10. Bajwa, A.A.; Mahajan, G.; Chauhan, B.S. Nonconventional weed management strategies for modern agriculture. *Weed Sci.* **2015**, *63*, 723–747. [CrossRef]
11. Gerhards, R.; Oebel, H. Practical experiences with a system of site-specific weed control in arable crops using real-time image analysis and GPS-controlled patch spraying. *Weed Res.* **2006**, *46*, 185–193. [CrossRef]
12. Weiss, U.; Biber, P.; Laible, S.; Bohlmann, K.; Zell, A. Plant species classification using a 3D LIDAR sensor and machine learning. *Proc. Int. Conf. Mach. Learn. Appl. ICMLA* **2010**, 339–345. [CrossRef]
13. Nørremark, M.; Griepentrog, H.W.; Nielsen, J.; Søgaard, H.T. The development and assessment of the accuracy of an autonomous GPS-based system for intra-row mechanical weed control in row crops. *Biosyst. Eng.* **2008**, *101*, 396–410. [CrossRef]
14. Reiser, D.; Martín-López, J.; Memic, E.; Vázquez-Arellano, M.; Brandner, S.; Griepentrog, H. 3D imaging with a sonar sensor and an automated 3-axes frame for selective spraying in controlled conditions. *J. Imaging* **2017**, *3*, 9. [CrossRef]
15. Tillett, N.D.; Hague, T.; Grundy, A.C.; Dedousis, A.P. Mechanical within-row weed control for transplanted crops using computer vision. *Biosyst. Eng.* **2008**, *99*, 171–178. [CrossRef]
16. Astrand, B.; Baerveldt, A.J. An agricultural mobile robot with vision-based perception for mechanical weed control. *Auton. Robots* **2002**, *13*, 21–35. [CrossRef]
17. Cloutier, D.C.; Van der Weide, R.Y.; Peruzzi, A.; Leblanc, M.L. Mechanical weed management. *Non Chem. Weed Manag.* **2007**, 111–134. [CrossRef]

18. Griepentrog, H.W.; Nørremark, M.; Nielsen, J. Autonomous intra-row rotor weeding based on GPS. In Proceedings of the 2006 CIGR World Congress Agricultural Engineering for a Better World, Bonn, Germany, 3–7 September 2006; pp. 2–6.
19. Lee, W.S.; Slaughter, D.C.; Giles, D.K. Robotic weed control system for tomatoes. *Precis. Agric.* **1999**, *1*, 95–113. [CrossRef]
20. Slaughter, D.C.; Giles, D.K.; Downey, D. Autonomous robotic weed control systems: A review. *Comput. Electron. Agric.* **2008**, *61*, 63–78. [CrossRef]
21. Ge, Z.; Wu, W.; Yu, Y.; Zhang, R. Design of mechanical arm for laser weeding robot. In Proceedings of the 2nd International Conference on Computer Science and Electronics Engineering (CSE 2013), Los Angeles, CA, USA, 1–2 July 2013; pp. 2340–2343.
22. Blackmore, B.S.; Griepentrog, H.W.; Fountas, S.; Gemtos, T.A. A specification for an autonomous crop production mechanization system. *Agric. Eng. Int. CIGR Ej.* **2007**, *IX*. [CrossRef]
23. Bechar, A.; Vigneault, C. Agricultural robots for field operations: Concepts and components. *Biosyst. Eng.* **2016**, *149*, 94–111. [CrossRef]
24. Anatis The Robotic in Support of the Environmental Friendly Agriculture! The Only Robot for Hoeing and Aid with Decision Making. Available online: http://www.trp.uk.com/trp/uploads/carrebrochures/023_Anatis%20-%20GB.pdf (accessed on 9 January 2019).
25. Zhang, C.; Li, N. *System Integration Design of Intra-Row Weeding Robot*; ASABE: Saint Joseph, MI, USA, 2013.
26. Contente, O.M.D.S.; Lau, J.N.P.N.; Morgado, J.F.M.; Santos, R.M.P.M. Dos vineyard skeletonization for autonomous robot navigation. In Proceedings of the 2015 IEEE International Conference on Autonomous Robot Systems and Competitions (ICARSC 2015), Villa Real, Portugal, 8–10 April 2015; pp. 50–55. [CrossRef]
27. Igawa, H.; Tanaka, T.; Kaneko, S.; Tada, T.; Suzuki, S. Visual and tactual recognition of trunk of grape for weeding robot in vineyards. In Proceedings of the 35th Annual Conference of the Industrial Electronics Society (IECON 2009), Porto, Portugal, 3–5 November 2009; pp. 4274–4279. [CrossRef]
28. Velicka, R.; Mockeviciene, R.; Marcinkeviciene, A.; Pupaliene, R.; Kriauciuniene, Z.; Butkeviciene, L.M.; Kosteckas, R.; Cekanauskas, S. The effect of non-chemical weed control on soil biological properties in a spring oilseed rape crop. *Zemdirb. Agric.* **2017**, *104*, 107–114. [CrossRef]
29. Sick AG Waldkirch Operating Instructions LMS1xx. Available online: https://mysick.com/saqqara/im0031331.pdf (accessed on 11 January 2017).
30. Vectornav VN-100 IMU/AHRS High-Performance Embedded Navigation. Available online: https://www.vectornav.com/docs/default-source/documentation/vn-100-documentation/PB-12-0002.pdf?sfvrsn=9f9fe6b9_16 (accessed on 1 December 2017).
31. Microsonic Datasheet pico+100/U. Available online: https://www.microsonic.de/DWD/_111327/pdf/1033/microsonic_pico+100_U.pdf (accessed on 9 January 2019).
32. Quigley, M.; Conley, K.; Gerkey, B.; Faust, J.; Foote, T.; Leibs, J.; Berger, E.; Wheeler, R.; Mg, A. ROS: An open-source robot operating system. In Proceedings of the 2009 ICRA Workshop on Open Source Software, Kobe, Japan, 12–17 May 2009; Volume III, p. 5.
33. Foote, T. Tf: The transform library. In Proceedings of the 2013 IEEE Conference On Tehnologies for Practical Robot Applications (TePRA), Woburn, MA, USA, 22–23 April 2013.
34. Fischler, M.A.; Bolles, R.C. Random sample consensus: A paradigm for model fitting with applications to image analysis and automated cartography. *Commun. ACM* **1981**, *24*, 381–395. [CrossRef]
35. Software, I. Open Source Software ImageJ. Available online: https://imagej.net/Welcome (accessed on 29 October 2018).

Monitoring of the Pesticide Droplet Deposition with a Novel Capacitance Sensor

Pei Wang [1,2,*,†], Wei Yu [2,†], Mingxiong Ou [1,2], Chen Gong [1,2] and Weidong Jia [1,2,*]

[1] Key Laboratory of Modern Agricultural Equipment and Technology, Ministry of Education of PRC, Jiangsu University, Zhenjiang 212300, China; myomx@ujs.edu.cn (M.O.); chengong@ujs.edu.cn (C.G.)
[2] Key Laboratory of Plant Protection Engineering, Ministry of Agriculture and Rural Affairs of PRC, Jiangsu University, Zhenjiang 212300, China; 2211616028@stmail.ujs.edu.cn
[*] Correspondence: wangpei@live.cn (P.W.); jiaweidong@ujs.edu.cn (W.J.).
[†] The first two authors Pei Wang and Wei Yu contributed same to this publication as the co-first authors. Their role orders in this publication should be regarded equally as the first authors.

Abstract: Rapid detection of spraying deposit can contribute to the precision application of plant protection products. In this study, a novel capacitor sensor system was implemented for measuring the spray deposit immediately after herbicide application. Herbicides with different formulations and nozzles in different mode types were included to test the impact on the capacitance of this system. The results showed that there was a linear relationship between the deposit mass and the digital voltage signals of the capacitance on the sensor surface with spray droplets. The linear models were similar for water and the spray mixtures with non-ionized herbicides usually in formulations of emulsifiable concentrates and suspension concentrates. However, the ionized herbicides in formulation of aqueous solutions presented a unique linear model. With this novel sensor, it is possible to monitor the deposit mass in real-time shortly after the pesticide application. This will contribute to the precision application of plant protection chemicals in the fields.

Keywords: capacitor sensor; deposit mass; pesticide droplets; formulations; ionization

1. Introduction

The usage of pesticides has increased significantly in the last two decades in China [1]. Meanwhile, its average application dose is 2.5 times of the global average level [2,3]. The over usage of pesticides will cause unnecessary invest for the farmers, leading to more residue of the pesticide ingredients in the food products and the soil, and damage the eco-system as well [4–6]. Furthermore, it can also induce the resistant property of the weeds, insects, and diseases, which makes the pest management strategies more complicated [7,8]. Thus, the effective approaches to monitor the droplet deposition doses during the pesticide application are in urgent demands.

To enhance the assessment efficiency of the deposition quality for the spray, several approaches have been introduced based on the image processing technology [9]. Various systems have been open for users to identify the pesticide deposition effect by analyzing the images of the water sensitive paper after the application, for instance, the Swath Kit [10], the USDA Image Analyzer [11], the Droplet Scan [12], the Deposit Scan [13], the Image J [14], and the Drop Vision-Ag [15]. Most of the new technologies are applied for measuring the coverage, droplet density, and or the droplet sizes. Few methodologies have been reported to evaluate the droplet deposition doses after pesticide application.

Conventionally, the droplet deposition dose is measured based on the elution procedure [16]. Pigments like the poinsettia and the methylene blue are usually selected to replace the pesticide for the spray preparation. After the application, plant leaves or the Petri dishes will be collected and the pigment ingredients will be eluted with deionized water from the leaf surfaces or the Petri dishes. Then the deposition dose can be calculated from the concentration of the eluent by the measurement with a spectrophotometer [17]. Salyani and Serdynski [18] reported a prototype sensor for the spray deposit monitoring based on the measurement of the electric capacitance of the parallel copper conductors with spray deposited between the copper gaps. Thus, it could indicate the mass information of the droplet deposit with the voltage signals, which enables the real-time and rapid measurements of the pesticide deposition doses. Recently, Zhang et al. [19] applied a similar system for droplet deposition evaluation of the aerial spraying. Meanwhile, a fringing capacitive sensor with similar principles was also applied for water content measurement in the field [20]. However, the prototype equipment was tested with the electrolyte solution with high conductivity, such as NaCl and NaOH, which is not a common feature of the typical spray mixture. As is known, besides the water solvent and the emulsion in water with the ingredients as ionic compound, most pesticide formulations with the organic ingredients are hardly ionized in the water mixture. Thus, further studies are required to verify the fitness and the measurement accuracy of this novel sensor before its application in the agricultural practice.

The objectives of this study are, firstly, to test the sensors application capability on the deposition dose measurement of spray mixtures with pesticide in different formulations; secondly, to evaluate the measuring accuracy of droplet deposition doses of spray mixture with several common pesticide formulations including the emulsifiable concentrates (EC), suspension concentrates (SC), and aqueous solutions (AS).

2. Materials and Methods

2.1. Implementation of the Droplet Deposit Sensing System

A leaf like capacitor (Yingtai Tech., Tianjin, China) was adopted for the implementation of the droplet deposit sensing system in this study. The sensor was designed based on a capacitor with 84 parallel coppers (Figure 1). The coppers were separated into two groups and connected respectively as two electrode plates of the capacitor. The whole structure of the capacitor was packaged (painted) with insulation material of ceramic. Lim et al. has presented that, the capacitance varies according to the dielectric constant of the media composition, the air or the spray, inside the gap of the electrodes [21]. The dielectric constant changes when the ratio of each component in the media composition varies. The dielectric constant can be calculated due to the equation

$$C_s = \frac{\varepsilon_a S_a + \varepsilon_s S_s}{d} \qquad (1)$$

where C_s is the capacitance of the capacitor with droplets depositing onside, ε_a is the dielectric constant of the air, ε_s is the dielectric constant of the spray mixture, S_a is the surface area of the electrode contacting to the air, S_s is the surface area of the electrode contacting to the spray droplet, and d is the distance between the copper electrodes.

When considering the total area of the electrodes (S), which should be the sum of S_a and S_S, the equation of the dielectric constant can be revised as

$$C_s = \frac{\varepsilon_a(S-S_s)+\varepsilon_s S_s}{d} = \frac{\varepsilon_a S}{d} + \frac{(\varepsilon_s-\varepsilon_a)S_s}{d} = C_0 + \frac{(\varepsilon_s-\varepsilon_a)S_s}{d} \qquad (2)$$

where, C_0 is a constant for each sensor.

Theoretically, the surface area of the electrode contacting to the spray droplet (S_S) can be calculated from the droplets' mass (m_s) and density (ρ_s). Then, the deposit mass of the spray can be calculated following the equations

$$m = \rho_s S_s d = \rho_s \frac{(C_s-C_0)d}{\varepsilon_s-\varepsilon_a} d = C_s \frac{\rho_s d^2}{\varepsilon_s-\varepsilon_a} - \frac{\rho_s C_0 d^2}{\varepsilon_s-\varepsilon_a} \qquad (3)$$

Thus, a linear model can be fitted to calculate the deposit mass of the spray according to the measurement of the capacitance of the capacitor with droplets depositing onside.

With the leaf like capacitor adopted in this study, it could provide the analog signals to indicate its capacitance. The sensor is implemented with an operational amplifier circuit as shown in Figure 1d. With a capacitor in fixed capacitance, the leaf like capacitance sensor will share different voltages when the dielectric constant inside it is changed. The capacitor was linked to an analog to digital converter (ADS1115, Texas Instruments, Dallas, TX, USA), which could convert the analog signal of the real time capacitance into the digital voltage signals. The digital signal was then processed by a 32-bit microcontroller (STM32F4 EXPLORER, ST Microelectronics, Geneva, Switzerland). The microcontroller could read the input digital signal and display the voltage information representing the analog signal on a thin film transistor-liquid crystal display (TFT screen). Linear models of deposit mass to the voltage signals were developed due to different herbicide spray in this research and installed in the microcontroller. Thus, the microcontroller could calculate the deposit results of each measurement and then display that on the TFT screen. All the measurement results were recorded with a trans-flash Card (SD card) which was mounted on the microcontroller processing board. Figure 1 presents the structure of the system.

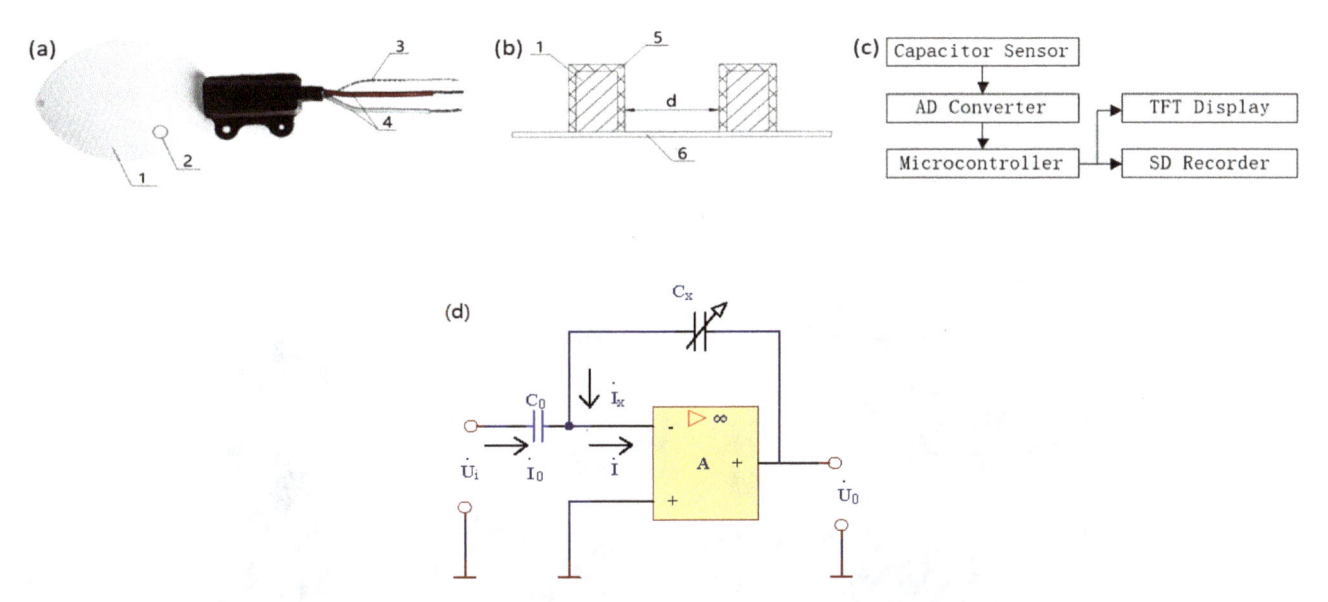

Figure 1. The structure of the leaf-like capacitor sensor and the deposit monitoring system. (**a**) Is the leaf-like sensor. (**b**) Is the electrode structure on the resin board. (**c**) Is the implementation diagram of the deposit monitoring system. (**d**) Is an electric schematic of the circuit of the leaf like sensor. 1 = electrodes, 2 = capacitor, 3 = data cable, 4 = power cable, 5 = insulating coating, 6 = resin board. The sensor is in 11.2 cm of length, 5.8 cm of width and 0.075 cm of thickness. The distance between each electrode is 1.58–1.78 mm. The width of the electrode is 0.59–0.79 mm and the thickness is 0.01–0.02 mm. The thickness of the insulating layer is about 2 μm. In (**d**), C_0 is a fixed capacitor, C_x is the leaf like sensor.

2.2. Experimental Setup and Processing

To develop the models of deposit mass to the voltage signals, experiments were conducted to measure the voltage signals of the leaf like sensor with different herbicide spray depositing onside. Herbicides in different formulation types were selected for the spray preparation with the suggested concentration on the product labels including the Yudasheng® (SC, 20% a.i. atrazine, Xinnong Guotai, Beijing, China), Lvlilai Butachlor® (EC, 50% a.i. butachlor, Lvlilai, Suzhou, China), Ruidefeng Lanhuoyan® (AS, 41% a.i. glyphosate isopropylamine salt, Noposion, Shenzhen, China), as well as the water for control. The sprays were prepared according to the details listed in Table 1. A hanging orbit sprayer (Figure 2) was used for the herbicide application. The application pressure was set at 0.3 MPa, while the application height was set at 50 cm above the sensor as is common in field conditions. Standard flat fan nozzles with different sizes were selected for the spraying (Lechler® ST 110-01/015/02/03, Lechler GmbH, Metzingen, Germany). Before spraying onto the sensors, the medium droplet sizes were measured for each nozzle with a laser particle size analyzer (Winner® 318B, Winner Particle Instrument, Ji'nan, China). During the testing experiment of the sensor, the moving velocity of the nozzles was fixed for three repeated measurements of each treatment and then varied randomly to obtain different deposit on the sensor surface. For each application, the signals from the sensor were recorded for 40 times over three seconds after the application. Mass of the sensor was measured with an electrical scale (YP102N, Sop-Top, Shanghai, China) before and after application. The deposit mass was calculated from the two mass measurements.

Table 1. Preparation of the sprays of the experiment

No	Formulation	Ingredient	Product Amount	Solution/Water Volume (mL)
1		water		5000
2	SC	atrazine	75 mL	5000
3	EC	butachlor	50 g	5000
4	AS	glyphosate isopropylamine salt	225 mL	5000

Figure 2. The hanging orbit sprayer for herbicide application in the experiment.

2.3. Data Analysis

The data analysis was processed with R Studio 0.98.490 [22]. The linear model was employed to fit the relationship between the voltage signals and the deposit mass of the spray. Analysis of variance (ANOVA) was carried out for the evaluation of differences between each treatment. Data were tested for normal distribution using the Shapiro-Wilk test ($p > 0.05$). Equality for heterogeneity of variances was tested using Levene's test for each treatment ($p > 0.05$).

3. Results and Discussion

3.1. Measurement Impact Analysis

To evaluate the impact of application factors as the nozzle types (droplet sizes) and the formulation types of the herbicides on the sensor's measurement results, experiments were conducted. The measuring results were presented in Figures 3 and 4.

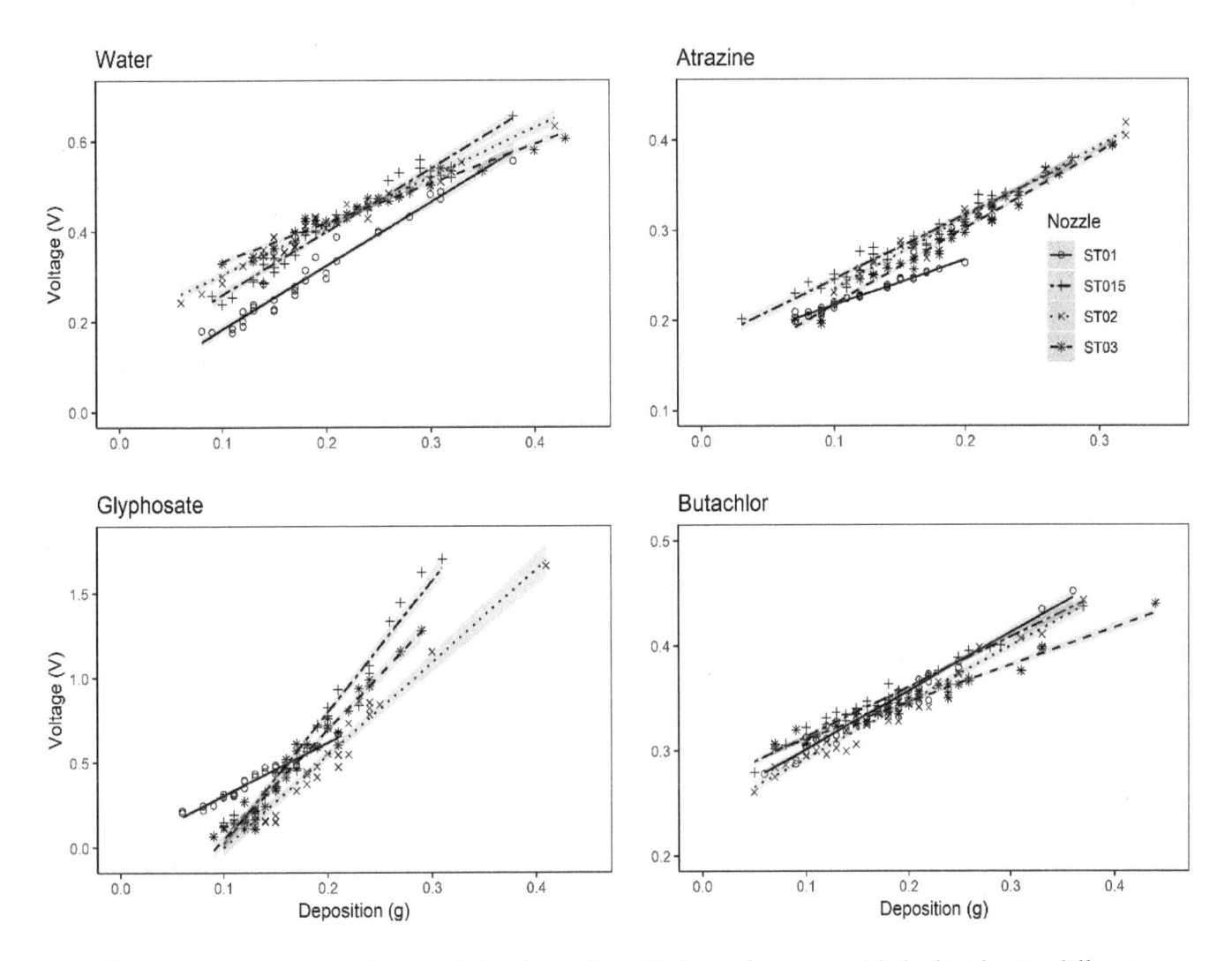

Figure 3. Linear regression models of the deposit data of sprays with herbicides in different formulations. The shadow area represents the standard error of the regressed linear models.

Linear models were employed for the regression of the signal voltage to deposit mass curves of each treatment. All the models were compared by ANOVA ($\alpha = 0.05$). It indicated that, when the nozzle type was fixed, there was no significant differences between the deposit to voltage curves of the spray mixture with EC (butachlor), SC (atrazine), and water. However, considering the AS (glyphosate), the voltage signal curve of the spray deposit was markedly different from the other groups. When the deposit is greater than 0.2 g on the leaf like sensor, the signal voltages were much higher of the sensor with AS droplets than the EC, SC, or water droplets. To explain this difference, the ionization property of the glyphosate isopropylamine salt should be concerned. The glyphosate is a weak acid herbicide. It usually exists as a salt compound and will divide into two ions with opposite charges (the cation and the anion) after being dissolved in the water, while the other herbicides with formulations like EC or SC usually contain organic ingredients which could not dissolve and ionize in the water. The concentration of ions would vary the dielectric constant of the spray mixture. As a result, the voltage signals of the electric capacity would be higher when the concentration of the glyphosate was increased.

In Figure 4, deposit curves of different nozzles were compared on each herbicide formulation. It presented that, for the spray mixture with same herbicide formulations, there was no significant differences of the deposit curves between the treatments with each nozzle type. Therefore, the spray mixture could be classified into two groups due to the ionization feature of the ingredients.

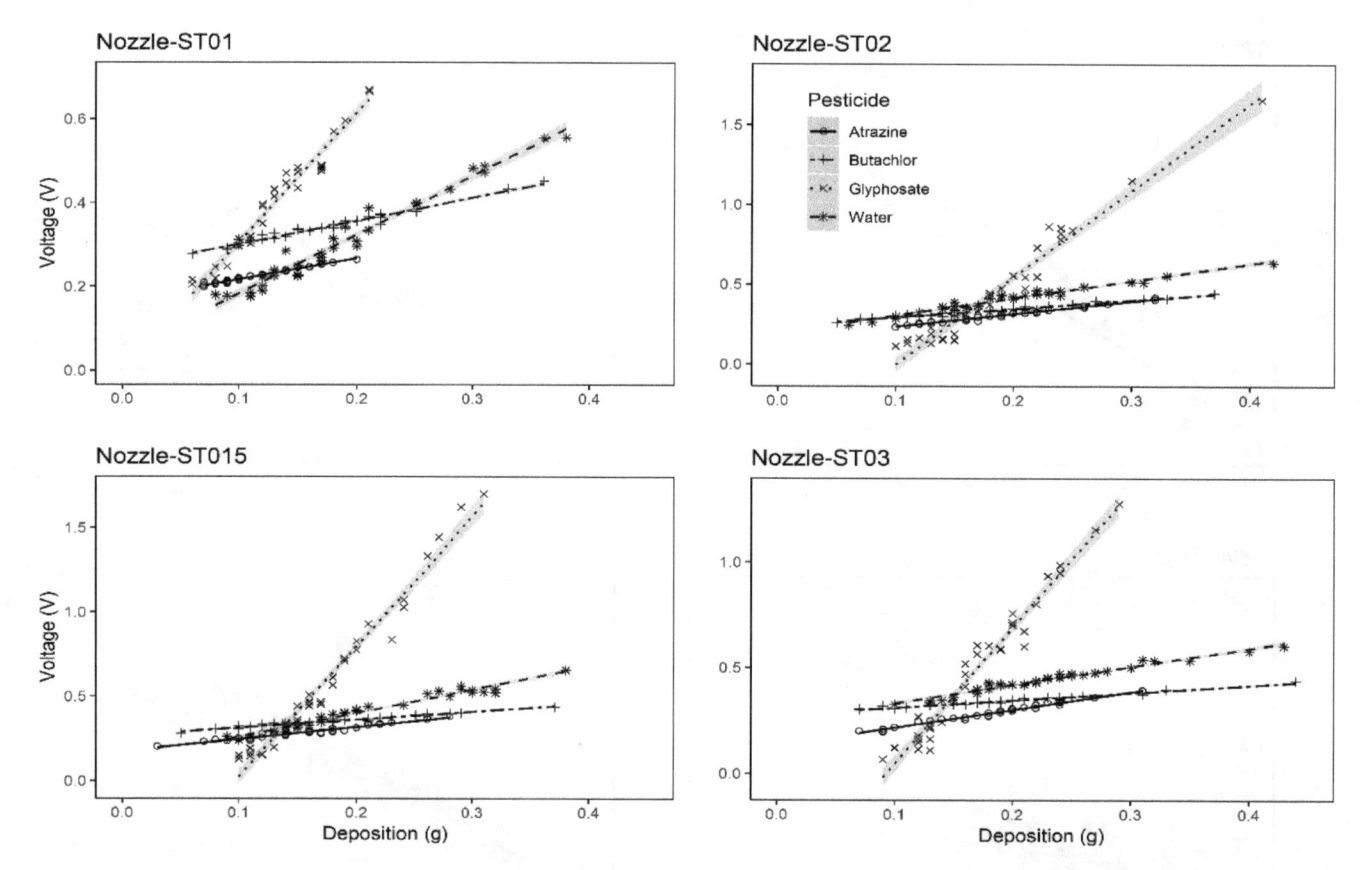

Figure 4. Linear regression models of the deposit data of sprays with droplet size spectrum generated from different nozzles. The shadow area represents the standard error of the regressed linear models.

3.2. Statistical Modeling

Considering the little differences of the deposit curves of water, SC and EC sprays, the data could be gathered for the modeling of the signal voltage to the deposit mass for all of the non-ionized herbicides, while the ionized herbicides should be tested for separated models. In this research, the glyphosate was taken as an example of the cases for ionized ingredients application. Since, the nozzle types did not show any impact on the signal voltages of the sensor measurement, the data of different herbicides but with the same nozzle type could be gathered for the modeling. Linear models were applied respectively for the regression of the two cases. The linear regression fitted well to the grouped data, which corresponding to the results of a former study by Zhang et al. [19].

With the linear model of the electric signals to the deposit of the non-ionized sprays, the algorithm could be programmed and installed in the microcontroller. Thus, the sensor was able to give precise information of the quantified spray deposit in the field. The dielectric constant of a dielectric material depends on the frequency of the applied electric field. The behavior of the dielectric constant is affected by three types of polarization including the orientation polarization, the ionic polarization and the electronic polarization. The ionized sprays may show all the three types of polarization, so that all polarization mechanisms are acting at lower frequencies till the water dipoles fail to follow the field alternations and so the orientation polar ceases to play, after which the polarization is dominated by ionic polarization. This continues until the ions' oscillations cannot follow the field alternation and it gets out the play at which the reaming will be the electronic polarization. When the frequency of the applied field equal to the resonance frequency of ionic movement, the dielectric constant show resonance behavior. The dielectric constant is proportional to density of the induced dipoles. Therefore, it is expected to increase with the salt concentration of the electrolyte [23]. Thus, the electric signal of ionized spray will be affected by the mount, concentration and other parameters of the electric charges in the mixture. Therefore, the model of one variable regression could not be used to simulate the application deposit with different ingredients. However, in this certain case of glyphosate application with the recommended doses on the product label, the linear regression model presented in Figure 5 could be adopted with the algorithm program similar to the one used for the deposit detection of the non-ionized sprays.

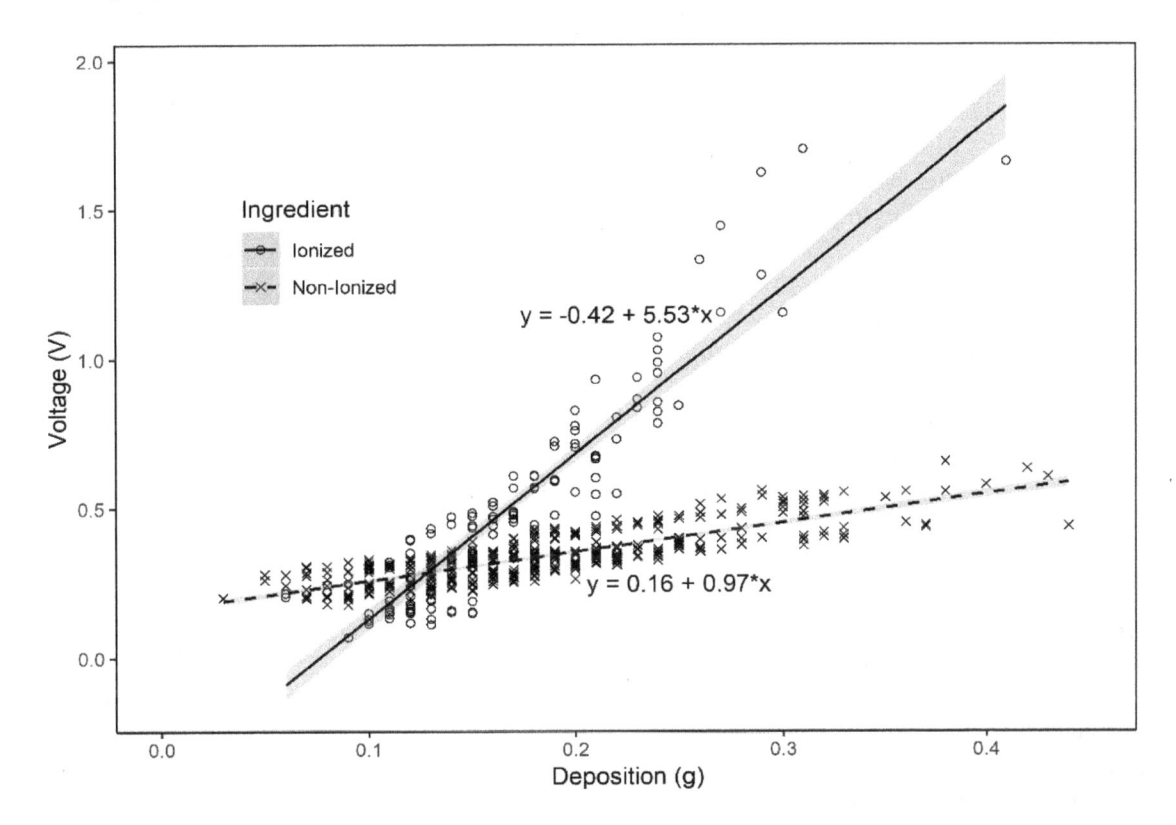

Figure 5. Linear regression models of the deposit data of sprays with ionized and non-ionized ingredients. The shadow area represents the standard error of the regressed linear models (ionized: $R^2 = 0.837$, non-ionized: $R^2 = 0.882$).

4. Conclusions

According to this study, we can conclude that the electric capacitor sensor has shown strong potential for its application on the real-time monitoring of the herbicide spraying deposit. The system and model implemented in this study can already be applied in the deposit measurement of the herbicide with non-ionized ingredients. For other water-soluble herbicides with ionized ingredients. However, further studies are required for the measuring model simulation, as the concentration and the valence of the ions will have impact on the dielectric constant of the spray mixture, which can directly affect the signal voltage representing the electric capacity of the sensor.

Author Contributions: Conceptualization, P.W.; Data curation, W.Y.; Formal analysis, P.W.; Funding acquisition, M.O. and W.J.; Investigation, W.Y.; Methodology, P.W., W.Y., M.O., and C.G.; Project administration, W.J.; Resources, C.G. and W.J.; Software, W.Y.; Supervision, P.W., M.O., and W.J.; Writing—original draft, P.W.; Writing—review & editing, P.W.

Acknowledgments: In addition, the authors would like to thank Huitao Zhou, Guangyang Mao, Shengnan Cao, Wanting Yang, and Shuai Zang for technical help in this study.

References

1. National Bureau of Statistics of China. 2018. Available online: http://data.stats.gov.cn/easyquery.htm?cn=C01 (accessed on 25 December 2018).

2. Zhang, C.; Shi, G.; Shen, J.; Hu, R. Productivity effect and overuse of pesticide in crop production in China. *J. Integr. Agric.* **2015**, *14*, 1903–1910. [CrossRef]
3. Zhang, Y.; McCarl, B.A.; Luan, Y.; Kleinwechter, U. Climate change effects on pesticide usage reduction efforts: A case study in China. *Mitig. Adapt. Strateg. Glob. Chang.* **2018**, *23*, 685–701. [CrossRef]
4. Bonny, S. Herbicide-tolerant transgenic soybean over 15 years of cultivation: Pesticide use, weed resistance, and some economic issues. The case of the USA. *Sustainability* **2011**, *3*, 1302–1322. [CrossRef]
5. Myers, J.P.; Antoniou, M.N.; Blumberg, B.; Carroll, L.; Colborn, T.; Everett, L.G.; Hansen, M.; Landrigan, P.J.; Lanphear, B.P.; Mesnage, R.; et al. Concerns over use of glyphosate-based herbicides and risks associated with exposures: A consensus statement. *Environ. Health* **2016**, *15*, 19. [CrossRef] [PubMed]
6. Budzinski, H.; Couderchet, M. Environmental and human health issues related to pesticides: From usage and environmental fate to impact. *Environ. Sci. Pollut. Res.* **2018**, *25*, 14277–14279. [CrossRef] [PubMed]
7. Denholm, I.; Rowland, M.W. Tactics for managing pesticide resistance in arthropods: Theory and practice. *Annu. Rev. Entomol.* **1992**, *37*, 91–112. [CrossRef] [PubMed]
8. Jutsum, A.R.; Heaney, S.P.; Perrin, B.M.; Wege, P.J. Pesticide resistance: Assessment of risk and the development and implementation of effective management strategies. *Pest Manag. Sci.* **1998**, *54*, 435–446. [CrossRef]
9. Ferguson, J.C.; Chechetto, R.G.; O'Donnell, C.C.; Fritz, B.K.; Hoffmann, W.C.; Coleman, C.E.; Chauhan, B.S.; Adkins, S.W.; Kruger, G.R.; Hewitt, A.J. Assessing a novel smartphone application–SnapCard, compared to five imaging systems to quantify droplet deposition on artificial collectors. *Comput. Electron. Agric.* **2016**, *128*, 193–198. [CrossRef]
10. Mierzejewski, K. *Aerial Spray Technology: Possibilities and Limitations for Control of Pear Thrips in Towards Understanding Thysanoptera General Technical Report NE-147*; USDA Forest Service: New England, USA, 1991.
11. Hoffmann, W.C.; Hewitt, A.J. Comparison of three imaging systems for water sensitive papers. *Appl. Eng. Agric.* **2005**, *21*, 961–964. [CrossRef]
12. Wolf, R.E. Assessing the ability of Droplet Scan to analyze spray droplets from a ground operated sprayer. *Appl. Eng. Agric.* **2003**, *19*, 525.
13. Zhu, H.; Salyani, M.; Fox, R.D. A portable scanning system for evaluation of spray deposit distribution. *Comput. Electron. Agric.* **2011**, *76*, 38–43. [CrossRef]
14. Rasband, W.S. *Image J*; US National Institutes of Health: Bethesda, MD, USA, 2014.
15. Leading Edge Associates. DropVision®AG. 2015. Available online: https://www.leateam.com/product/129/1144/ (accessed on 25 December 2018).
16. Hewitt, A.J. Droplet size and agricultural spraying, Part I: Atomization, spray transport, deposition, drift, and droplet size measurement techniques. *At. Spray* **1997**, *7*, 235–244. [CrossRef]
17. Hewitt, A.J. Tracer and collector systems for field deposition research. *Aspects Appl. Biol.* **2010**, *99*, 283–289.
18. Salyani, M.; Serdynski, J. Development of a sensor for spray deposition assessment. *Trans. ASAE* **1990**, *33*, 1464–1468. [CrossRef]
19. Zhang, R.; Chen, L.; Lan, Y.; Xu, G.; Kan, J.; Zhang, D. Development of a deposit sensing system for aerial spraying application. *Trans. CSAM* **2014**, *45*, 123–127.
20. Da Costa, E.F.; De Oliveira, N.E.; Morais, F.J.O.; Carvalhaes-Dias, P.; Duarte, L.F.C.; Cabot, A.; Siqueira Dias, J.A. A self-powered and autonomous fringing field capacitive sensor integrated into a micro sprinkler spinner to measure soil water content. *Sensors* **2017**, *17*, 575. [CrossRef] [PubMed]
21. Lim, L.G.; Pao, W.K.; Hamid, N.H.; Tang, T.B. Design of helical capacitance sensor for holdup measurement in two-phase stratified flow: A sinusoidal function approach. *Sensors* **2016**, *16*, 1032. [CrossRef] [PubMed]
22. R Development Core Team. *R: A Language and Environment for Statistical Computing*; R Foundation for Statistical Computing: Vienna, Austria, 2013; ISBN 3-900051-07-0.
23. Föll, H. Chapter 3: Dielectrics. Available online: https://www.tf.uni-kiel.de/matwis/amat/elmat_en/kap_3/backbone/r3_3_5.html (accessed on 16 January 2019).

Contribution to Trees Health Assessment using Infrared Thermography

Rui Pitarma *, João Crisóstomo and Maria Eduarda Ferreira

Research Unit for Inland Development, Polytechnic of Guarda, Avenida Francisco Sá Carneiro 50, 6300-559 Guarda, Portugal
* Correspondence: rpitarma@ipg.pt

Abstract: Trees are essential natural resources for ecosystem balance, regional development, and urban greening. Preserving trees has become a crucial challenge for society. It is common for the use of invasive or even destructive techniques for health diagnosis of these living structures, and interventions after visual inspection. Therefore, the dissemination and implementation of increasingly less aggressive techniques for inspection, analysis and monitoring techniques are essential. The latest high-definition thermal cameras record thermal images of high resolution and sensitivity. Infrared thermography (IRT) is a promising technique for the inspection of trees because the tissue of the sap is practically on the surface of the living structure. The thermograms allow the identification of deteriorated tissues and to differentiate them from healthy tissues, and make an observation of the tree as a functional whole body. The aim of this study is to present, based on differences in the temperatures field given by the thermal images, a qualitative analysis of the status of two different arboreal species, *Quercus pyrenaica* Willd and *Olea europaea* L. The results show the IRT as an expeditious, non-invasive and promising technique for tree inspection, providing results that are not possible to reach by other methods and much less by a visual inspection. The work represents a contribution to make IRT a tree decision-making tool on the health status of trees.

Keywords: trees inspection; trees monitoring; infrared thermography; IRT; VTA; sustainability

1. Introduction

The tree is an essential natural resource for ecosystem balance. It regulates nature, climate and urban greening. Trees play a key role in local biodiversity as they release oxygen and reduce global warming. Trees regulate climate by mitigating urban heat islands. Trees behave as barriers for noise pollution and wind. They provide moisture to the atmosphere, which favours precipitation. They facilitate water infiltration into the soil contributing to the formation and maintenance of groundwater aquifers. They are also essential for soil building as their roots fix the soil preventing erosion. All trees are of importance, whether by their age, type, size and shape. Some of them are classified as remarkable or even monumental trees, and the law protects them [1]. Therefore, it is fundamental to act for the preservation and sustainability of these living beings. This is a current social challenge [2–5]. It is decisive to move from the anthropocentric conception of trees to ecocentric conception, that is, all living beings including humankind are interrelated and interdependent to keep the ecosphere equilibrium.

Contrary to what people thought for many years, trees react to physical and biological damages that lead to deterioration. The compartmentalization of decay in trees (CODIT) is a good example of this [6]. Therefore, trees must be cared for in a way that enhances their own defence systems [6]. In order to verify the health status of trees especially when more detailed information is required, invasive and even destructive techniques [7] are used. Unfortunately, they interfere with the structural integrity of this living being. Therefore, it is urgent to disseminate and implement non-aggressive inspection techniques that preserve biological integrity and functionality [3,4]. The basic rule must always be to start with less invasive techniques and only if required, to utilise the most aggressive ones, so that the damage produced in the tree [4,5,8] is minimized.

Infrared thermography (IRT) is a non-invasive non-contact technique that relies on the detection of body heat emission [9–11]. It measures continuously surface temperature in real-time [9]. Aerial surveillance of the canopies to detect the distribution and spread of forest damage was the first use of IRT in trees [4]. Later, this technique was applied to tree bark (branches and trunks). Now, it allows the evaluation of some types of damage in the lower trunk and to estimate the presumable cause that affects roots (the root system) [4]. The latest thermal cameras capture thermal images of high resolution and sensitivity. It has been demonstrated that IRT in conjunction with visual inspection when applied to assessment, and monitoring of tree health provides reliable data [2–4,12]. The IRT capabilities for tree inspection have not been exploited sufficiently yet, since the elaborate sap conductive tissues are practically on the surface of the living structure. The reading of tree surface temperature reveals specific variations when there is deterioration and voids inside the tree. The IRT observes the tree as a functional whole body and thermograms provide information to differentiate deteriorated tissues from healthy tissues. Other methods do not have this assessment capacity, much less the naked eye inspection. Other non-invasive methods survey the body by points, and then extrapolation applies to have an idea of the functional whole body. Through the naked eye, only advanced deterioration is detected, then a corrective measure is hardly effective, and the solution is tree felling.

Thermography allows the early detection of damage, while it is still not visually noticeable. Even more, it monitors the progress of pathology. The IRT is expeditious and non-invasive [2–5,12]. It is, therefore, a powerful, fast and efficient tool to detect changes in the integrity of trees and branches, identifying if one of them should be removed. The differences in the thermal patterns of the tree surface indicate the deteriorated areas of the tree. The greater the differences in the thermal patterns of the trunk and branches, the worse the tree health condition [2–5].

The methods different from IRT that acquire data for diagnosis are time-consuming, and require more hand labour, especially if the part of the tree to be examined cannot be reached from the ground [2–4]. Some techniques, such as the resistograph, require perforation, and these holes may become pathways for pathogens [2–4]. Other methods use X-rays or γ. A short summary of the other main methods for detecting tree deterioration is presented in Table 1. The ionising radiation used in some methods conveys the perception that they are not safe for the health of living beings [3]. Moreover, when IRT is compared to more sophisticated techniques, such as X-ray and γ, as well as tomographic acoustic techniques such as Picus and ArborSonic 3D, or even nuclear magnetic resonance [7,13–15], IRT is the only one able to evaluate the health condition and functionality of tree tissues. That is, the assessment of the structural integrity to detect voids and deterioration inside the tree is possible because the tree is analysed as a functional whole body [4,5]. IRT analyses the tree as a whole, in a holistic way, while other techniques provide information only on specified points, and the whole is obtained by extrapolation after a series of investigations [2–4]. The authors raise concerns about avoiding errors when selecting the emissivity value for temperature reading calibration. However, more parameters are required for IRT quantitative readings. It is necessary to assess the value of reflected (or reflective) apparent temperature, which varies according to the angle between the camera and the tree surface, and the direction of radiation from the environment and sunrays. In some studies, the tree is cut into logs, and then the logs are perforated to simulate deterioration after the logs are sealed at both sides to maintain the water content. When observing the logs in thermograms, it is difficult to detect the holes and the larvae introduced inside them. Even with the same water content, the temperature of the logs does not behave like the tree because there is no flow of the sap, as this happens when the tree structure is alive [16,17]. Some studies analyse trees of the same species affected by several pathologies. The main aim was to look for similar temperature patterns on the surface all along the trees applying IRT [18]. The statistical analysis showed that there was no correlation [18].

Table 1. A brief summary of the other main methods for detecting tree deterioration (adapted from [19]).

Method	Principle	Key Highlights
Increment borer	Visual inspection	Measures growth rate, age and soundness. Invasive method (it may itself be a decay factor) and needs experience on decay causes.
Boroscope	Remote visual inspection	Visual analysis from inside. Same disadvantages of the "Increment borer".
Resistograph	Penetration resistance	Fast, easy to execute and interpret. Does not detect early to intermediate decay stages; requires comparison with known patterns.
Shigometer	Electrical resistivity	Detects deterioration in early stages. Information is limited by the probe length.
Fractometer	Strength and stiffness	Small device and easy to carry. Portable compression meter has depth limited.
Stress wave velocity	Single-path acoustic wave velocity	Quickly performed; defines the location and extent of internal decay. Difficult to determine early stages of decay.
Electrical resistance	Electrical.	Effective to detect advanced decay stages.
Stress wave tomography	Multiple path acoustic wave velocity	Detects internal decay; accurately locates the anomalies; sensitive to early stages of decay. High cost and difficult to operate.
Electromagnetic tomography	Electromagnetic wave permittivity	Higher frequencies provide better resolution but penetration depth decreases. High cost and difficult to operate.
Nuclear magnetic resonance (NMR)	Magnetic properties	Non-ionising radiation; very detailed images that facilitate the analysis. High cost and difficult to operate.
Electronic nose	Odour	High levels of accuracy and reliability. Difficult to determine early stages of decay.
Gamma-ray computed Tomography	Gamma-ray transmissivity	Reliable and non-invasive. Ionizing radiation; high cost, and difficult to carry and operate.

Considering the above, IRT is a well-established technique in a range of fields, such as industrial and building maintenance. Despite its merits, it is still a relatively recent technique in assessing tree health [4,5] and remains residually implemented in agriculture. Due to the relatively scarce studies in this field [5], more research to ensure its potential and applicability on a large scale is required. Thus, it is crucial more studies focus on contributing to show and analyse the complexity of the technique applied to trees, the new and atypical aspects of the problem and respond to some knowledge gaps. Accordingly, the present paper intends to detail some relevant features related to the applicability of the technology, thus contributing to turning the IRT technique as a decision-making tool to assess the health status of trees. To accomplish this, two sample trees, one of species *Quercus pyrenaica* Willd and another of species *Olea europaea* L., are qualitatively analysed from their thermal images.

2. Materials and Methods

2.1. Background

The fundamental aim of thermographic surveying is the heat transfer process shown in Figure 1. When the IRT camera aims at the target, it receives radiation from the object itself, radiation reflected on the surface of the object coming from the emissions of neighbouring bodies, and radiation emitted by the atmosphere. In fact, the atmosphere results in interfering in some way with radiation that arrives at the camera [20]. The total power of the radiation arriving at the IRT (W_{tot}) camera is equal to the sum of the emission of the body ($\varepsilon\,\tau\,W_{obj}$), the emission reflected in the object coming from sources of the near ambient (($1-\varepsilon$) $\tau\,W_{refl}$), and the emission of the atmosphere itself (($1-\tau$) W_{atm}). That is:

$$W_{tot} = \varepsilon\tau W_{obj} + (1-\varepsilon)\tau W_{refl} + (1-\tau)W_{atm}, \qquad (1)$$

where:
- ε Emissivity
- τ Coefficient of atmosphere transmission
- W_{obj} Energy radiation emitted by the object
- W_{refl} Reflected energy from surrounding bodies
- W_{atm} Atmospheric energy

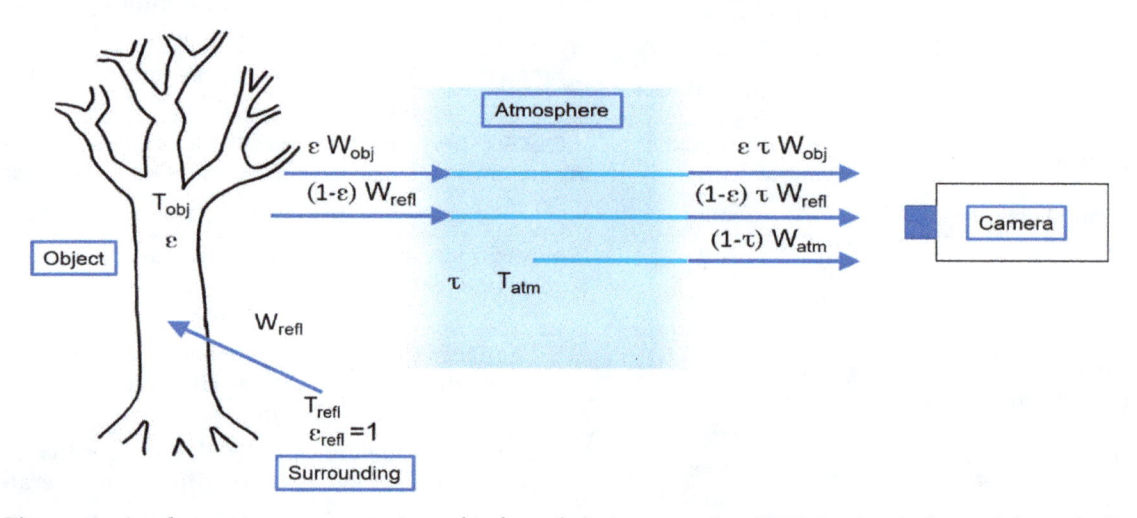

Figure 1. A schematic representation of infrared thermography (IRT) basics (adapted from [19]).

The thermal regime of a tree is determined by the heating of the surface by solar radiation and the transport, by conduction, of sensible heat to the interior. During the day, the surface heats up, generating a sensible heat flow towards the tree (a relatively slow process). The transportation of the elaborated sap promotes the convective transport of heat and ensures more tree thermal uniformity (a faster process of energy exchange). At night, the surface cooling by the emission of radiation (long waves) reverses the flow direction that changes from the interior to the surface leading to the cooling of the entire structure. The sap circulation causes the temperature gradient to vary along the trunk, and trees with more water available have better sap circulation [16]. This characteristic allows the differentiation between functional and dysfunctional tissue since the transport is done through the functional tissue. This feature allows health and vitality to be verified by IRT [16].

The density, the specific heat and the thermal conductivity play an important role in these processes. High specific heat leads to a smaller thermal stimulation for the same amount of heat. Thermal conductivity expresses the ability of the material to allow the heat to pass through it. It is commonly used in steady-state heat transfer analysis. Thermal diffusivity is much more relevant for transient heat transfer processes because it shows how well the heat could diffuse through the material. Thus, the diffusivity is a more important variable for the thermal characterization of a body than the conductivity (k), because it expresses how quickly the body adjusts completely to the temperature of its environment. In fact, the material thermal diffusivity (α) expresses how much easier the heat moves through its volume, that depends on the velocity conduction of the heat (k) and the amount of heat needed to increase temperature (ρC_p). It is given by the following equation:

$$\alpha = \frac{k}{\rho C_p}, \qquad (2)$$

where k is the thermal conductivity, ρ is the volumetric mass density, and C_p is the material specific heat. For instance, low diffusivity material delays the transfer of external temperature variations into the interior. Thus, it derives from the above, that damaged wood zones or unhealthy tissues have different thermal properties, leading to different heating and cooling times in relation to healthy zones. It is this principle that is explored in tree health analysis using IRT technology.

As mentioned above, wood (deteriorated/non-live tissues) and trees (functional tissues) have significantly different thermal characteristics, and as a consequence, different temperatures. This fact can be seen in Figure 2, that shows the difference in surface temperatures between a tree and a wooden stake supporting it [21]. The analysed tree was a specimen *Prunus domestica* L. (common name plum-tree) and the thermogram was obtained in the summer (air temperature 22.5 °C, relative humidity 55%), 4 h after the sunset, with an emissivity of 0.97 and a rainbow colour pallet. The wooden stake displayed a lower temperature as compared to the tree (except its lower part). The lower part shows slightly higher temperature values than the lower trunk because the tree was irrigated before the observation. The tree keeps a balanced relationship with the environment temperature, so the tree temperatures usually are lower than the atmospheric temperature when the sun heating effect is over [5]. Therefore, in Figure 2, the tree shows higher temperatures in the healthier parts and lower temperatures in the deteriorated parts, as well as in parts where the tree was recently wet with water [22]. The greater the differences in the thermal pattern of trunk and branches, the worse the condition of the tree health [2–5].

Figure 2. A tree (*Prunus domestica* L.) and a wooden stake supporting it: (**a**) photograph; (**b**) IRT thermogram (passive mode). The temperature values (°C): Sp1 = 20.0; Sp2 = 20.5; Sp3 = 21.5; Sp4 = 20.5; Sp5 = 19.5. (Adapted from [21]).

Equipment

A thermohygrometer, FLIR MR 176 (Figure 3A) was used to measure the atmospheric temperature and relative humidity. The thermographic camera used in this study is the FLIR T1030sc (Figure 3B).

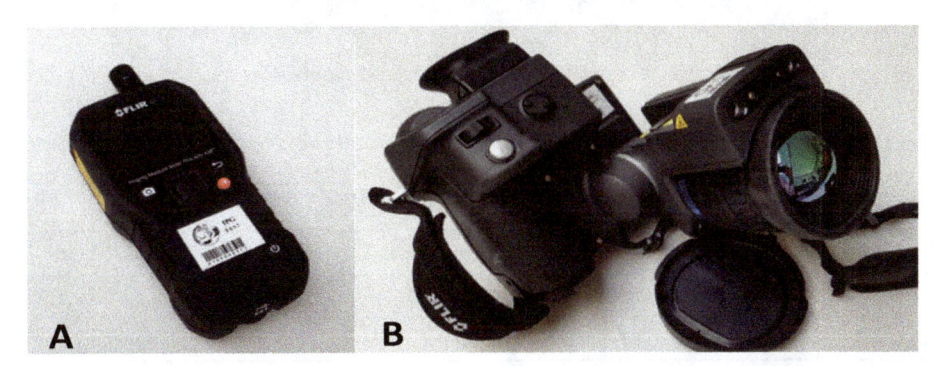

Figure 3. (**A**) Humidity meter FLIR MR 176; (**B**) Thermal camera FLIR T1030sc.

It is a HD imaging and measurement camera recording 1024 × 768 pixels. It has a focal plane array (FPA) detector type, uncooled microbolometer with a detector pitch of 17 μm. The T1030sc operates in a spectral range of 7.5–14 μm and offers a thermal sensitivity < 20 mK, at 30 °C (86 °F). The accuracy, at 25 °C, is +1 °C or +1%, for body temperatures from 5 °C to 150 °C. The camera was equipped with 28° lens, FOV 28° × 21° (36 mm) and number F 1.2. The thermal camera is equipped with a compass and GPS functions to locate the data on the thermogram [20]. The minimum focusing distance is 0.4 m for the standard 28° lens. The focus is manual, automatic or continuous. The continuous digital zoom from 1 to 8× can be used. It has a tactile screen of 800 × 480 pixels that facilitates the introduction of instructions. The software used to treat the thermograms were FLIR Tools + and FLIR ResearchIR Max 4 [23,24].

2.2. Sample Trees

Two trees were analysed: *Quercus pyrenaica* Willd (Figure 4), and *Olea europaea* L. (Figure 5). Table 2 shows their general characteristics. The common name of species *Quercus pyrenaica* Willd is Pyrenean oak. It is an autochthonous species of the Iberian Peninsula, from the mountain regions of continental climates [25]. This deciduous tree sprouts in acidic substrates preferably of granitic and schist origin. The immature subjects are marcescent with the capacity to sprout at the base of the trunk, as well as from the roots. The trunk is straight with opaque grey bark cracked in the form of panels. The specimen under study (Figure 4) is approximately 10 m in height. It consists of three trunks that arise from the base. The diameter at breast height (DBH) measured at 1.3 m in height, from south to north (south at the left of the photo) was: 0.20 m, 0.17 m and 0.17 m, respectively. The second tree is a perennial of the species *Olea europaea* L. It is native to the Mediterranean region (Southern Europe, North Africa and the Middle East) and the common name is olive (Figure 5). The olive tree is of exceptional longevity, being able to surpass 1500 years of age, and reaches a height of up to 8 m. It has greyish bark and branches. The specimen studied is approximately 3.5 m high. The branching begins at 1.75 m from the base, and the DBH measured at 1.30 m height is 0.35 m. Both specimens are located in the interior region of Portugal.

Figure 4. A photograph of the *Quercus pyrenaica* Willd tree taken at 2 p.m. under direct sunlight.

Table 2. Characteristics of sample trees.

Scientific Name	Common Name	Height (m)	HGB (m)	DBH (m)	Picture
Quercus pyrenaica Willd	Pyrenean oak	10.0	1.50	0.20/0.17/0.17	Figure 4
Olea europaea L.	Olive	3.5	1.75	0.35	Figure 5

HGB: Height from the ground (collar) to where branches begin (bifurcation); DBH: diameter at breast height.

Figure 5. A photograph of the *Olea europaea* L. tree taken at 5:30 p.m. under direct sunlight.

3. Methodology

The tree inspection was carried out applying qualitative IRT at the passive mode, that is, the heat source was the environment through solar radiation. The heat flows from the hotter to the colder zones of the tree. The irregularities of the thermal pattern on the tree surface can indicate the existence of defects, voids and deteriorated tissue. When it is found, deterioration as voids and defects occur in both the thermal properties of the constituents and the heat transfer process [2–5,7].

Thermograms were recorded at different times along the day, from when direct solar radiation was hitting the trees until after sunset. Besides the thermograms, photographs were taken to support the visual inspection and thermogram interpretation. The thermograms were processed running FLIR software [23,24]. The authors marked the spots at different locations over the tree trunk. The thermal patterns were analysed in order to correlate the temperature distribution with tree health. The atmospheric temperature and the relative humidity were measured at the time of the recording pf each thermogram, as well as the observation distance (between the thermal camera and the tree).

From a qualitative approach, although important, the value of emissivity is not preponderant. In fact, what is at issue is not the exact value of the measured temperature, but the temperature differences among the spots. The temperature values are comparable because the camera measured them with the same emissivity and reflected temperature.

Nevertheless, the emissivity value introduced was the one appropriate to the surface of the element under study. As the wood and bark are little reflective materials, the emissivity is high. Then, a high emissivity value was set in the camera. Thus, for both specimens, the emissivity value was 0.95. For each thermogram, the atmospheric temperature at the time of the capture was set in the camera. As it was a qualitative analysis, this study did not consider the reflected temperature. Further, it did not rain more than one week before surveying, then the thermograms did not register noises due to the humidity factor.

From the qualitative analysis, a more quantitative approach was taken, in particular, as regards to the comparison of recorded temperature values. It is important to understand thermal pattern differences among the trunks of similar diameter. The thermal patterns resulted from differences in the temperature along the surface. Naturally, this process introduces systematic errors that affect all points equally in each thermogram, whereby do not affect the result.

4. Results and Discussion

Figure 6 shows the thermogram of the *Quercus pyrenaica* Willd tree recorded from the same point of view and at the same time as the photo shown in Figure 4 (on 21 January 2018, 2 p.m., winter in the northern hemisphere). The southern part of the tree (left side of the thermogram) is exposed to sunlight. Figure 7 shows another thermogram of the same tree at the same day but it was taken at

9:45 p.m. (4 h after sunset) from the same point of view. Table 3 shows the observation conditions, and the parameters set in the camera to capture the thermograms. Table 4 contains the temperatures of the spots marked in the thermograms.

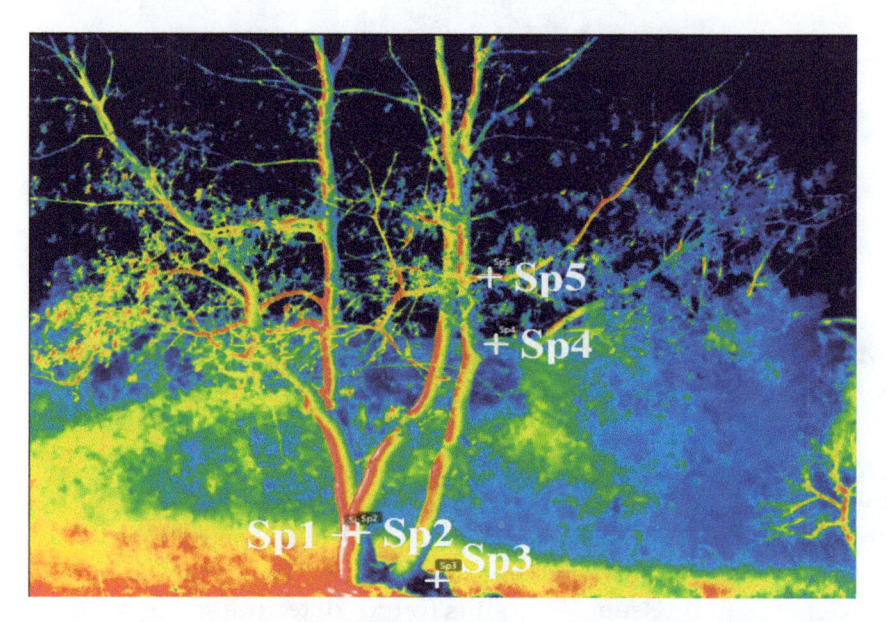

Figure 6. Thermogram *of Quercus pyrenaica* Willd tree taken at 2 p.m. under direct sunlight. (Spot's size out of scale for better visualization).

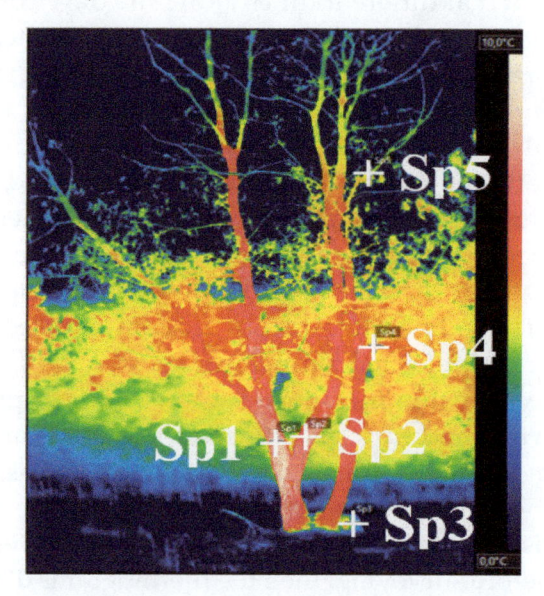

Figure 7. Thermogram of the *Quercus pyrenaica* Willd tree taken at 9:45 p.m., 4 h after sunset. (Spot's size out of scale for better visualization).

Table 3. The ambient conditions and parameters for the thermograms of *Quercus pyrenaica* (Figures 6 and 7).

Thermogram Image	Tree (Specimen)	Air Temperature (°C)	Relative Humidity (%)	Emissivity	Reflected Temperature (°C)	Observation Distance (m)	Colour Palette	Temperature Range (°C)	Daytime
Figure 6	Quercus pyrenaica	20	50	0.95	20	5	Rainbow	10–35	sunlight
Figure 7	Quercus pyrenaica	10	70	0.95	10	5	Rainbow	0–10	night

Contribution to Trees Health Assessment using Infrared Thermography 67

Table 4. The temperature (°C) of spots on the *Quercus pyrenaica* thermograms (Figures 6 and 7).

Thermogram Image	Sp1	Sp2	Sp3	Sp4	Sp5
Figure 6	32.5	18.0	12.0	15.0	23.5
Figure 7	8.5	8.0	3.0	6.0	4.5

Figure 6 shows areas under direct sun exposure (left zone of the tree) and areas in the shadow (right of the tree). On the left of the trunk is Sp1 under direct sun exposure at 32.5 °C, while Sp2 is in the shadow at 18 °C. The Sp3, on the bottom right, with 12 °C, point out an old cut trunk. Sp4 is in the shadow at 15 °C on a branch of larger diameter than the branch where Sp5 is located. Sp5 is under direct sun exposure at 23.5 °C.

Figure 7 shows the thermogram of the same tree captured 4 h after sunset. It can be seen that Sp1 is on the left trunk at 8.5 °C and Sp2 is on the central trunk at 8 °C. Sp3 is at 3 °C on the stump. Sp4 at 6 °C is located on a branch of larger diameter than the branch of Sp5. Sp5 is at 4.5 °C. The observation conditions and the parameters assumed to capture the thermogram are in Table 3. Table 4 shows the spot temperatures in the thermogram.

Figure 8 shows a photograph (A) and a thermogram (B) of the tree *Quercus pyrenaica* Willd taken at the same time. The capture angle (west) is different from previous images. A red arrow (B) shows a stump that remained from an old cut trunk identified as Sp3 in the thermograms of Figures 6 and 7.

Figure 8. Photo (**A**) and thermogram (**B**) of the *Quercus pyrenaica* Willd tree. The white arrow shows an enlarged image of the CODIT development wall; the green arrow shows the same wall of CODIT on the thermogram; the red arrow indicates a stump. CODIT: compartmentalization of decay in trees.

Figure 9 shows the thermogram of the tree of species *Olea europaea* L. described in Figure 5 and Table 2. Table 5 shows the environmental conditions during the observation, and the parameters assumed. The spot temperatures in Figure 9 are detailed in Table 6. The thermogram was recorded after sunset on a winter day (on 21 January 2018, at 5:50 p.m.; sunset at 5:35 p.m.). The atmospheric temperature was 18.0 °C and the relative humidity was 50%. The atmospheric temperature was 18.0 °C and the relative humidity was 50%. The thermogram was taken a distance of 5 m, using an emissivity of 0.95. A rainbow colour palette and a temperature range of 5.0 °C to 20.0 °C were used.

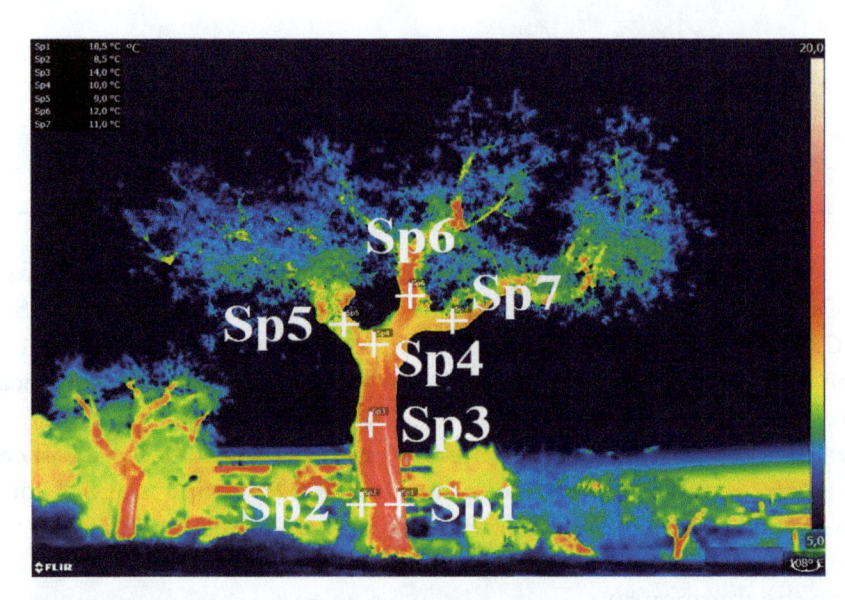

Figure 9. Thermogram of the *Olea europaea* L. tree recorded in passive mode at 5:50 p.m., 15 min after sunset. (Spot's size out of scale for better visualization).

Table 5. The ambient conditions and parameters assumed for recording thermogram of *Olea europaea* L.

Thermogram Image	Tree (Specimen)	Air Temperature (°C)	Relative Humidity (%)	Emissivity	Reflected Temperature (°C)	Observation Distance (m)	Colour Palette	Temperature Range (°C)	Daytime
Figure 9	*Olea europaea* L.	18	50	0.95	18	5	Rainbow	5–20	after sunset

Table 6. The temperature (°C) of the spots on the *Olea europaea* L. thermogram (Figure 9).

Thermogram Image	Sp1	Sp2	Sp3	Sp4	Sp5	Sp6	Sp7
Figure 9	18.5	8.5	14.0	10.0	9.0	12.0	11.0

The *Olea europaea* L. tree shows a large crack identified as Sp1 in the thermogram of Figure 9. Sp1 temperature is slightly higher than the atmospheric temperature. This spot represents a zone of the tree that has been exposed to the sun for a longer time. Sp2 temperature is much lower than Sp1 temperature—the difference between them is 10 °C. The differences in the temperature at the same diameter trunk is an indicator of possible deterioration. Then, it is possible that the Sp2 zone has deteriorated. However, this zone was in the shadow for a longer time. In addition, Sp1 is in a crack area. Thus, the comparison of these two spots is not conclusive, but it reveals a clue of possible deterioration to take into account.

In the same Figure 9, Sp2 is at a much lower temperature than Sp3. These spots are comparable because they are located on trunks of similar calibre. On the other hand, Sp2 was in the shadow for some time. Thus, the comparison is not conclusive, but it reveals a new clue to take into account for possible deterioration. Then, this study analysed the temperature of Sp2 in relation to the atmospheric temperature. The Sp2 temperature of 8.5 °C is much lower than the atmospheric temperature of 18 °C. This condition strengthens the possibility of deterioration at Sp2. Note that Sp3 is in the middle of the main trunk receiving sunlight. Although it already lost temperature, it was at 14 °C.

Sp4 in Figure 9 is at 10 °C, even exposed to direct sunlight, and it is on the same trunk as Sp3, which is at 14 °C. The two of them are located in trunks of the same diameter. The difference of 4 °C between the two points is a strong indicator of deterioration in Sp4. When comparing the temperature of Sp4 with the atmospheric temperature, the difference is 8 °C, which is relevant since Sp4 was exposed to direct solar radiation until just before the thermogram was recorded. This is another reason that indicates deterioration. Sp4 is approximately at the opening of an orifice. Then, it was expected the temperature to be higher than the registered temperature, and still higher than the values of the boundary zones, as is the case at the slot of Sp1. A probable cause for unhealthy conditions in that area is the hole at Sp4 that stores water from precipitation.

Sp5, Sp6 and Sp7 (Figure 9) are on branches of approximately the same diameter. They are at lower temperatures than the main trunk, because they are of smaller calibre. The trunks of smaller calibre heat and cool faster than larger ones. Smaller trunks have a larger surface per volume unit, so they heat and cool faster. The thermal inertia of the trunks of larger diameter is greater because they have more mass. Thus, from the point of view of health, Sp5, Sp6 and Sp7 should be at approximately the same temperature if subjected to the same solar exposure. Sp7 should have a higher temperature since it received directly more sunlight. This did not happen. Sp6 temperature is 1 °C higher than Sp7, which is not significant. The most disturbing is Sp5, because it is at 9 ° C. This value is too low for a healthy area even if exposed to less time in the sun. Sp5 is at a much lower temperature than the atmospheric temperature, which is 18 °C. This is a strong indicator of possible deterioration.

Sp2, Sp4 and Sp5 conditions (Figure 9) strongly indicate possible deterioration. This strong evidence was confirmed with observations of an enormous interior cavity linked to the crack (Figure 10).

Figure 10. Felling the tree has shown the presence and size of the concealed damage (red arrow) detected by the IRT method.

For doubtful cases, more thermography analysis would be recommended, e.g., at night, and if doubt remained, these spots become flags for the application of other diagnostic techniques and methods. Note that, these other diagnostic techniques do not need to be applied to the whole tree, but only in the identified points.

Epiphytic vegetation, such as lichens, is distributed more or less homogeneously throughout the tree and should not be responsible for the differences in the thermal pattern presented.

In IRT, as in all existing techniques, there are several conditions to take into account in order to achieve reliable results on tree analysis. That is: Exposure to sunlight and shadow; thermal contrast between the environment and the object targeted; the absence of water like rainfall; vegetation covers such as mosses and lichens; typical bark patterns of each species; the thermal comparison between the trunks of different calibre within the same tree [2–4,9].

Thermographic observations to trees are often made against the sun, or to tree surfaces when they are under the incidence of solar radiation. Thermographic recordings of trees when the sun is in front of the camera and sunlight directly hitting the tree surface introduces recording noise. In these cases, the temperature differences are more influenced by reading errors (lens, light exposure and reflections) than by material properties and defects. The shaded or less-illuminated areas (lower direct solar exposure) exhibit lower temperature values than expected, suggesting that the tree has deteriorated [2–4]. The IRT application requires a strong thermal contrast between the object to observe the environment and the objects that surround it. There must be a significant difference between the radiative power of the environment and the object analysed [9].

The thermograms and photographs presented are illustrative of relevant aspects to be taken into account in the observation and thermographic analysis of tree salubrity. It can be seen that on the thermogram of Figure 6, Sp1 and Sp5 are exposed to direct sunlight, and they are at a higher temperature than the atmospheric temperature, whereas the spots in the shadow are at a lower temperature than the atmospheric temperature. On the thermogram of Figure 7, all temperatures are lower than the atmospheric temperature. The colour patterns of thermograms of Figures 6 and 7 appear different. The various pattern colours correspond to different temperatures in the thermogram as shown in Table 4. The temperature differences by themselves do not indicate deterioration, as they can result from sunlight exposure and shadows. Figure 7 shows a general homogeneous temperature distribution, in which the different shades of Sp3 in Figure 7 is on a tree stump. Even its diameter is larger than the other trunks and it is at the lowest temperature and considered lifeless, and therefore not functional. It is probably because it does not circulate sap and its water content is much lower than its surroundings. In the upper part of the tree, the branches are of smaller diameter, as in Sp5. The branches heat and cool faster. Therefore, at 2 p.m. (under the action of sun exposure) Sp3 is at a higher temperature than Sp4, and at 9:45 p.m. (night), the opposite is registered.

Under closer inspection, there are subtle differences of the temperature in the thermograms, which result from the bark pattern, pruning wounds, and epiphytic vegetation such as mosses and lichens. They appear as spots of slightly lower temperature than the adjacent ones. In any of these cases, these small thermal contrasts are not deterioration signals. Another case that leads to some doubts is the self-defence process (CODIT) that trees develop.

It is important to point out that thermographic inspection requires photographs because they are visual inspection tools (VTA) that facilitate the interpretation of thermograms. By improving the interpretation of the results, it is often possible to avoid invasive methods [8].

Finally, it is relevant to highlight the main merits and weaknesses of the IRT technique. IRT has an enormous capacity to analyse the trees as a whole and differentiate functional tissue from dysfunctional tissue. This is crucial for the inspection of the vitality and health status, representing a fast, economical, nondestructive and environmentally friendly monitoring tool. However, as in other non-invasive methods, the main limitation is that IRT does not identify specifically the damage detected, i.e., does not identify the pathology, nor its causative agent. Nor can it also give precise indications of the magnitude of the damage. More studies are needed to optimise the technology and training, in order to make the system even more efficient and reliable.

5. Conclusions

In this study, the general principles of the IRT methodology for health status of trees was presented. A qualitative IRT approach of two sample trees, namely species *Quercus pyrenaica* Willd and *Olea europaea* L., was used. Several details were highlighted to describe the IRT analysis applied to trees. The results show the IRT is a non-invasive, sustainable and expedited technique with high potential for tree inspection. It allows for the early diagnosis of damage, even those that are not yet visually noticeable, which is relevant to advanced tree maintenance. As in any other technique, its correct application requires a deep and multidisciplinary knowledge of the phenomena and a high familiarisation with the technique. The study intends, therefore, that thermography and other related methodologies for tree diagnosis result in interventions that privilege sustainability to benefit the economy and nature.

6. Recommendations

IRT is a well-established technique in many fields. However, it is relatively new in agriculture where it remains residually implemented. Most farmers are ready to accept technology if it is profitable, less complex and makes their life easier [26]. For this to happen, more detailed studies to establish a solid application basis, which can guide practitioners needs to be created. In fact, reliable guidelines are not available to describe the acceptable protocol and parameters tailored to adopt for a more straightforward approach for many tree health problems. Several challenges need to be addressed: The complexity of the technique; the new and atypical aspects of the problem; several knowledge gaps related to the technical issues and applicability specificities. Thus, more research is required to ensure its potential and applicability on a large scale. The latest high resolution and sensitivity thermal cameras can also contribute to overcoming these challenges.

Author Contributions: R.P. and J.C. conceptualize the study, developed the methodology and performed the analysis. R.P. wrote the paper. M.E.F. supervised and revised the Biology details. The investigation supervision and project administration are made by R.P.

References

1. ICNF. Instituto da Conservação da Natureza e das Florestas, Árvores Monumentais de Portugal Árvores Monumentais de Portugal. Available online: http://www2.icnf.pt/portal/florestas/aip/aip-monum-pt (accessed on 14 March 2019).
2. Catena, G. A new application of thermography. *Atti della Fondazione Giorgio Ronchi* **1990**, 947–952.
3. Catena, A. Thermography Reveals Hidden Tree Decay. *Arboric. J.* **2003**, *27*, 27–42. [CrossRef]
4. Catena, A.; Catena, G. Overview of Thermal Imaging For Tree Assessment. *Arboric. J.* **2008**, *30*, 259–270. [CrossRef]
5. Bellett-Travers, M.; Morris, S. The Relationship between Surface Temperature and Radial Wood Thickness of Twelve Trees Harvested in Nottinghamshire. *Arboric. J.* **2010**, *33*, 15–26. [CrossRef]
6. Shigo, A.L.; Marx, H.G. *Compartmentalization of Decay in Trees*; USDA Forest Service: Washington, DC, USA, 1977.
7. Goh, C.L.; Abdul Rahim, R.; Fazalul Rahiman, M.H.; Mohamad Talib, M.T.; Tee, Z.C. Sensing wood decay in standing trees: A review. *Sens. Actuators A Phys.* **2018**, *269*, 276–282. [CrossRef]
8. Mattheck, C.; Breloer, H. Field guide for visual tree assessment (Vta). *Arboric. J.* **1994**, *18*, 1–23. [CrossRef]
9. Holst, G.C. *Common Sense Approach to Thermal Imaging*; JCD Publishing: Winter Park, FL, USA; SPIE Optical Engineering Press: Bellingham, WA, USA, 2000; ISBN 978-0-9640000-7-0.
10. Pitarma, R.; Crisóstomo, J.; Jorge, L. Analysis of Materials Emissivity Based on Image Software. In *New Advances in Information Systems and Technologies*; Rocha, Á., Correia, A.M., Adeli, H., Reis, L.P., Mendonça Teixeira, M., Eds.; Springer International Publishing: Cham, Switzerland, 2016; Volume 444, pp. 749–757. ISBN 978-3-319-31231-6.
11. Crisóstomo, J.; Pitarma, R. The Importance of Emissivity on Monitoring and Conservation of Wooden Structures Using Infrared Thermography. In *Advances in Structural Health Monitoring [Working Title]*; Hassan, M., Ed.; IntechOpen: London, UK, 2019.
12. FLIR. Assessing Tree Health with Infrared, Application Story. Available online: http://www.flirmedia.com/MMC/THG/Brochures/T559239/T559239_EN.pdf (accessed on 28 April 2019).
13. Leong, E.-C.; Burcham, D.C.; Fong, Y.-K. A purposeful classification of tree decay detection tools. *Arboric. J.* **2012**, *34*, 91–115. [CrossRef]
14. Johnstone, D.; Moore, G.; Tausz, M.; Nicolas, M. The measurement of wood decay in landscape trees. *Arboric. Urban For.* **2010**, *36*, 121–127.
15. Rudnicki, M.; Wang, X.; Ross, R.J.; Allison, R.; Perzynski, K. *Measuring Wood Quality in Standing Trees: A Review*; U.S. Department of Agriculture, Forest Service, Forest Products Laboratory: Madison, WI, USA, 2017; p. 13.
16. Burcham, D.C.; Leong, E.-C.; Fong, Y.-K. Passive infrared camera measurements demonstrate modest effect of mechanically induced internal voids on Dracaena fragrans stem temperature. *Urban For. Urban Green.* **2012**, *11*, 169–178. [CrossRef]
17. Hoffmann, N.; Schröder, T.; Schlüter, F.; Meinlschmidt, P. Potenzial von Infrarotthermographie zur Detektion von Insektenstadien und -schäden in Jungbäumen. *J. für Kulturpflanzen* **2013**, *65*, 2013.
18. Burcham, D.C.; Leong, E.-C.; Fong, Y.-K.; Tan, P.Y. An Evaluation of Internal Defects and Their Effect on Trunk Surface Temperature in *Casuarina equisetifolia* L. (Casuarinaceae). *Arboric. Urban For.* **2012**, *38*, 277–286.
19. Vidal, D.; Pitarma, R. Infrared Thermography Applied to Tree Health Assessment: A Review. *Agriculture* **2019**, *9*, 156. [CrossRef]
20. FLIR. *FLIR Systems T10xx Series User's Guide*; FLIR Systems, Inc.: Wilsonville, OR, USA, 2016.

21. Ferreira, M.; Crisóstomo, J.; Pitarma, R. Infrared Thermography Technology to Support Science Teaching—Meaningful Learning about Trees with University Students. In Proceedings of the 13th International Technology, Education and Development Conference (INTED2019), Valencia, Spain, 11–13 March 2019.
22. Meola, Carosena Origin and Theory of Infrared Thermography. In *Infrared Thermography Recent Advances and Future Trends*; Meola, C. (Ed.) Bentham Science Publishers: New York, NY, USA, 2012; pp. 3–28. ISBN 978-1-60805-143-4.
23. FLIR. *FLIR Tools+ User's Guide*; FLIR Systems, Inc.: Wilsonville, OR, USA, 2016.
24. FLIR. *FLIR ResearchIR 4 User's Guide*; FLIR Systems, Inc.: Wilsonville, OR, USA, 2016.
25. Humphries, C.J.; Press, J.R.; Sutton, D.A. *Árvores de Portugal e Europa*; Fapas: Porto, Portugal, 2005; ISBN 978-972-95951-2-7.
26. Sharma, S.; Kaushik, A. Views of Irish Farmers on Smart Farming Technologies: An Observational Study. *AgriEngineering* **2019**, *1*, 164–187.

A Non-Invasive Method based on Computer Vision for Grapevine Cluster Compactness Assessment using a Mobile Sensing Platform under Field Conditions

Fernando Palacios [1,2], Maria P. Diago [1,2] and Javier Tardaguila [1,2,*]

[1] Televitis Research Group, University of La Rioja, 26006 Logroño (La Rioja), Spain
[2] Instituto de Ciencias de la Vid y del Vino, University of La Rioja, CSIC, Gobierno de La Rioja, 26007 Logroño, Spain
[*] Correspondence: javier.tardaguila@unirioja.es.

Abstract: Grapevine cluster compactness affects grape composition, fungal disease incidence, and wine quality. Thus far, cluster compactness assessment has been based on visual inspection performed by trained evaluators with very scarce application in the wine industry. The goal of this work was to develop a new, non-invasive method based on the combination of computer vision and machine learning technology for cluster compactness assessment under field conditions from on-the-go red, green, blue (RGB) image acquisition. A mobile sensing platform was used to automatically capture RGB images of grapevine canopies and fruiting zones at night using artificial illumination. Likewise, a set of 195 clusters of four red grapevine varieties of three commercial vineyards were photographed during several years one week prior to harvest. After image acquisition, cluster compactness was evaluated by a group of 15 experts in the laboratory following the International Organization of Vine and Wine (OIV) 204 standard as a reference method. The developed algorithm comprises several steps, including an initial, semi-supervised image segmentation, followed by automated cluster detection and automated compactness estimation using a Gaussian process regression model. Calibration (95 clusters were used as a training set and 100 clusters as the test set) and leave-one-out cross-validation models (LOOCV; performed on the whole 195 clusters set) were elaborated. For these, determination coefficient (R^2) of 0.68 and a root mean squared error (RMSE) of 0.96 were obtained on the test set between the image-based compactness estimated values and the average of the evaluators' ratings (in the range from 1–9). Additionally, the leave-one-out cross-validation yielded a R^2 of 0.70 and an RMSE of 1.11. The results show that the newly developed computer vision based method could be commercially applied by the wine industry for efficient cluster compactness estimation from RGB on-the-go image acquisition platforms in commercial vineyards.

Keywords: image analysis; cluster morphology; RGB; machine learning; non-invasive sensing technologies; proximal sensing; precision viticulture

1. Introduction

Grapevine cluster compactness is a key attribute related to grape composition, fruit health status, and wine quality [1,2]. Compactness defines the density of the cluster by the degree of the aggregation of its berries. Highly compacted winegrape clusters can be affected to a greater extent by fungal diseases, such as powdery mildew [3] and botrytis [4], than loose ones [5].

The most prevalent method for assessing cluster compactness was developed by the International Organization of Vine and Wine (OIV) [6] and has been applied in several research studies [7,8]. This OIV method procures cluster compactness assessment by visual inspection in five different classes. This compactness class takes into account several morphological features of the berries and pedicels, which are visually appraised by trained experts. This method and others designed to evaluate compactness on specific varieties [9–11] tend to be inaccurate due to the intrinsic subjectivity of the evaluation linked to the evaluator's opinion. Moreover, these visual inspection methods are laborious and time-consuming, as they may also require the manual measurement of specific cluster morphological parameters. Therefore, alternative methods for objectively and accurately assessing cluster compactness are needed for wine industry applications.

Computer vision and image processing technology enables low-cost, automated information extraction and its analysis from images taken using a digital camera. This technology is being used in viticulture to estimate key parameters such as vine pruning weight [12,13], the number of flowers per inflorescence [14,15], canopy features [16], or yield [17,18], as well as to provide relevant information to grape harvesting robots [19,20].

Automated cluster compactness estimation by computer vision methods was recently attempted by Cubero et al. [21] and Chen et al. [22]. The former involved the automated extraction of image descriptors from red, green, blue (RGB) cluster images taken from different cluster views under laboratory conditions. From these descriptors, a partial least squares (PLS) calibration model was developed to predict their associated OIV compactness rating. In the approach followed by Chen et al. [22], a multi-perspective imaging system was developed, which made use of different mirror reflections that facilitated the simultaneous acquisition of images from multiple views from a single shot. Additionally, the system also included a weighing sensor for cluster mass measurement. Then, a set of image descriptors and features derived from the data provided by the sensing system were automatically extracted and used to calibrate several models. Of these, the PLS model achieved the best results.

Previously developed computer vision methods for cluster compactness assessment provided accurate and objective compactness estimation only working under controlled laboratory conditions, which requires the destructive collection of clusters in the vineyard. This is a laborious and time-consuming practice that precludes the appraisal of cluster compactness as a standard grape quality parameter prior to harvest, thus limiting its industrial applicability. Moreover, to the best of our knowledge, there is no commercial method available to assess grapevine cluster compactness under field conditions in an automated way.

The purpose of this work was to develop a new, non-invasive, and proximal method based on computer vision and machine learning technology for assessing grapevine cluster compactness from on-the-go RGB image acquisition in commercial vineyards.

2. Materials and Methods

2.1. Experimental Layout

The trials were carried out during seasons 2016, 2017, and 2018 in three commercial vineyards planted with four different red grapevine varieties (*Vitis vinifera* L). The vines were trained onto a vertical shoot positioned (VSP) trellis system and were partially defoliated at fruit set.

An overall set of 195 red grape clusters involving five distinct datasets were labeled in the field prior to image acquisition in three commercial vineyards.

- Vineyard site #1: Located in Logroño (lat. 42°27′42.3″N; long. 2°25′40.4″W; La Rioja, Spain) with 2.8 m row spacing and 1.2 m vine spacing, where a set of 95 Tempranillo clusters were imaged and sampled during season 2016, denoted as T16.
- Vineyard site #2: Located in Logroño (lat. 42°28′34.2″N; long. 2°29′10.0″W; La Rioja, Spain) with 2.5 m row spacing and 1 m vine spacing, where a set of 25 Grenache clusters were imaged and sampled during season 2017, denoted as G17.
- Vineyard site #3: Located in Vergalijo (lat. 42°27′46.0″ N; long. 1°48′13.1″ W; Navarra, Spain) with 2 m row spacing and 1 m vine spacing, where three sets of 75 clusters of Syrah, Cabernet

Sauvignon, and Tempranillo (25 per grapevine variety) were imaged and sampled during season 2018, denoted as S18, CS18, and T18, respectively.

The data were divided into a training set, formed by 95 clusters of T16 dataset, and an external validation test set, formed by 100 clusters of G17, S18, CS18, and T18 datasets, in order to test the system performance on new or additional varieties and vineyards.

2.2. Image Acquisition

Vineyard canopy images were taken on-the-go at a speed of 5 Km/h one week prior to harvest using a mobile sensing platform developed at the University of La Rioja. Image acquisition was performed at night using an artificial illumination system mounted onto the mobile platform in order to obtain homogeneity on the illumination of the vines and to separate the vine under evaluation from the vines of the opposite row. An all-terrain vehicle (ATV) (Trail Boss 330, Polaris Industries, Medina, Minnesota, USA) was modified to incorporate all components as described in the work of Diago et al. [23] (Figure 1).

Figure 1. Mobile sensing platform for on-the-go image acquisition: a modified all-terrain vehicle (ATV) incorporating a red, green, blue (RGB) camera, Global Positioning System (GPS), and an artificial illumination system mounted on an adaptable structure.

Additionally, some elements were modified:

- RGB camera: a mirrorless Sony α7II RGB camera (Sony Corp., Tokyo, Japan) mounting a full-frame complementary metal oxide semiconductor (CMOS) sensor (35 mm and 24.3 MP resolution) and equipped with a Zeiss 24/70 mm lens was used for image acquisition in vineyard sites #1 and #2, while a Canon EOS 5D Mark IV RGB camera (Canon Inc. Tokyo, Japan) mounting a full-frame CMOS sensor (35 mm and 30.4 MP) equipped with a Canon EF 35 mm F/2 IS USM lens was used in vineyard site #3.
- Industrial computer: A Nuvo-3100VTC industrial computer was used for image storage and camera parameters setting for the Canon EOS 5D Mark IV using custom software developed, while the parameters of the Sony α7II camera were set in the camera itself and the storage in a Secure Digital (SD)-card.

The camera was positioned at a distance of 1.5 m from the canopy. The camera parameters were manually set at the beginning of the experiment in each vineyard.

2.3. Reference Measurements of Cluster Compactness

After image acquisition, the labeled clusters were manually collected, and their compactness was visually evaluated in a laboratory at the University of La Rioja by a panel composed of 15 experts following the OIV 204 standard [6]. In this reference method, each cluster was classified in one of five discrete classes (Figure 2) ranging from 1, the loosest clusters, to 9, the most compact clusters. In the visual assessment, several aspects related to the morphology of the cluster, such as berries' mobility, pedicels' visibility, and berries' deformation by pressure, were taken into consideration. The average of the evaluators' ratings was used as the reference compactness value for each cluster.

Figure 2. Examples of clusters with different compactness ratings according to the International Organization of Vine and Wine (OIV) 204 standard: class 1 (**a**) very loose clusters; class 3 (**b**) loose clusters; class 5 (**c**) medium compact clusters; class 7 (**d**); compact clusters; class 9 (**e**) very compact clusters.

2.4. Image Processing

Image processing comprised several steps that can be summarized as semi-supervised image segmentation followed by cluster detection and compactness estimation for each detected cluster. While cluster detection and compactness estimation were fully automated, the image segmentation step required the intervention of the user for each dataset. The algorithm was developed and tested using Matlab R2017b (Mathworks, Natick, MA, USA). The flowchart of the algorithm process for a new set of images is described in Figure 3.

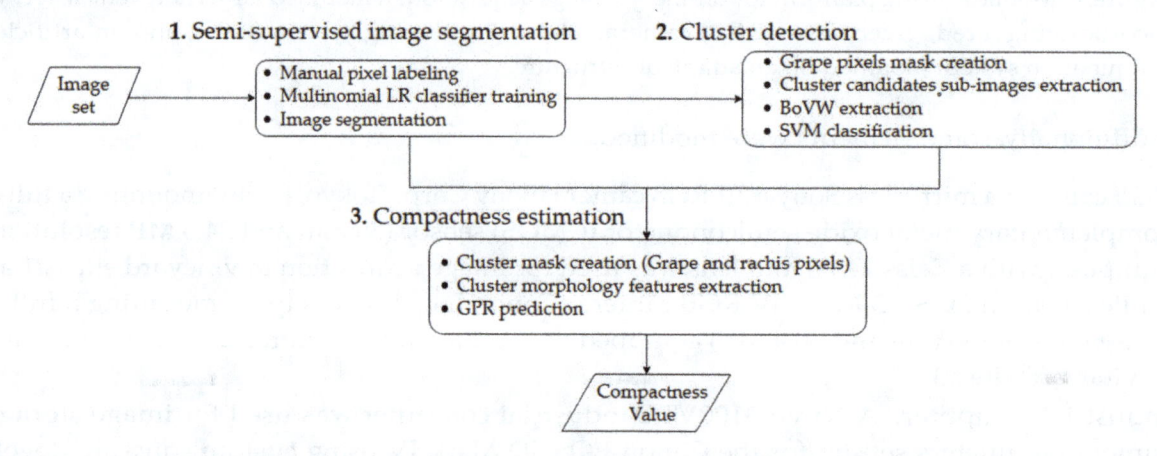

Figure 3. Flow-chart of the full algorithm for a new set of images. First, a set of pixel labeling was required to train a multinomial logistic regression model to segment the whole set. Second, the cluster candidates were extracted and filtered using a bag of visual words model. Finally, compactness features were extracted, and the estimation was performed on each cluster by the Gaussian process regression model.

To ensure consistency in the analysis of the complete algorithm, the classifier used at each step was trained using the training set and evaluated with the output obtained by the classifier of the previous step for the test set, except for the initial image segmentation, where a model was trained on each individual dataset.

2.4.1. Semi-Supervised Image Segmentation

For the proper compactness estimation of every cluster visible in the image, a previous detection of the clusters and their main elements (grape and rachis) was needed.

Most of the red winegrape pixels are easily distinguishable from pixels of other vine elements by their color. Hence, an initial pixelwise color-based segmentation was performed on every image. For this approach, seven classes representing the elements of the grapevines potentially present in their images were defined: "grape", "rachis", "trunk", "shoot", "leaf", "gap", and "trellis". For extracting cluster candidates, only groups of pixels belonging to the first class ("grape") were used, but rachis identification was also relevant for compactness assessment. In summary, an image segmentation considering the seven classes described above as a first step eliminated the necessity of further color segmentation.

A set of 3500 pixels were manually labeled (500 pixels per class), and color features were extracted considering a combination of two color spaces: RGB and CIE L*a*b* (CIELAB) [24]. In this approach, a pixel p was mathematically represented as in Equation (1):

$$p = [R_p, G_p, B_p, L_p, a_p, b_p] \qquad (1)$$

where R_p, G_p, B_p and L_p, a_p, b_p represent the values of p for the three channels of the *RGB* and the *CIELAB* color spaces, respectively. The function rgb2lab from Matlab R2017b was used for the RGB to CIELAB color space conversion.

A multinomial logistic regression was trained with the set described above in order to obtain a pixel wise color based classifier. This classifier, which is a generalization of logistic regression for multiclass problems [25], predicts the probability of each possible outcome for an observation as the relative probability of belonging to each class over belonging to another one chosen arbitrarily as the reference class. Assuming that n is the reference class of a set of $\{1,\ldots,n\}$ classes, the output of the classifier for p is $[\pi_1,\ldots,\pi_n]$, where π_i represents the probability of p belonging to the class i for $i = 1,\ldots,n-1$ [Equation (2)], and π_n represents the probability for the reference class [Equation (3)].

$$\pi_i = \frac{e^{\beta_{i,0} + \sum_{j=1}^{k} \beta_{i,j} x_j}}{1 + \sum_{l=1}^{n-1} e^{\beta_{l,0} + \sum_{j=1}^{k} \beta_{l,j} x_j}} \qquad (2)$$

$$\pi_n = \frac{1}{1 + \sum_{l=1}^{n-1} e^{\beta_{l,0} + \sum_{j=1}^{k} \beta_{l,j} x_j}} = 1 - \sum_{l=1}^{n-1} \pi_l \qquad (3)$$

where k is the number of predictor variables, x_j the j-th predictor variable, and $\beta_{i,j}$ is the estimated coefficient for the i-th class.

The segmentation was performed by assigning to each pixel the class with the highest probability (Figure 4).

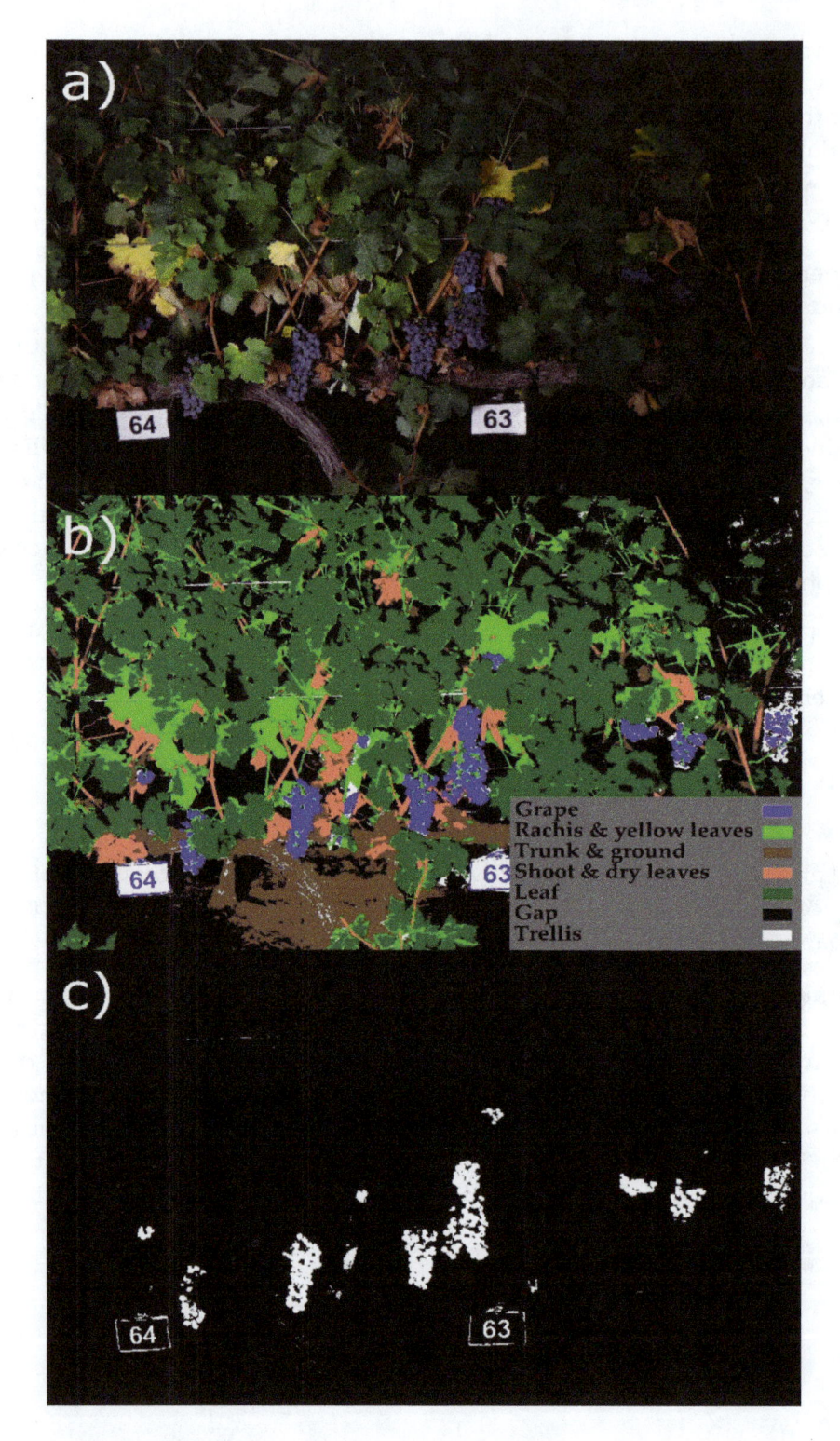

Figure 4. Initial semi supervised segmentation. A set of pixels were manually labeled on the set of the original images (**a**) into seven predefined classes ("grape", "rachis", 'trunk", "shoot", "leaf", "gap", or "trellis"), and the multinomial logistic regression model performed the segmentation on the whole set of images. (**b**) Some pixels of elements without a predefined class were misclassified (e.g., yellow leaves identified as "rachis", dry leaves identified as "shoot", or ground identified as "trunk"). The pixels classified as "grape" and marked in white (**c**) were used for identifying cluster candidates.

2.4.2. Cluster Detection

Using the initial segmentation, a mask of cluster candidates was then generated by selecting those pixels assigned to the "grape" class (Figure 4c).

While the initial color segmentation allowed filtering most of the non-cluster elements presented in the image, a second filtering step was needed to remove those non-cluster groups of pixels with similar color to the "grape" class, which can form objects with different shapes and sizes. This second filtering can be summarized as:

- A morphological opening (morphological erosion operation followed by a dilation) of the clusters' candidates mask using a circular kernel with a radius of three pixels.
- An extraction of a sub-image per minimal bounding box that contains a connected component (groups of connected pixels) in the clusters' candidates mask.
- An extraction of features for each sub-image that represents the information contained on it. For this, the bag-of-visual-words (BoVW) was employed.
- A classification of "cluster" vs. "non-cluster" sub-images.

The bag-of-visual-words (BoVW) model is a concept derived from document classification for image classification and object categorization [26]. In this model, images are treated as documents formed by local features denominated "visual words". These words are grouped to form a "vocabulary" or "codebook". Then, every image is represented by the number of occurrences of every "codeword" in the codebook. In this work, local features were 64-length speeded up robust features (SURF) descriptor vectors [27] clustered by a k-means algorithm [28].

Given a set of n training sub-images, Tr, represented as $Tr = \{tr_1, \ldots, tr_n\}$ and their class $Y = \{y_1, \ldots, y_n\}$ manually labeled into "cluster" (total and partial clusters) and "non-cluster", the process adopted for the training set is the following:

1. To extract SURF points for every sub-image.
2. Cluster SURF points applying k-means. The set of cluster centroids would form the codebook of k codewords.
3. Extraction of the bag-of-words per sub-image:

    a. To assign each SURF point of the image to the nearest centroid of the codebook.
    b. To calculate the histogram by counting the number of SURF points assigned to each centroid.

Then, tr_i had a feature vector $x_i \in \mathbb{R}^k$ that was used to train a support vector machine (SVM) classifier [29]. This is a machine learning algorithm for supervised learning classification or regression tasks that transforms input data into a high-dimensional feature space using a kernel function and finds the hyperplane that maximizes the distance to the nearest training data point of any class. With the classifier trained, only step 3 was performed for new sets of cluster candidates sub-images, and the resulting feature vectors were used to classify each sub-image into "cluster" or "non-cluster" classes, preserving only cluster sub-images for further analysis (Figure 5).

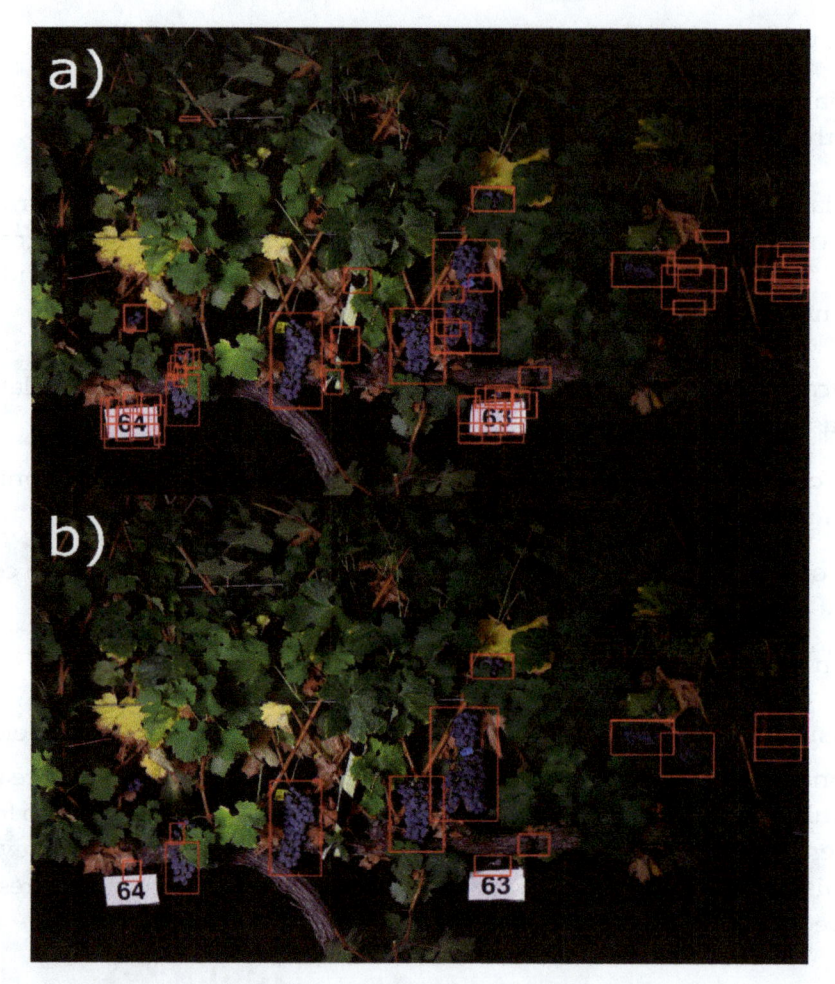

Figure 5. Cluster candidates' extraction and filtering. (**a**) Bounding boxes were extracted from connected components of grape pixels and (**b**) filtered using the bag-of-visual-words (BOVW) model to estimate the cluster compactness of the final non-filtered regions.

2.4.3. Cluster Compactness Estimation

This step involved the extraction of a set of features from the cluster morphology that were related to its compactness. For that purpose, the segmented pixels corresponding to the initial segmentation mask of the sub-images classified as clusters in the previous step had to be extracted. For a given detected cluster sub-image, the next procedure was followed:

- A new mask using only pixels of "grape" and "rachis" classes was created.
- A morphological opening on "grape" pixels using a circular kernel with a radius of two pixels was applied.
- A morphological opening on "rachis" pixels using a circular kernel with a radius of two pixels was also applied.
- A mask containing only the largest connected component formed by "grape" and "rachis" pixels, denoted as mask "A", was created.
- A mask containing the convex hulls of each "grape" pixel's connected component (that can represent several grouped berries on compact clusters or isolated grapes on loose clusters), denoted as mask "B", was created.
- The final mask was created containing "grape" pixels and "rachis" pixels that were in mask "A" and inside the region of the convex hull of mask "B". Those "rachis" pixels in mask "A" that were outside of the convex hull of mask "B" and connected at least two connected components of "grape" pixels were included as well.

The features to estimate the cluster compactness were extracted from the last mask containing only "grape" and "rachis" pixels (Figure 6). These features were the following:

- Ratio of the area of the convex hull body of the cluster corresponding to holes (AH)
- Ratio of the clusters area corresponding to berries (AB)
- Ratio of the area corresponding to "rachis" (AR)
- Average width at 25 ± 5% of the length of the cluster (W25)
- Average width at 50 ± 5% of the length of the cluster (W50)
- Average width at 75 ± 5% of the length of the cluster (W75)
- Ratio between "rachis" and "grape" pixels (RatioRG)
- Roundness of "grape" pixels (RDGrape): $\frac{4.0 \times \pi \times A_{Grape}}{P_{Grape}^2}$
- Compactness shape factor of "grape" pixels (CSFGrape): $\frac{P_{Grape}^2}{A_{Grape}}$
- Ratio between the maximum width and the length of the cluster (AS)
- Ratio between W75 and W25 (RatioW75_W25)
- Proportion of the "rachis" pixels "inside" the cluster (RR_{in})
- Proportion of the "rachis" pixels "outside" the cluster (RR_{out})
- Ratio of the area of the cluster over the mean area of the clusters of its set (R_{AoM})

Where A_{Grape} and P_{Grape} correspond to the area and the perimeter, considering only grape pixels.

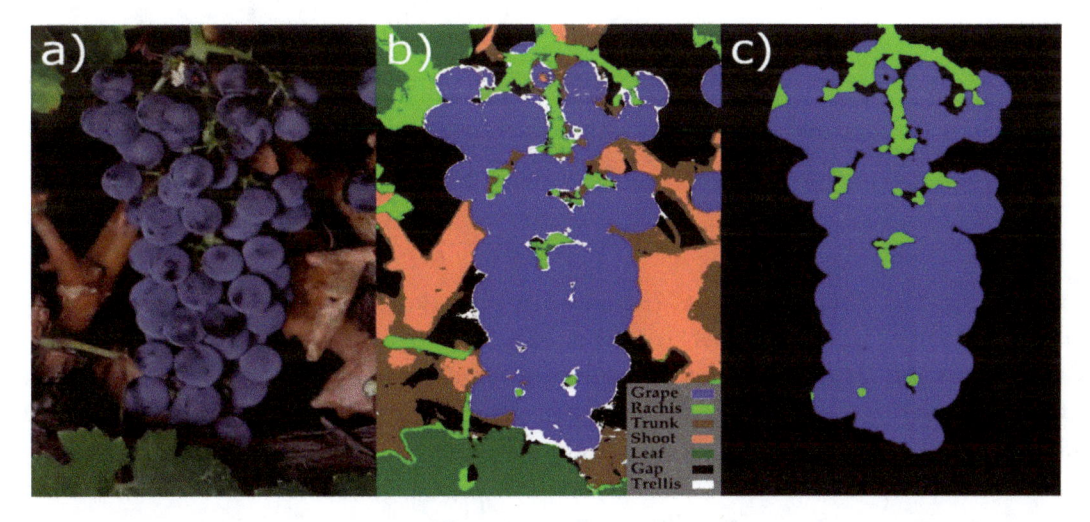

Figure 6. Extraction of the clusters' final masks for compactness estimation. (**a**) Extracted cluster sub-image, (**b**) its corresponding segmentation, and (**c**) a cluster mask obtained using "grape" and "rachis" pixels and morphological operations.

While some of these features were already addressed by Cubero et al. [21] and Chen et al. [22], they had to be adapted to the new environmental situation of field conditions, while others were specifically designed for this study. Features based on clusters' widths and lengths (W25, W50, W75) required a prior rotation of the cluster mask along the longest axis to match the width of the cluster with the horizontal axis, thus the whole set of clusters could have a similar orientation. The features RR_{in} and RR_{out} were calculated as the proportion of "rachis" pixels completely surrounded by "grape" pixels, and the rest of the "rachis" pixels that were not, respectively. The feature R_{AoM} provided a measure about the size of the cluster over the average of the clusters on its set and added robustness by incorporating images of clusters taken at different distances.

The compactness estimation was performed by a Gaussian process regression (GPR) model trained with the data extracted from n clusters $\{\{x_i, y_i\}\}_{i=1}^{n}$ where $x_i \in \mathbb{R}^{14}$ represents the 14-feature vector, and $y_i \in [1,9]$ represents the average of the ratings of the evaluators for the i-th cluster.

Gaussian process regression models are probabilistic kernel-based machine learning models that use a Bayesian approach to solve regression problems estimating uncertainty at predictions [30]. A Gaussian process regression model is described in Equation (4):

$$g(x) = f(x) + h(x)^T \beta \qquad (4)$$

where $h(x)$ is a vector of basis functions, β is the coefficient of $h(x)$, and $f(x) \sim GP(0, k(x, x'))$ is a zero mean Gaussian process with a $k(x, x')$ covariate function.

2.4.4. Performance Evaluation Metrics

The results obtained for each step were analyzed using a set of metrics corresponding to classification tasks in the case of multinomial logistic regression and support vector machine and regression for the Gaussian process regression. The metrics chosen for classification performance are commonly used for evaluating results of binary classifiers, where a sample can be identified as positive class or negative class. For multinomial logistic regression, the positive and the negative classes were the class under evaluation and the rest of them (e.g., "grape" class vs. "non-grape" classes), while for support vector machine, the "cluster" and the "non-cluster" classes were considered, respectively. The metrics calculated were sensitivity [Equation (5)], specificity [Equation (6)], F1 score [Equation (7)], and intersect over union [IoU; Equation (8)].

$$Sensitivity = \frac{TP}{TP + FN} \qquad (5)$$

$$Specificity = \frac{TN}{TN + FP} \qquad (6)$$

$$F_1 = 2 \times \frac{Precision \times Sensitivity}{Precision + Sensitivity} \qquad (7)$$

$$IoU = \frac{TP}{TP + FP + FN} \qquad (8)$$

where TP represents the "true positives" (number of positive samples correctly classified as positive class), FP represents the "false positives" (number of negative samples incorrectly classified as positive class), TN represents the "true negatives" (number of negative samples correctly classified as negative class), and FN represents the "false negatives" (number of positive samples incorrectly classified as negative class). *Precision* was defined as in Equation (9):

$$Precision = \frac{TP}{TP + FP} \qquad (9)$$

The area under the receiver operating characteristic (ROC) curve (AUC) [31] was also considered. For regression, the determination coefficient (R^2) and the root mean squared error (RMSE) were selected.

2.4.5. Hyperparameter's Optimization Procedure

Support vector machine and Gaussian process regression are two machine learning algorithms that have a set of hyperparameters that are not learned from the data and need to be set before the training. The most traditional hyperparameter selection method is a brute-force grid search of the best subset of hyperparameters combined with a manually predefined set of values established for each hyperparameter in order to optimize a performance metric. Instead, in this work, a Bayesian optimization algorithm [32] was used for finding the best hyperparameters set, which proved to outperform other optimization algorithms [33]. This algorithm finds the best hyperparameter set that optimizes an objective function (in this context, a performance metric of the machine learning algorithm) using a Gaussian process trained with the objective function evaluations. The Gaussian process is updated with the result of each evaluation of the objective function, and an acquisition function is used to determine the next point to be evaluated in a bounded domain, i.e., the next set of hyperparameters.

The functions fitcsvm and fitrgp of Matlab R2017b were used to train the support vector machine and the Gaussian process regression models, respectively, selecting their hyperparameters with Bayesian optimization. A Gaussian process with automatic relevance determination (ARD) Matérn 5/2 kernel model and the expected-improvement-plus acquisition function were used for this purpose. The ranges considered for each hyperparameter and the final values are shown in Table 1.

Table 1. Hyperparameter range considered for each classifier and the used final values.

	Kernel Function (fixed)	Optimized Hyperparameters Range		Final Values	
		Box Constraint	Kernel scale	Box Constraint	Kernel scale
SVM	Radial basis function (RBF)	$[10^{-3}, 10^3]$	$[10^{-3}, 10^3]$	1.4654	24.628
		Sigma	Kernel scale	Sigma	Kernel scale
GPR	Exponential	$[10^{-4}, 22.5184]$	$[0.1216, 121.6122]$	0.83194	91.5821

SVM: support vector machine; GPR: Gaussian process regression.

3. Results and Discussion

3.1. Initial Segmentation Performance

The initial segmentation process was a key step towards the accurate assessment of cluster compactness. Therefore, a different segmentation model was applied to each grapevine variety and vineyard to avoid errors associated with slight differences in color and illumination from the images captured from one vineyard to another, which would occur if applying a unique segmentation model. Likewise, five sets of 3500 pixels each were manually labeled (500 pixels per class) for each variety and vineyard, which were used to train five distinct multinomial logistic regression models.

As shown in Table 2, overall, the five models achieved good results in terms of sensitivity, specificity, F1 score, AUC, and IoU when applied to their specific set of images. With regard to the most relevant classes ("rachis" and "grape"), similar and equally good values were obtained for specificity for all sets, while more variable outcomes were obtained for the remaining metrics. In general terms, the T18 model yielded the best results for these two relevant classes, closely followed by the S18 model, in this case only for the "grape" class, and slightly outperformed by the CS18 model for the "rachis" class in terms of sensitivity and AUC. More modest results were obtained for models G17 and T16 (Table 1). Particularly, model T16 yielded values under 0.9 in sensitivity and F1 score metrics for the "rachis" class and in IoU for both "rachis" and "grape" classes.

Table 2. Performance results of each multinomial logistic regression model for segmenting images in their respective dataset using a 10-fold stratified cross validation on the manually labeled pixel sets in terms of sensitivity, specificity, F1 score, area under the receiver operating characteristic (ROC) curve (AUC) and intersect over union (IoU) metrics. Their average was compared with the performance of a single segmentation with all datasets combined using a 5-fold and a 10-fold stratified cross validation.

Vineyard Canopy Class	T16	G17	Dataset S18	CS18	T18	Average	5-Fold CV	10-Fold CV
			Sensitivity					
Trellis	0.9420	0.8760	0.9500	0.9200	0.9760	0.9328	0.5284	0.6700
Gap	0.9740	0.9900	0.9920	0.9960	0.9940	0.9892	0.9868	0.9880
Leaf	0.9760	0.9340	0.9060	0.9680	0.9700	0.9508	0.7476	0.8980
Shoot	0.9440	0.9520	0.9880	1.0000	0.9900	0.9748	0.8992	0.9208
Rachis	0.8640	0.9020	0.9220	0.9660	0.9580	0.9224	0.6912	0.7964
Trunk	0.9020	0.9560	0.9540	0.9660	0.9980	0.9552	0.3620	0.6992
Grape	0.9320	0.9720	0.9700	0.9540	0.9800	0.9616	0.7652	0.8732

Table 2. *Cont.*

Vineyard Canopy Class	T16	G17	Dataset S18	CS18	T18	Average	5-Fold CV	10-Fold CV
			Specificity					
Trellis	0.9897	0.9883	0.9930	0.9883	0.9970	0.9913	0.9554	0.9648
Gap	0.9950	0.9987	0.9967	0.9990	0.9977	0.9974	0.9955	0.9963
Leaf	0.9960	0.9893	0.9897	0.9960	0.9947	0.9931	0.9796	0.9829
Shoot	0.9900	0.9963	0.9983	0.9997	0.9987	0.9966	0.9335	0.9789
Rachis	0.9813	0.9850	0.9810	0.9927	0.9947	0.9869	0.9181	0.9673
Trunk	0.9803	0.9820	0.9923	0.9933	0.9987	0.9893	0.9151	0.9423
Grape	0.9900	0.9907	0.9960	0.9927	0.9963	0.9931	0.9661	0.9751
			F1 Score					
Trellis	0.9401	0.9003	0.9538	0.9246	0.9789	0.9396	0.5884	0.7123
Gap	0.9721	0.9910	0.9861	0.9950	0.9900	0.9868	0.9801	0.9831
Leaf	0.9760	0.9349	0.9207	0.9719	0.9690	0.9545	0.7996	0.8978
Shoot	0.9421	0.9645	0.9890	0.9990	0.9910	0.9771	0.7826	0.8996
Rachis	0.8745	0.9056	0.9057	0.9612	0.9628	0.9220	0.6334	0.7993
Trunk	0.8931	0.9264	0.9540	0.9631	0.9950	0.9463	0.3869	0.6836
Grape	0.9357	0.9586	0.9729	0.9550	0.9790	0.9602	0.7773	0.8634
			AUC					
Trellis	0.9658	0.9322	0.9715	0.9542	0.9865	0.9620	0.7419	0.8174
Gap	0.9845	0.9943	0.9943	0.9975	0.9958	0.9933	0.9912	0.9922
Leaf	0.9860	0.9617	0.9478	0.9820	0.9823	0.9720	0.8636	0.9405
Shoot	0.9670	0.9742	0.9932	0.9998	0.9943	0.9857	0.9164	0.9499
Rachis	0.9227	0.9435	0.9515	0.9793	0.9763	0.9547	0.8047	0.8818
Trunk	0.9412	0.9690	0.9732	0.9797	0.9983	0.9723	0.6386	0.8207
Grape	0.9610	0.9813	0.9830	0.9733	0.9882	0.9774	0.8656	0.9241
			IoU					
Trellis	0.8870	0.8187	0.9117	0.8598	0.9587	0.8872	0.4169	0.5532
Gap	0.9456	0.9821	0.9725	0.9901	0.9803	0.9741	0.9610	0.9667
Leaf	0.9531	0.8778	0.8531	0.9453	0.9399	0.9139	0.6661	0.8146
Shoot	0.8906	0.9315	0.9782	0.9980	0.9821	0.9561	0.6428	0.8175
Rachis	0.7770	0.8275	0.8276	0.9253	0.9283	0.8571	0.4635	0.6657
Trunk	0.8068	0.8628	0.9120	0.9288	0.9901	0.9001	0.2399	0.5193
Grape	0.8792	0.9205	0.9473	0.9138	0.9589	0.9239	0.6358	0.7596

To compare the segmentation performance between individual models on each dataset versus a unique segmentation model, two additional cross-validation methods were applied: a five-fold cross-validation, where at each iteration, the training fold was formed by four datasets and the test fold by the remaining dataset, and a ten-fold stratified cross-validation, where at each iteration, the training and the test folds comprised data at an equal proportion of each dataset and class. The comparison of the results for these two validation methods revealed that better results were obtained for all metrics when the training and the test contained data from the same dataset (ten-fold CV). Comparing both methods with the average of the results obtained by the individual models indicated that applying individual segmentation models produced a substantial improvement in all metrics (except for specificity, for which only a slight increase was recorded) and for all classes (with the exception of the "gap" class, for which similar results were obtained using the three methods). The increase in these metrics for the relevant classes ("grape" and "rachis") and the importance of this step highlights the need for applying individual models for each dataset. Differences between performances could be related to differences in color tonality of the vine elements segmented (e.g., different green tonalities for leaves) between grape varieties and vineyards.

The results show that, given a set of images taken on a vineyard, a multinomial logistic regression model trained with a small subset of pixels manually labeled from the images can be applied to effectively segment vine images in the predefined classes using color information. Also, the pixels needed for compactness estimation ("grape" and "rachis" pixels) can be extracted.

3.2. Cluster Detection Performance

A support vector machine was trained with 600 sub-images manually labeled into 300 "cluster" (total and partial clusters) and 300 "non-cluster" sub-images automatically extracted from the segmentation performed on the T16 set.

The classifier was validated against a set of 800 sub-images automatically extracted from the segmentation performed on sets G17, S18, CS18, and T18 (200 sub-images per set) and manually labeled into 400 "cluster" (total and partial clusters) and 400 "non-cluster" sub-images (100 sub-images of each class per set). A set of k values in a range from 10–200 was chosen for the k-means algorithm, and the performance of the classifier for the "cluster" class was evaluated (Table 3). The model trained with $k = 100$ yielded the best results for all metrics. Similar results were obtained for all k values in terms of sensitivity and F1 score, while for specificity and AUC, the model trained with $k = 10$ performed poorer than the rest of models. The model that yielded the best results ($k = 100$) showed similar values in sensitivity, specificity, and F1 (between 0.76 and 0.8), with specificity being slightly superior, while a higher value was obtained for AUC.

Table 3. Performance results of the support vector machine classifier for cluster detection validated with the external set and performing a 5-fold cross validation on the whole data for several k values of k-means tested in terms of sensitivity, specificity, F1 score, and AUC metrics.

		Sensitivity	Specificity	F1 Score	AUC
Test Set	$k = 10$	0.760	0.660	0.724	0.751
	$k = 50$	0.738	0.770	0.750	0.828
	$k = 100$	0.765	0.795	0.777	0.865
	$k = 150$	0.750	0.770	0.758	0.841
	$k = 200$	0.720	0.765	0.737	0.848
5-Fold CV	$k = 10$	0.811	0.678	0.761	0.821
	$k = 50$	0.804	0.788	0.798	0.884
	$k = 100$	0.821	0.781	0.805	0.903
	$k = 150$	0.790	0.805	0.796	0.888
	$k = 200$	0.799	0.813	0.804	0.902

CV: cross validation.

These results could be improved by incorporating new data, as is visualized in Table 3. A five-fold cross validation (each fold being a different set) was performed to train new support vector machine classifiers. The results show an improvement over all metrics for all k tested values, with the exception of specificity for the $k = 100$ model, whose value was slightly diminished. This model still achieved the best results for all metrics except specificity. The $k = 100$ model was chosen as the final model, as it yielded the best results with the test set in all metrics as well as the best results in almost all metrics at the cross validation.

The performance of the support vector machine proves that this classifier trained with the bag-of-visual-words representation of "cluster" and "non-cluster" sub-images can be applied to classify new sub-images from new datasets previously unknown to the classifier, and therefore, it can be used to filter non-cluster pixel groups before compactness estimation.

3.3. Cluster Compactness Estimation

A Gaussian process regression model was trained on the set of features extracted from the clusters of the T16 set (95 clusters) and validated with the automatically detected clusters of G17, S18, CS18, and T18 sets (100 clusters). A coefficient of determination (R^2) of 0.68 and an RMSE of 0.96 were achieved (Figure 7). This is a remarkable result, considering that the four test sets were totally unknown to the classifier and were taken on different vineyards. Also, S18, CS18, and T18 test sets were photographed with a different RGB camera than the one used to photograph the T16 training set. Even more relevant is that three of the four sets (75 clusters of 100) were formed by varieties different from the one used for training (Tempranillo was used for training, while Grenache, Syrah, Cabernet Sauvignon, and Tempranillo clusters were included in the test set). This outcome paves a way to cluster compactness estimation of winegrape varieties without the requirement of representing variety in the training data—in contrast to the work presented by Cubero et al. [21]—and paves a way to real context application on new varieties and vineyards without the necessity of including specific data from those varieties and vineyards, which would require retrieving new clusters and assessing their compactness by trained experts.

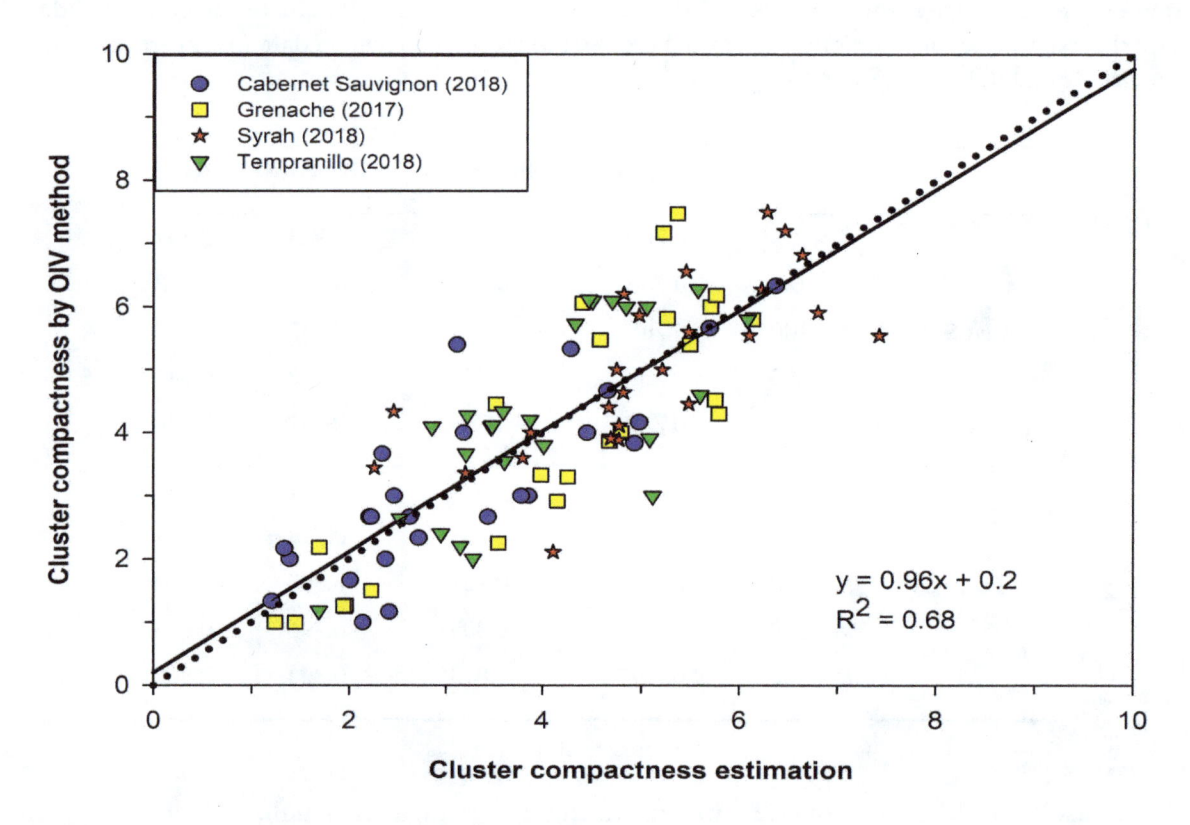

Figure 7. Performance of the GPR model on the test set (100 clusters); correlation between the cluster compactness estimation performed by the model and the OIV ratings (reference method) evaluated visually by the panel of experts.

On the other hand, when leave-one-out cross validation over all datasets (195 clusters) was performed, a coefficient of determination (R^2) of 0.70 and an RMSE of 1.11 were obtained (Figure 8). It can be observed that the algorithm had an accurate performance along most of the compactness range but tended to slightly underestimate highly compacted clusters with an OIV rating close to 9. A feasible reason for this could be that, since this very high compact class was mainly characterized by the deformation of the berries due to the pressure among berries, this feature was difficult to extract by image analysis.

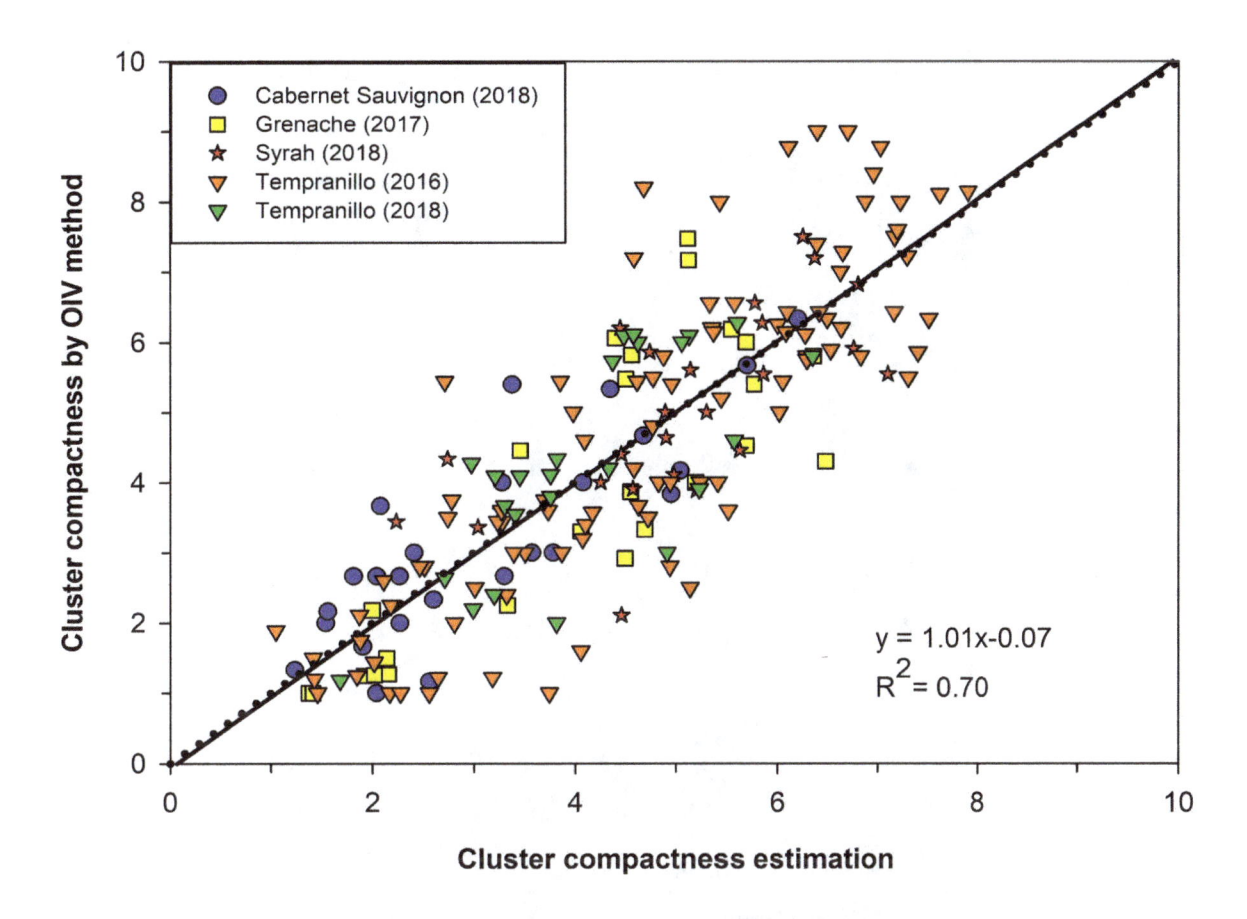

Figure 8. Performance of the GPR model performing leave-one-out cross validation (LOOCV) on the whole data (195 clusters); correlation between the cluster compactness estimation performed by the model and the OIV ratings (reference method) evaluated visually by the panel of experts.

The accuracy of compactness estimation is also meant to be highly influenced by the results obtained in previous steps, i.e., a high misclassification rate in the initial segmentation of cluster and "rachis" pixels would produce wrong shapes in the final cluster mask and a poor feature extraction for the cluster. Also, for estimating the compactness of a given cluster, this has to be previously detected by the BOVW model.

The cluster compactness estimation using the developed methodology in this work could be limited by some experimental in-field conditions, as follows:

- Occlusion of the cluster: the estimation was only performed on the visible region of the cluster. Therefore, a high level of occlusion of the cluster could increase the estimation error. An example of a cluster partially occluded by leaves is shown in Figure 9a and the final mask extracted for compactness estimation in Figure 9b, where the cluster mask presents an anomalous shape that would lead to incorrect compactness estimation.

- Cluster overlapping: highly overlapped clusters would be identified as one, and therefore a unique estimation would be obtained, associated with the set of overlapped clusters. An example is illustrated in Figure 9c, where a set of clusters are overlapped, and in the extracted mask (Figure 9d), the clusters cannot be separated from each other for proper individual compactness estimation.

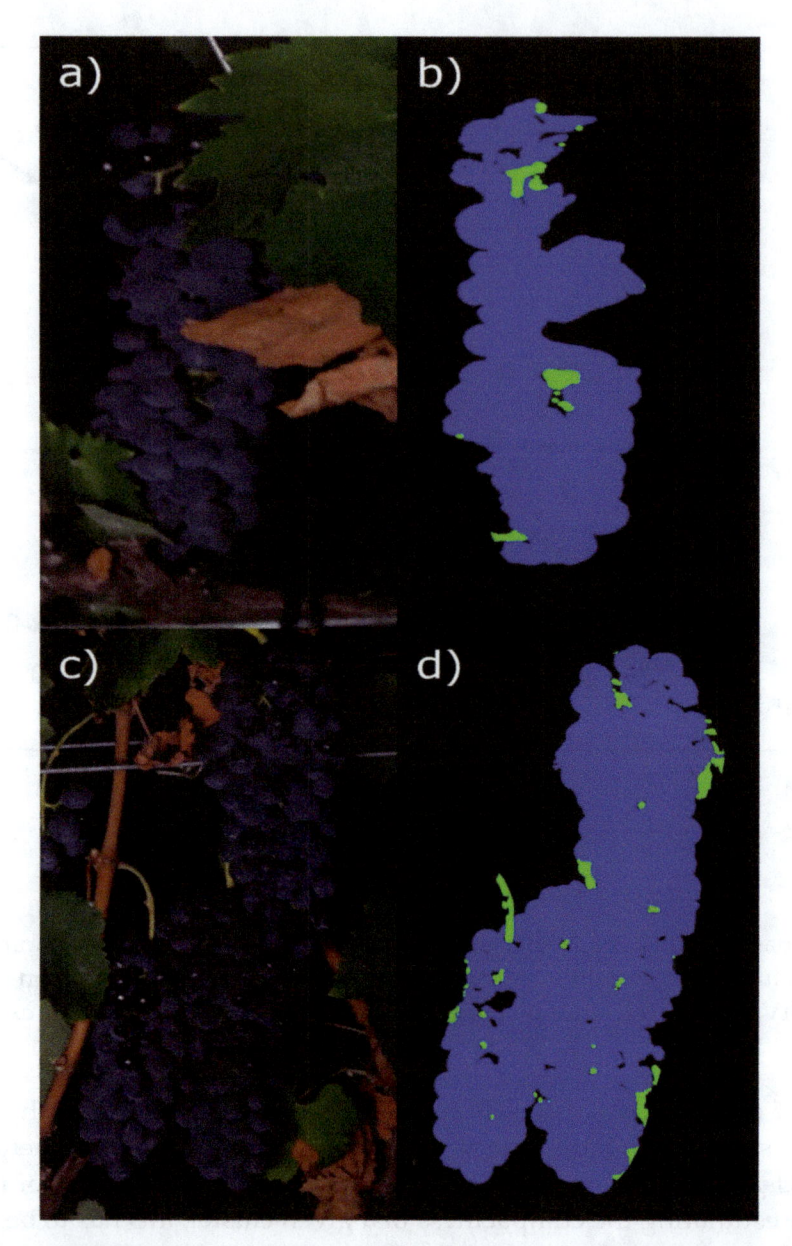

Figure 9. Examples of cluster occlusions and overlapping in commercial vineyards limiting compactness estimation. Cluster partially occluded by leaves (**a**), multiple overlapped clusters (**c**), and final segmented masks used for compactness estimation (**b,d**).

At the current state of the system, the occlusion problem could be overcome by defoliating the side of the vineyard to be photographed. For cluster overlapping, those groups of overlapped clusters could be isolated in the field, or the separation between them could be manually labeled on the images with a clearly different color than "rachis" and "grape" colors (e.g., "trunk" color).

3.4. Commercial Applicability

The developed system can be efficiently used to estimate the cluster compactness in commercial vineyards. The image acquisition carried out by a mobile sensing platform allows the user to take a high number of images in extensive vineyards, which can be automatically geo-referenced. Therefore, the geo-referenced compactness estimations could be used to generate a map that illustrates the spatial variability in cluster compactness, which could be very useful to delineate zones according to cluster compactness and to identify those with similar values. This information could be highly relevant for sorting grapes before harvest, as cluster compactness is often linked to grape quality and health status.

The non-invasive nature of the system could also enable an early identification of very compact clusters before harvest in order to establish strategies against fungal diseases, such as botrytis.

It is also remarkable that the absence of features non-extracted directly from image analysis in the model opens the possibility of a direct application of the algorithm in new vineyards and varieties, in contrast to previous works. Cubero et al [21] introduced, in the PLS model, the cluster winegrape variety as a feature, which requires collecting additional clusters of the variety whose compactness is going to be estimated, evaluating its compactness following the OIV method, and re-training the model. Chen et al. [22] introduced features derived from the cluster mass measured by a weighing sensor, which requires prior harvesting of clusters.

3.5. Future Work

While the current state of the system is capable of estimating compactness in commercial vineyards under uncontrolled field conditions, some improvements still can be made. The algorithm works properly only for red winegrape varieties due to the initial segmentation step, which uses only color information. In this regard, any white grape pixel could be easily misclassified as leaves or "rachis" pixels. A more robust segmentation algorithm could be developed to overcome this problem, combining color and texture information or recurring deep learning techniques. Also, these solutions could help to develop a fully automated system.

The compactness estimation model would also benefit from a more advanced image analysis algorithm capable of extracting features representing the deformation of berries (to increase the accuracy in detecting highly compact clusters), the degree of occlusion of the cluster (to avoid estimations on highly occluded clusters), and to separate overlapped clusters (to enable an individualized estimation on each cluster of the overlapped set).

4. Conclusions

The results of this work show that the developed system was able to estimate winegrape cluster compactness using RGB computer vision on-the-go (at 5 km/h using a mobile sensing platform) and machine learning technology under field conditions. This system enabled a semi-automated, non-invasive method for compactness estimation of a high number of red grapevine clusters under field conditions with low time-consumption. It could be applied to determine the spatial variability of cluster compactness in commercial vineyards, which could be used as new quality input to drive decisions on harvest classification or differential fungicide spraying, for example. The developed methodology constitutes a new tool to improve decision making in precision viticulture, which could be helpful for the wine industry.

Author Contributions: M.P.D. and J.T. conceived and designed the experiments. F.P. developed the algorithm and validated the results. F.P., M.P.D. and J.T. wrote the paper.

Acknowledgments: Authors would like to thank Ignacio Barrio, Diego Collado, Eugenio Moreda and Saúl Río for their help collecting field data.

References

1. Hed, B.; Ngugi, H.K.; Travis, J.W. Relationship between cluster compactness and bunch rot in Vignoles grapes. *Plant Dis.* **2009**, *93*, 1195–1201. [CrossRef] [PubMed]
2. Tello, J.; Marcos, J.I. Evaluation of indexes for the quantitative and objective estimation of grapevine bunch compactness. *Vitis* **2014**, *53*, 9–16.
3. Austin, C.N.; Wilcox, W.F. Effects of sunlight exposure on grapevine powdery mildew development. *Phytopathology* **2012**, *102*, 857–866. [CrossRef] [PubMed]

4. Vail, M.; Marois, J. Grape cluster architecture and the susceptibility of berries to Botrytis cinerea. *Phytopathology* **1991**, *81*, 188–191. [CrossRef]
5. Molitor, D.; Behr, M.; Hoffmann, L.; Evers, D. Research note: Benefits and drawbacks of pre-bloom applications of gibberellic acid (GA3) for stem elongation in Sauvignon blanc. *S. Afr. J. Enol. Vitic.* **2012**, *33*, 198–202. [CrossRef]
6. OIV. OIV Descriptor list for grape varieties and Vitis species. *OIV* **2009**, *18*, 178. Available online: http://www.oiv.int/public/medias/2274/code-2e-edition-finale.pdf (accessed on 2 September 2019).
7. Palliotti, A.; Gatti, M.; Poni, S. Early leaf removal to improve vineyard efficiency: gas exchange, source-to-sink balance, and reserve storage responses. *Am. J. Enol. Vitic.* **2011**, *62*, 219–228. [CrossRef]
8. Tardaguila, J.; Blanco, J.; Poni, S.; Diago, M. Mechanical yield regulation in winegrapes: Comparison of early defoliation and crop thinning. *Aust. J. Grape Wine Res.* **2012**, *18*, 344–352. [CrossRef]
9. Zabadal, T.J.; Bukovac, M.J. Effect of CPPU on fruit development of selected seedless and seeded grape cultivars. *HortScience* **2006**, *41*, 154–157. [CrossRef]
10. Evers, D.; Molitor, D.; Rothmeier, M.; Behr, M.; Fischer, S.; Hoffmann, L. Efficiency of different strategies for the control of grey mold on grapes including gibberellic acid (Gibb3), leaf removal and/or botrycide treatments. *OENO One* **2010**, *44*, 151–159. [CrossRef]
11. Tello, J.; Aguirrezábal, R.; Hernáiz, S.; Larreina, B.; Montemayor, M.I.; Vaquero, E.; Ibáñez, J. Multicultivar and multivariate study of the natural variation for grapevine bunch compactness. *Aust. J. Grape Wine Res.* **2015**, *21*, 277–289. [CrossRef]
12. Kicherer, A.; Klodt, M.; Sharifzadeh, S.; Cremers, D.; Töpfer, R.; Herzog, K. Automatic image-based determination of pruning mass as a determinant for yield potential in grapevine management and breeding. *Aust. J. Grape Wine Res.* **2017**, *23*, 120–124. [CrossRef]
13. Millan, B.; Diago, M.P.; Aquino, A.; Palacios, F.; Tardaguila, J. Vineyard pruning weight assessment by machine vision: towards an on-the-go measurement system. *OENO One* **2019**, *53*. [CrossRef]
14. Aquino, A.; Millan, B.; Gutiérrez, S.; Tardáguila, J. Grapevine flower estimation by applying artificial vision techniques on images with uncontrolled scene and multi-model analysis. *Comput. Electron. Agric.* **2015**, *119*, 92–104. [CrossRef]
15. Liu, S.; Li, X.; Wu, H.; Xin, B.; Petrie, P.R.; Whitty, M. A robust automated flower estimation system for grape vines. *Biosystems Eng.* **2018**, *172*, 110–123. [CrossRef]
16. Diago, M.P.; Krasnow, M.; Bubola, M.; Millan, B.; Tardaguila, J. Assessment of vineyard canopy porosity using machine vision. *Am. J. Enol. Vitic.* **2016**, *67*, 229–238. [CrossRef]
17. Nuske, S.; Wilshusen, K.; Achar, S.; Yoder, L.; Narasimhan, S.; Singh, S. Automated visual yield estimation in vineyards. *J. Field Rob.* **2014**, *31*, 837–860. [CrossRef]
18. Millan, B.; Velasco-Forero, S.; Aquino, A.; Tardaguila, J. On-the-Go Grapevine Yield Estimation Using Image Analysis and Boolean Model. *J. Sens.* **2018**, *2018*. [CrossRef]
19. Luo, L.; Tang, Y.; Zou, X.; Ye, M.; Feng, W.; Li, G. Vision-based extraction of spatial information in grape clusters for harvesting robots. *Biosystems Eng.* **2016**, *151*, 90–104. [CrossRef]
20. Luo, L.; Tang, Y.; Lu, Q.; Chen, X.; Zhang, P.; Zou, X. A vision methodology for harvesting robot to detect cutting points on peduncles of double overlapping grape clusters in a vineyard. *Comput. Ind.* **2018**, *99*, 130–139. [CrossRef]
21. Cubero, S.; Diago, M.P.; Blasco, J.; Tardáguila, J.; Prats-Montalbán, J.M.; Ibáñez, J.; Tello, J.; Aleixos, N. A new method for assessment of bunch compactness using automated image analysis. *Aust. J. Grape Wine Res.* **2015**, *21*, 101–109. [CrossRef]
22. Chen, X.; Ding, H.; Yuan, L.-M.; Cai, J.-R.; Chen, X.; Lin, Y. New approach of simultaneous, multi-perspective imaging for quantitative assessment of the compactness of grape bunches. *Aust. J. Grape Wine Res.* **2018**, *24*, 413–420. [CrossRef]
23. Diago, M.P.; Aquino, A.; Millan, B.; Palacios, F.; Tardaguila, J. On-the-go assessment of vineyard canopy porosity, bunch and leaf exposure by image analysis. *Aust. J. Grape Wine Res.* **2019**, *25*, 363–374. [CrossRef]
24. Luo, M.R. CIELAB. In *Encyclopedia of Color Science and Technology*; Springer: Berlin/Heidelberg, Germany, 2014; pp. 1–7.
25. Dobson, A.J.; Barnett, A. *An introduction to generalized linear models*; Chapman and Hall/CRC: New York, NY, USA, 2008.

26. Csurka, G.; Dance, C.; Fan, L.; Willamowski, J.; Bray, C. Visual categorization with bags of keypoints. In Proceedings of the Workshop on statistical learning in computer vision, ECCV, Prague, Czech Republic, 15 May 2004; pp. 1–2.
27. Bay, H.; Tuytelaars, T.; Van Gool, L. Surf: Speeded up robust features. *Springer* **2006**, *3951*, 404–417.
28. Lloyd, S. Least squares quantization in PCM. *IEEE Trans. Inf. Theory* **1982**, *28*, 129–137. [CrossRef]
29. Cortes, C.; Vapnik, V. Support-vector networks. *Mach. Learn.* **1995**, *20*, 273–297. [CrossRef]
30. Williams, C.K.; Rasmussen, C.E. *Gaussian Processes for Machine Learning*; MIT Press: Cambridge, MA, USA, 2006.
31. Bradley, A.P. The use of the area under the ROC curve in the evaluation of machine learning algorithms. *Pattern Recognit.* **1997**, *30*, 1145–1159. [CrossRef]
32. Mockus, J.; Tiesis, V.; Zilinskas, A. The application of Bayesian methods for seeking the extremum. In *Towards Global Optimization*; Elsevier: Amsterdam, The Netherlands, 2014; pp. 117–129.
33. Jones, D.R. A Taxonomy of Global Optimization Methods Based on Response Surfaces. *J. Glob. Optim.* **2001**, *21*, 345–383. [CrossRef]

Utilisation of Ground and Airborne Optical Sensors for Nitrogen Level Identification and Yield Prediction in Wheat

Christoph W. Zecha [1,*], **Gerassimos G. Peteinatos** [2], **Johanna Link** [1] **and Wilhelm Claupein** [1]

[1] Department of Agronomy (340a), Institute of Crop Science, University of Hohenheim, Fruwirthstraße 23, 70599 Stuttgart, Germany; johannalink@gmx.de (J.L.); claupein@uni-hohenheim.de (W.C.)

[2] Department of Weed Science (360b), Institute of Phytomedicine, University of Hohenheim, Otto-Sander-Straße 5, 70599 Stuttgart, Germany; G.Peteinatos@uni-hohenheim.de

* Correspondence: zechachristoph+sensing@gmail.com

Abstract: A healthy crop growth ensures a good biomass development for optimal yield amounts and qualities. This can only be achieved with sufficient knowledge about field conditions. In this study we investigated the performance of optical sensors in large field trails, to predict yield and biomass characteristics. This publication investigated how information fusion can support farming decisions. We present the results of four site-year studies with one fluorescence sensor and two spectrometers mounted on a ground sensor platform, and one spectrometer built into a fixed-wing unmanned aerial vehicle (UAV). The measurements have been carried out in three winter wheat fields (*Triticum aestivum* L.) with different Nitrogen (N) levels. The sensor raw data have been processed and converted to features (indices and ratios) that correlate with field information and biological parameters. The aerial spectrometer indices showed correlations with the ground truth data only for site-year 2. FERARI (Fluorescence Excitation Ratio Anthocyanin Relative Index) and SFR (Simple Fluorescence Ratio) from the Multiplex® Research fluorometer (MP) in 2012 showed significant correlations with yield (Adj. $r^2 \leq 0.63$), and the NDVI (Normalised Difference Vegetation Index) and OSAVI (Optimized Soil-Adjusted Vegetation Index) of the FieldSpec HandHeld sensor (FS) even higher correlations with an Adj. $r^2 \leq 0.67$. Concerning the available N (N_{avail}), the REIP (Red-Edge Inflection Point) and CropSpec indices from the FS sensor had a high correlation (Adj. $r^2 \leq 0.86$), while the MP ratio SFR was slightly lower (Adj. $r^2 \leq 0.67$). Concerning the biomass weight, the REIP and SAVI indices had an Adj. $r^2 \leq 0.78$, and the FERARI and SFR ratios an Adj. $r^2 \leq 0.85$. The indices of the HandySpec Field® spectrometer gave a lower significance level than the FS sensor, and lower correlations (Adj. $r^2 \leq 0.64$) over all field measurements. The features of MP and FS sensor have been used to create a feature fusion model. A developed linear model for site-year 4 has been used for evaluating the rest of the data sets. The used model did not correlate on a significant de novo level but by changing only one parameter, it resulted in a significant correlation. The data analysis reveals that by increasing mixed features from different sensors in a model, the higher and more robust the r^2 values became. New advanced algorithms, in combination with existent map overlay approaches, have the potential of complete and weighted decision fusion, to ensure the maximum yield for each specific field condition.

Keywords: precision farming; sensor fusion; remote sensing; fluorescence; reflectance; spectrometry; nitrogen fertilisation; wheat; yield

1. Introduction

Agricultural systems using Precision Farming (PF) technologies have already been introduced in the market. The range varies from entry level guidance to data acquisition systems integrated into the farm management software. Most of these systems gather tractor-implement information, or perform tailor made applications [1]. The more intensive the crop production system is, the more advanced the technology adaptation on farms is [2]. This serves the goal of higher yields and better crop quality, with the support of sensor systems. The increasing number of available sensors, along with the high diversity of sensor technologies, e.g., imaging sensors, multi- and hyperspectral optical sensors, fluorometers, etc., has increased the possibility for integrating these sensor systems into the daily farm operation. Each sensor has advantages and disadvantages, and can provide important information concerning the field status [2–4]. Yet each sensor type has limitations to overcome. By merging the data of different sensors and sensor types, their limitations can be reduced, since data can be complementary or more informative [5]. In that sense, data fusion approaches are necessary, achieving better results by merging numerous sensor data deriving from the field and comparing them with ground truth data like yield or biomass.

Hall and Llinas [6] defined data fusion as "the integration of information from multiple sources to produce specific and comprehensive unified data about an entity". Brooks and Iyengar [5] classified four categories for sensor data fusion: (1) redundant; (2) complementary; (3) coordinated; or (4) independent fusion. Dasarathy [7] defined three levels: (I) raw data fusion; (II) feature fusion with feature extraction; and (III) decision fusion, which includes inter alia weighted decision methods [8]. Many different terms are used in literature to describe and discuss "fusion" concerning data. Dasarathy [9] also decided to use "information fusion" instead, as the overall term. In all cases, fusion of the sensor information can improve our knowledge of the field conditions [6].

For agricultural applications many sensors have been proposed. Several research studies based on spectral data are available, e.g., using data mining techniques with a genetic algorithm for nitrogen (N) status and grain yield estimation [10], or acquiring multispectral aerial images for the detection of wheat crop and weeds [11]. They are often based on measurements with one single sensor. There is a lack of information, of how informative different sensors and combination of sensors are, in the variability presented at the field level. Peteinatos et al. [12] measured stress levels in outdoor wheat pots with three optical sensors. Yet there is work to be done, connecting ground data with aerial data, even more in real field conditions. Using mobile platforms for data acquisition offers the possibility of system automation with fusion approaches. The advantage of ground platforms is their ability of carrying higher loads and more equipment than it would be possible with Unmanned Aerial Vehicles (UAV) [13].

In the current paper, the investigated research fields were planted with winter wheat utilising different N levels. These fields were examined with a fluorescence sensor and spectrometers, one spectrometer installed on an UAV, the other two spectrometers and the fluorescence sensor on a ground platform. The aim of this research was to test research sensors on field trails close to normal, practical farming conditions. This publication will discuss redundant and complementary fusion approaches, on a raw data and feature fusion level. It investigates the questions; (i) how the used research sensors perform in a large field; (ii) which of the calculated features are statistically significant for assessments of wheat yield, biomass and the available N for the plant; and (iii) how information fusion can support farming decisions.

2. Material and Methods

2.1. Experimental Site

The investigations have been made at the facility Ihinger Hof in Renningen (Germany), an institution of the University of Hohenheim, Stuttgart in South-West Germany. The location of Ihinger Hof (N 48°44′41″, E 8°55′26″) has a mean annual precipitation of 690 mm (710 mm in 2011 and 727 mm in 2012), and an average annual temperature of 7.9 °C. The measurements about four site-years have been carried out with winter wheat (*Triticum aestivum* L., cv. Toras and Schamane) on the experimental fields "Inneres Täle" in 2011 respectively site-year (1), "Riech" in 2011 and 2012

(2) + (3), as well as "Lammwirt" in 2012 (4). The term "site-year" is a combination of two factors site and year, according to Beres et al. [14], where site relates to an individual field of a farm.

On site-year 1 and 4, the N levels ranged from 60 to 180 kg N·ha^{-1} in five distinct levels. Additionally a dosage of 170 kg N·ha^{-1}, that is usually applied on this farm, were used as the conventional application level. Figure 1 represents the N levels of site-year 1; site-year 4 had the same levels. Nitrogen was distributed in three fertiliser applications in the early growing periods (Zadoks' Scale (Z) 27–Z 47) [15] with a pneumatic fertiliser spreader and a tractor with an automatic steering system and GPS Real-Time Kinematic (RTK) precision (approx. ±2.5 cm). The first N application of 60 kg N·ha^{-1} has been distributed equally over the whole field. The second application had 0–80 kg N·ha^{-1} based on the treatment; and the third application has been carried out with 0–40 kg·N ha^{-1} to reach the planned total amount of N for the respective N level.

For site-year 2 + 3, the fertilisation was applied in repeating rows over the whole field: (1) control; (2) APOLLO model output [16]; and (3) Yara N-sensor control. The field design in this case was different compared to site-year 1 and 4, however it provided the required randomisation for the data analysis, with N levels from 60–170 kg N·ha^{-1} in eight distinct levels (see Figure 2).

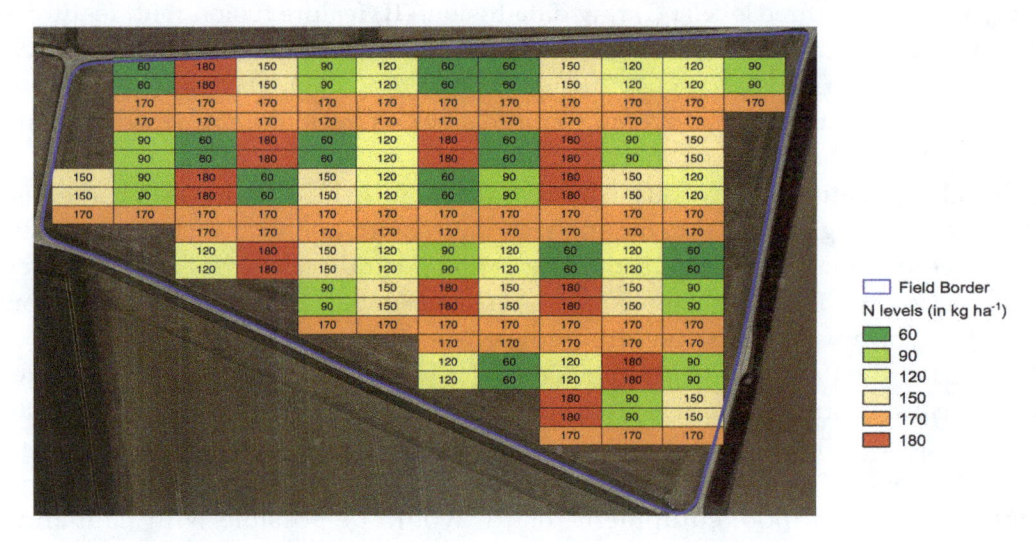

Figure 1. The colored plots reflect the different N levels in kg N·ha^{-1} for site-year 1, as shown in the legend. Each plot has a size of 36 m × 12 m (L × W).

Figure 2. The colored plots reflect the different N levels for site-years 2 and 3, as shown in the legend. Each plot has a size of 36 m × 12 m (L × W).

Table 1 gives an overview of the site characteristics for the research fields. The research fields at the location Ihinger Hof have a high, natural field variability with soil types reaching from pure clay to silty loam. All fields of the location Ihinger Hof were investigated in the year 2009 on their electrical soil conductivity with an EM38 sensor (Geonics Limited, Mississauga, ON, Canada).

Table 1. Site characteristics for winter wheat (*Triticum aestivum* L.): plant density (No.·m^{-2}), seeding and harvest dates, and electrical soil conductivity (mS), of the three research fields Inneres Täle (IT), Riech (RI) and Lammwirt (LW) at experimental site Ihinger Hof, Renningen (Germany). C = Corn, WW = Winter wheat, OSR = Oilseed rape, T = Toras, S = Schamane, SD = Standard deviation.

Site-Year	Site	Year	Previous Crop	Variety	Plant Density	Seeding Date	Harvesting Date	Soil Conductivity Min	Mean	Max	SD
1	IT	2011	C	T	340	27 November 2010	11 August 2011	19.68	38.19	56.98	4.93
2	RI	2011	WW	S	300	14 October 2010	4 August 2011	14.48	31.82	69.53	6.01
3	RI	2012	WW	S	300	17 October 2011	31 July 2012	14.48	31.82	69.53	6.01
4	LW	2012	OSR	T	300	14 October 2011	1 August 2012	52.49	64.48	85.57	6.38

On all four site-years, biomass (BM) samples have been collected over the whole field at three growing stages: stem elongation (approx. Z 35), flowering (approx. Z 61), and before harvest (approx. Z 93). To determine the N content in the soil (N_{min}-method), the samples were analysed on three soil depths: (1) 0–30 cm; (2) 30–60 cm; and (3) 60–90 cm. This took place at the end of tillering (Z 29) and after the harvest. The BM samples have been analysed for grains per ear, the number of tillers, the protein content and the BM weight. The wheat fields have been harvested with a standard New Holland combine harvester, equipped with a header of 6 m cutting width and a GPS receiver with RTK precision to geo-reference the yield data. The laboratory analysis and the yield logging are considered in the current manuscript as the ground truth data, with which the sensor data will be compared. The available N for the plant (N_{avail}), used in this manuscript, is defined as the sum of N_{min} and applied N until the respective sensor measurement date. N_{avail} is a simplified form to express the N supply for the plants in field, as atmospheric entries and mineralisation may provide additional N during the growing season. In spring, soil samples over the whole field have been taken and after harvesting. Table 2 gives an overview of the measurement dates for the ground and UAV mounted sensors for site-year 1. A similar frequency of the field sampling applies to the rest of the site-years.

Table 2. Exemplary for the other site-years, the overview shows the dates for site-year 1 (2011) regarding ground and aerial sampling in the different growing stages (Z). A = aerial spectrometer, G = ground spectrometer.

Z	Spectrometer	Fluorescence Sensor
30	G: 28 April 2011	28 April 2011
37	A + G: 20 May 2011	20 May 2011
75	G: 16 June 2011	16 June 2011
77	G: 28 June 2011	28 June 2011
85	A: 4 July 2011, G: 6 July 2011	6 July 2011

2.2. Measurement Set-Up

The sensor measurements derive from data of three sensors, two spectrometer devices, FieldSpec Handheld (FS—Analytical Spectral Devices, Boulder, CO, USA), HandySpec Field® (HS—tec5 AG, Oberursel, Germany), and the fluorescence sensor Multiplex® Research (MP—Force-A, Orsay, France). The ground sensors (Table 3) were mounted on a rebuilt self propelled Hege 76 multi-equipment carrier (Wintersteiger AG, Ried, Austria), the so called Hohenheim multi-sensor platform "Sensicle"; for more information and image see Keller et al. [3] and Zecha et al. [4]. The sensors mounted to the Sensicle have been adjusted at every measurement date at a specific height for each sensor relative to the canopy (Table 3). The spectrometers are passive sensors, highly dependent on the sun illumination. On the other hand, the MP fluorometer is insensitive to the ambient light conditions due to its light-emitting diodes (LEDs) used for signal excitation. More information about the sensors can be found in Table 3.

The Monolithic Miniature-Spectrometer (MMS) 1 NIR enhanced (Carl Zeiss Jena GmbH, Jena, Germany & tec5 AG, Oberursel, Germany) has been selected due to the compact dimension, the low weight of only 500 g, and the high spectral resolution [17]. It has similar technical properties like

Table 3. Used sensor devices and sensor details. BGF = Blue-Green Fluorescence; RF = Red Fluorescence; FRF = Far-Red Fluorescence.

Type	Manufacturer	Sensor Model	Wavelength Range	Spectral Resolution	Footprint	Classification
Spectrometry	Analytical Spectral Devices	FieldSpec Handheld	325–1075 nm	1 nm	2.74 m^2	Passive, Ground
	tec5 AG	HandySpec Field®	360–1000 nm	10 nm	0.44 m^2	Passive; Ground
	Carl Zeiss Jena GmbH & tec5 AG	MMS1 NIR enhanced	310–1110 nm	3.3 nm	50.27 m^2	Passive, Aerial
Fluorescence	Force-A	Multiplex® Research	BGF, RF and FRF	–	0.005 m^2	Active, Ground

the HS sensor mounted on the Sensicle ground platform (Table 3). It was mounted in the centre of a fixed-wing UAV pointing with the detector to the ground and set to a flight altitude of 100 m above ground; for more information and images see Link et al. [17].

2.3. Information Fusion and Statistical Data Analysis

The ground sensor software for triggering the measurements has been developed by the respective sensor hardware companies. The data logging software for the aerial spectrometer has been developed in C++ for Windows mobile 5 on a Personal Digital Assistant (PDA) [17]. The sensor raw data have been processed and converted to features (indices and ratios) that correlate with field information and biological parameters. This has been done using `Unix-Shell` and `awk` scripts on Ubuntu 12.04 Long Term Support, in combination with the statistical software R [18]. For the spectral data, several indices were derived, allowing a comparison with other sensor data. Common plant characteristics like the chlorophyll content, are commonly used to determine the presence of stress or correlate with the field biomass [19,20]. In the current measurements, the following indices were calculated:

Red-Edge Inflection Point [21]

$$REIP = 700 + 40 \times \frac{(R_{670} + R_{780})/2 - R_{700}}{R_{740} - R_{700}} \quad (1)$$

Normalised Difference Vegetation Index [22]

$$NDVI = \frac{(R_{780} - R_{680})}{(R_{780} + R_{680})} \quad (2)$$

CropSpec [23]

$$CropSpec = (\frac{R_{808}}{R_{735}} - 1) \times 100 \quad (3)$$

Hyperspectral Vegetation Index e.g., [24]

$$HVI = \frac{R_{750}}{R_{700}} \quad (4)$$

Optimised Soil-Adjusted Vegetation Index [25,26]—factor L varies between 0 and 1

$$OSAVI/SAVI = \frac{(R_{800} - R_{670})}{(R_{800} + R_{670} + L)} \times (1 + L) \quad (5)$$

For our analysis, a specific L value (canopy background adjustment factor) was used, 0.16 for OSAVI and 0.20 for SAVI. Concerning the Multiplex® Research fluorescence sensor, the following signals and ratios were used, as described in Cerovic et al. [27] and Ghozlen et al. [28]. The index denotes the fluorescence type while the subindex denotes the wavelength excitation of the LEDs:

$$BGF_{UV/G} = \text{Yellow Fluorescence}$$
$$RF_{UV/G} = \text{Red Fluorescence}$$
$$FRF_{R/G} = \text{Far-Red Fluorescence}$$

Anthocyanins

$$ANTH = \log\left(\frac{FRF_R}{FRF_G}\right) \qquad (6)$$

Flavonols

$$FLAV = \log(FER_{RUV}) \qquad (7)$$

Fluorescence Excitation Ratio Anthocyanin Relative Index

$$FERARI = \log\left(\frac{5000}{FRF_R}\right) \qquad (8)$$

Simple Fluorescence or Chlorophyll Ratio

$$SFR_{R/G} = \frac{FRF_{R/G}}{RF_{R/G}} \qquad (9)$$

The geographic information system (GIS) Quantum GIS [29] has been used for data visualisation and for merging the geo-referenced features in form of indices, signals and ratios with the field design. Linear regression, analysis of variance (ANOVA) and branch-and-bound algorithm have been employed to the sensor data features with the aid of R. After post-processing the data, all features (independent variables—IDV) have been intensively analysed and correlated against the ground truth data (dependent variables—DV).

3. Results

3.1. Field Conditions

The average yield amounts of all site-years per N level are presented in Table 4.

Table 4. Average grain yield (t·ha^{-1}) for winter wheat (*Triticum aestivum* L.) with 14% grain moisture content at the different N levels (kg·ha^{-1}) of the four site-years.

Site-Year	\multicolumn{6}{c}{Yield for N Levels}
	60	90	120	150	170	180
1	7.2	6.7	6.7	7.1	7.3	7.3
2	4.9	-	4.9	5.4	5.3	5.9
3	6.5	-	6.3	6.6	6.9	6.6
4	6.2	6.5	7.5	7.8	7.9	7.9

For the six N levels of site-year 1, yield showed an increasing amount with more N, except for the N level of 60 kg·ha^{-1} (Table 4). The 60 kg·ha^{-1} plots had a similar yield than the plots at a higher N level. On site-year 4, the average yield increased with the N levels until the level of 150 kg·ha^{-1}. For the treatments of 150–180 kg·ha^{-1}, the yield remained on a similar amount. For site-years 2 and 3, these fields have been fertilised with a different strategy, so the average yield amounts per N level are not directly comparable to site-years 1 and 4. As shown in Table 1, all fields used in this research have a high deviation for the electrical soil conductivity with values ranging from 14.48 to 85.57 Milli-Siemens (mS). The field belonging to site-year 2 and 3 has the highest range of all investigated fields.

3.2. Regression Analysis

The feature extraction of the sensor raw data, presented as wavelength indices and fluorescence ratios, have been taken as independent variables (IDV) for the linear regression. As dependent variables (DV) for the following results have been chosen: (1) wheat yield; (2) BM weight; (3) leaf area index (LAI); and (4) available N (N_{avail}). The data analysis was carried out separately for each measurement date, to better observe the changes in correlation over time.

The linear regression results of the aerial sensor MMS1 data with the IDV's of site-years 2 and 3 are shown in Table 5. These results were not significant for DV's (1) and (3) of site-years 2 and 3, whereas DV LAI showed low correlations for site-year 2. The correlations with the DV's BM and LAI could not be measured for site-year 3. The number of valid UAV data fitting to the design layout of site-years 1 and 4 was too low for a significant data analysis.

Table 5. Linear regression analysis of UAV MMS1 sensor indices for site-years 2 and 3. Z = Zadoks' Scale, DV = Dependent Variable, IDV = Independent Variable, RMSE = Root mean square error, BM = Biomass, LAI = Leaf Area Index, HVI = Hyperspectral Vegetation Index, OSAVI = Optimized Soil-Adjusted Vegetation Index, REIP = Red-Edge Inflection Point, PVR = Plant Vigor Ratio, TCARI = Transformed Chlorophyll Absorption Reflectance Index.

Season	Z	DV	IDV	Adj. r^2	RMSE	*p*-Value
2011	73	LAI	HVI	0.15	0.31	0.0144
	73		NDVI	0.24	0.30	0.0020
	73		OSAVI	0.23	0.30	0.0024
	73		REIP	0.18	0.31	0.0077
	66	BM Weight	HVI	0.13	1521.64	0.0000
	73		HVI	0.21	852.72	0.0039
	34		NDVI	0.22	794.57	0.0246
	73		NDVI	0.23	838.76	0.0022
	34		OSAVI	0.22	793.42	0.0240
	73		OSAVI	0.23	842.12	0.0025
	34		PVR	0.19	812.01	0.0374
	73		REIP	0.17	875.04	0.0095
2012	61	LAI	REIP	0.09	0.47	0.0153
	85		NDVIg	0.08	0.49	0.0286
	85		HNDVI	0.08	0.49	0.0304
	85		OSAVI	0.07	0.49	0.0315
	85		NDVI	0.07	0.49	0.0332
	61	BM Weight	TCARI	0.08	117.17	0.0229

Tables 6 and 7 present the correlation results of site-year 4 for the ground sensors MP and FS, only for Adj. r^2 values > 0.46. The FERARI and SFR ratios are significant with yield and BM weight for end of heading and flowering growing stages onwards; the Yellow Fluorescence (BGF) correlates already at the beginning of stem elongation. For N_{avail}, the SFR ratio and the RF signal show significant results (Table 6). The calculated indices HVI, NDVI, OSAVI of the FS sensor show correlations with yield over several measurements of the growing season. The CropSpec and REIP indices highly correlate with N_{avail} for the end of heading stage and further on, HVI and NDVI on the other hand have a lower correlation (Table 7).

Table 6. Linear regression analysis of signals and ratios from Multiplex® Research fluorescence sensor for site-year 4. Z = Zadoks' Scale, DV = Dependent Variable, IDV = Independent Variable, RMSE = Root mean square error, BM = Biomass, FERARI = Fluorescence Excitation Ratio Anthocyanin Relative Index, SFR = Simple Fluorescence Ratio, BGF = Yellow Fluorescence. Significance level: p-value < 0.001.

Z	DV	IDV	Adj. r^2	RMSE
66		FERARI	0.48	0.75
85		FERARI	0.49	0.74
59		SFR_G	0.53	0.71
85	Yield	SFR_G	0.61	0.65
59		SFR_R	0.56	0.69
85		SFR_R	0.63	0.64
59	N_{avail}	SFR_G	0.67	16.52
66		RF_{UV}	0.63	26.77
31		BGF_G	0.46	97.66
59		BGF_G	0.61	83.32
85		BGF_G	0.73	69.21
91		BGF_G	0.74	68.11
85	BM Weight	FERARI	0.83	55.25
85		FLAV	0.62	81.98
85		FRF_R	0.86	50.40
85		RF_G	0.83	54.91
85		SFR_R	0.85	52.14

Table 7. Linear regression analysis of indices from spectrometer FieldSpec HandHeld for site-year 4. Z = Zadoks' Scale, DV = Dependent Variable, IDV = Independent Variable, RMSE = Root mean square error, BM = Biomass. Significance level: p-value < 0.001.

Z	DV	IDV	Adj. r^2	RMSE
59		HVI	0.56	0.68
66		HVI	0.54	0.70
85		HVI	0.59	0.65
66		NDVI	0.56	0.70
85	Yield	NDVI	0.63	0.62
59		OSAVI	0.55	0.69
85		OSAVI	0.67	0.58
85		REIP	0.54	0.69
85		CropSpec	0.53	0.70
59		CropSpec	0.79	13.12
66		CropSpec	0.68	24.57
85		CropSpec	0.78	20.56
59		HVI	0.69	15.98
66		HVI	0.67	25.08
85	N_{avail}	HVI	0.68	24.78
66		NDVI	0.62	27.15
85		NDVI	0.62	26.91
59		REIP	0.86	10.76
66		REIP	0.76	21.40
85		REIP	0.83	18.04
31	BM Weight	REIP	0.78	67.07
85		SAVI	0.75	67.22

3.3. Data Validation

Basis of the data validation were the results of Zecha et al. [4], in which four mixed correlation models were presented, based on measurements with the same spectral and fluorescence sensors. From this research, Model 4 is proposed by the authors for cross-validation with the sensor data of site-years 1–3. The parameters A_x, B_x, C_x and D_x denominate the modelled coefficients of the linear regression—Model 4 as shown in [4]:

$$y = A_4 + B_4 \times CropSpec + C_4 \times HVI + D_4 \times RF_{UV} \qquad (10)$$

By cross-validating the above model with the data from site-years 1–3, the correlations were low. Using a Find-Best-Model-Algorithm in R for the data sets of all site-years, the following yield model has been discovered:

$$Yield_{predicted} = A_x + B_x \times FERARI + C_x \times HVI + D_x \times RF_{UV} \qquad (11)$$

Table 8 highlights the corresponding correlations for Model 4 of site-year 4 and for the new model $Yield_{predicted}$, employed to the data sets of all site-years.

Table 8. Adj. r^2 values of Model 4 and model $Yield_{predicted}$ for the corresponding FS indices and MP signal, grouped by months and site-years. n.a. = not available, n.s. = not significant. p-values for Model 4 < 0.05. p-values for $Yield_{predicted}$ < 0.001.

Model	Site-Year	April	May	June	July
Model 4	4	0.52	0.52 + 0.79	0.75 + 0.77	0.20
	1	0.18	0.24	0.40 + 0.47	0.47
$Yield_{predicted}$	2	0.17	n.a.	0.32	0.37
	3	n.s.	0.31	0.40	0.39
	4	0.26 + 0.40	0.50 + 0.71	0.64 + 0.74	0.66

4. Discussion

This study describes the performance of the used optical sensors, and their ability of wheat yield, biomass and N_{avail} assessment. Based on the yield amounts, the crop development had a steady growth for all site-years, despite of an irregular high yield amount of site-year 1 at the field plots with an N level of 60 kg·ha^{-1}. The reason for this irregularity may be caused by the previous season in 2010. There, corn was planted which can have positive effects on the organic humus content of the field, e.g., Singh Brar et al. [30]. For site-years 1 and 4, the yield at N levels between 150 an 180 kg N·ha^{-1} had no increasing effect on the grain quantity or quality [4]. A lower N level can be recommended for the fertiliser management of these fields for the cultivation of wheat. The total average yield of site-year 3 was 26.7% higher than on site-year 2, which is an indication of more BM in the field, that is able to produce more grain.

The UAV MMS1 spectrometer has similar technical properties like the HS spectrometer, however, the results of both sensors are on a different prediction level for the IDV's in the presented research design. The analysis with the chosen DV's yield and N_{avail} for the MMS1 spectrometer data did not show any correlations. For site-year 2, there are low correlations for BM Weight and LAI; they were not repeatable for site-year 3 (Table 5). Reasons for the low or non existent correlations, based on the findings of Link et al. [17], are (1) a limited path accuracy with the consequence of outlaying data points not fitting to the research field design; (2) height inaccuracy of the UAV; (3) a short flight time of 15 min which required several flight missions to cover the entire research field; (4) that data post processing relies on accurate data from the autopilot system for pitch and roll correction of each data point, and on the control measurement of the MMS1 sensor at the start of the UAV. As the sensor in this setup only could be configured for continuous measurements, a lot of the logged data were of no use as they included the necessary flight turns and the surface measurements on the flight to the research field, Changing light conditions during the following flight mission affected the measurement precision in each design plot; and (5) the sensor footprint of 50.27 m^2 with an overlapping factor of 0.33 [17], covering a larger area at each measurement than the ground sensors were able to acquire (Table 3). As a consequence, the MMS1 data had a higher averaged value than the ground sensor data, which results in a lower resolution and a lower detection accuracy. However, this may be sufficient depending on other investigation purposes, ensuring a stable flight altitude and an integrated fusion approach on a raw data or feature fusion level. Other aerial platform approaches, like an electric multicopter, may lead to better results due to its better flight stability and easier point to point navigation behaviour. Geipel et al. [31] took the same MMS1 spectrometer like in the presented manuscript and mounted it to a hexacopter. With the same ground-truth information via sampling the above-ground BM they were able to measure higher correlations with BM and grain yield, taking into account a data acquisition system for all involved sensors [32].

The MP fluorometer was able to detect significant correlations with grain yield (Adj. r^2 of 0.48–0.63), notably in the ratios SFR with green and red excitation as well as in FERARI. They are linked to the chlorophyll content of the crop [33,34]. The correlations with the available N are high and reach Adj. r^2 values of 0.63–0.67 at a later growing stage (Z 59 and Z 66) with the RF signal and the SFR ratio. The highest correlations are with BM related properties. The correlations with the BM weight range from an Adj. r^2 of 0.46 at the early growing stage (Z 31), up to an Adj. r^2 of 0.86 at ripening (Z 85) and senescence (Z 91) stages. Fluorescence sensors for agricultural usage on tractors or other mobile platforms are barely in use. Their required contact with the crop canopy is one of the reasons why most of the used agricultural sensors are based on spectral characteristics [35]. However, due to the active LED emission source of the MP sensor, it provides a profound, reliable and repeatable technology especially for measurements on the field with changing illumination. Hyperspectral line scanners do not require close contact with the crop canopy and use sun induced fluorescence, however their field application is still in discussion and used on a research level [36].

Spectral sensors are already well adopted at large modern farms, and are able to fuse the measured data with previously gathered data sets via a map overlay approach [2,37,38]. Also in scientific research spectral sensors have a high acceptance, as more than 90% of the spectral information on crop canopy is contained in the red and near infrared (NIR) spectral bands [39,40]. For the FS indices HVI, NDVI, OSAVI, REIP and CropSpec, the correlations with yield increased, starting at heading stage (Z 51) to a high level of an Adj. $r^2 = 0.67$ at ripening stage (Z 85). Especially the indices CropSpec and REIP correlate very high with N_{avail} and provide an Adj. r^2 up to 0.86. For the BM characteristics, REIP, SAVI and CropSpec have high Adj. r^2 values > 0.63 already from stem elongation stage onwards (Z 30). The r^2 values of the HandySpec sensor data analysis was at a lower level than the ones from the FieldSpec sensor. They conclude in a maximum correlation of an Adj. $r^2 \leq 0.64$ at a significance level <0.05, with the presented DV's and IDV's.

For research, the high correlations of the MP fluorometer and the FS sensor can be merged on a feature fusion level. This has been done by Zecha et al. [4] and in the presented manuscript with a data post-processing method. The developed Model 4 from site-year 4 has been applied to the data sets of site-years 1–3. Model 4 did not correlate on a significant level with the gathered sensor data in these three site-years. However, a similar combination of indices and ratios (model $Yield_{predicted}$) resulted in significant correlations for all four site-years, by changing only one parameter (FERARI with CropSpec). By this change, the Adj. r^2 was between 0.32 and 0.74 two months before harvest for all site-years. The data analysis reveals that the more mixed indices and ratios are in a model, the higher and more robust the Adj. r^2 values became, like RF_{UV} and HVI, combined with index CropSpec or ratio FERARI, in the investigated linear models.

This model has a potential to continue working. Three out of the four parameters are exactly the same, providing results for the other three site-years. On the other hand, the ability of the presented model, predicting wheat yield by using unknown or different data, has not yet been validated, e.g., with machine learning methods proposed by Peña et al. [41] or as comparison with the linear models of Mortensen et al. [42] estimating above-ground biomass and N-uptake through aerial images. Future work needs to be done to train and test the real capabilities of this model, and to prove if it works.

5. Conclusions

(i) The used aerial data collection system, as a combination of a fixed wing UAV and the MMS1 spectrometer, cannot be recommended for multispectral data acquisition like it has been done in the presented setup. A limited path accuracy, a short flight time of approx. 15 min including take-off, flight turns and landing, the MMS1 sensor setup in continuous measurement mode, independent sensor data logging and the related huge post-processing efforts, and the footprint along with an overlap of 30% make it unfavorable for a qualitative data analysis and feature correlation with ground truth data. For the aerial data acquisition, the authors recommend an integrated data acquisition system with all sensors connected via a sensor data infrastructure.

(ii) Two ground sensors mounted to the Sensicle platform, the fluorometer Multiplex® Research (MP) and the FieldSpec HandHeld (FS) spectrometer, had high correlations with wheat yield, available nitrogen and the sampled biomass characteristics from the field plots. The HandySpec Field® (HS) spectrometer had lower significant correlations in all site-year than the FS sensor. The usage of the three ground sensors in continuous measurement mode is most reliable for the fluorometer MP. With an internal GPS sensor and an active LED source, measurement starts with one click and data storage on a SD card. The FS and HS spectrometer require an additional device for measurement triggering, and do rely on an external GPS receiver. The raw data post-processing cannot be handled without scripts, converting the raw data in features like indices, while calculating them with the white reference measurements, taken at the start of each continuous measurement series. The ability of the presented model, predicting wheat yield by using unknown or future data, has not yet been validated. Recommending the developed model for a general performance, further model training and model testing need to take place.

(iii) An enhanced algorithm during the raw data calculation of the spectrometer, taking into account the ambient solar radiation during each continuous measurement mission, may improve the correlations and make the developed model more robust to apply it in earlier growing stages with high correlations. Advanced algorithms considering the factors (1) ambient solar radiation; (2) electrical soil conductivity; (3) aerial images with feature extraction; or (4) soil scoring may result in better yield predictions by providing the right decision for each spot in a field. In combination with the existent map overlay approaches of today's spectral sensor systems, these complete and weighted decision can save field inputs and ensure the perfect crop development to reach the maximum yield for the specific field. Once, the field data collection and analysis process can be accomplished with sensors and software in an convenient way also for a farmer, the adoption of sensor technology in agriculture will increase.

Author Contributions: C.W.Z. collected the field data, made the statistical analysis and wrote the manuscript. G.G.P. contributed to the sensor data analysis and reviewed the document. J.L. proposed the field trial design, supported with the aerial data collection and the publication review. W.C. proposed the idea for the study and supported with editorial contributions.

Acknowledgments: A special thanks to all colleagues within the SENGIS team, namely Johanna Link, Martin Weis, Gerassimos Peteinatos, Markus Jackenkroll, Martina Keller and Jakob Geipel, for all their support. Moreover, the authors acknowledge Andrea Richter for their support in the field.

References

1. Auernhammer, H. Precision farming—The environmental challenge. *Comput. Electron. Agric.* **2001**, *30*, 31–43. [CrossRef]
2. Adamchuk, V.I.; Rossel, R.A.V.; Sudduth, K.A.; Lammers, P.S. Sensor Fusion for Precision Agriculture. In *Sensor Fusion—Foundation and Applications*; Thomas, C., Ed.; InTech: Rijeka, Croatia, 2011; Chapter 2. [CrossRef]

3. Keller, M.; Zecha, C.; Weis, M.; Link, J.; Gerhards, R.; Claupein, W. Competence centre SenGIS-exploring methods for georeferenced multi-sensor data acquisition, storage, handling and analysis. In Proceedings of the 8th European Conference on Precision Agriculture (ECPA), Prague, Czech Republic, 11–14 July 2011; Czech Centre for Science and Society: Ampthill, UK; Prague, Czech Republic, 2011; pp. 491–500.
4. Zecha, C.W.; Link, J.; Claupein, W. Fluorescence and reflectance sensor comparison in winter wheat. *Agriculture* **2017**, *7*, 78. [CrossRef]
5. Brooks, R.R.; Iyengar, S. *Multi-Sensor Fusion: Fundamentals and Applications with Software*; Prentice-Hall, Inc.: Upper Saddle River, NJ, USA, 1998.
6. Hall, D.L.; Llinas, J. A challenge for the data fusion community I: Research imperatives for improved processing. In Proceedings of the 7th National Symposium on Sensor Fusion, Albuquerque, NM, USA, 16–18 March 1994.
7. Dasarathy, B. Sensor fusion potential exploitation-innovative architectures and illustrative applications. *Proc. IEEE* **1997**, *85*, 24–38. [CrossRef]
8. Elmenreich, W. *An Introduction to Sensor Fusion*; Technical Report; Vienna University of Technology, Department of Computer Engineering: Vienna, Austria, 2002.
9. Dasarathy, B.V. Information Fusion—What, where, why, when, and how? *Inf. Fusion* **2001**, *2*, 75–76. [CrossRef]
10. Thorp, K.R.; Wang, G.; Bronson, K.F.; Badaruddin, M.; Mon, J. Hyperspectral data mining to identify relevant canopy spectral features for estimating durum wheat growth, nitrogen status, and grain yield. *Comput. Electron. Agric.* **2017**, *136*, 1–12. [CrossRef]
11. Mesas-Carrascosa, F.J.; Torres-Sánchez, J.; Clavero-Rumbao, I.; García-Ferrer, A.; Peña, J.M.; Borra-Serrano, I.; López-Granados, F. Assessing optimal flight parameters for generating accurate multispectral orthomosaics by UAV to support site-specific crop management. *Remote Sens.* **2015**, *7*, 12793–12814. [CrossRef]
12. Peteinatos, G.G.; Korsaeth, A.; Berge, T.W.; Gerhards, R. Using optical sensors to identify water deprivation, nitrogen shortage, weed presence and fungal infection in wheat. *Agriculture* **2016**, *6*, 24. [CrossRef]
13. Zecha, C.W.; Link, J.; Claupein, W. Mobile sensor platforms: Categorisation and research applications in precision farming. *J. Sens. Sens. Syst.* **2013**, *2*, 51–72. [CrossRef]
14. Beres, B.L.; Turkington, T.K.; Kutcher, H.R.; Irvine, B.; Johnson, E.N.; O'Donovan, J.T.; Harker, K.N.; Holzapfel, C.B.; Mohr, R.; Peng, G.; et al. Winter wheat cropping system response to seed treatments, seed size, and sowing density. *Agron. J.* **2016**, *108*, 1101–1111. [CrossRef]
15. Zadoks, J.C.; Chang, T.T.; Konzak, C.F. A decimal code for the growth stages of cereals. *Weed Res.* **1974**, *14*, 415–421. [CrossRef]
16. Link, J.; Graeff, S.; Batchelor, W.D.; Claupein, W. Evaluating the economic and environmental impact of environmental compensation payment policy under uniform and variable-rate nitrogen management. *Agric. Syst.* **2006**, *91*, 135–153. [CrossRef]
17. Link, J.; Senner, D.; Claupein, W. Developing and evaluating an aerial sensor platform (ASP) to collect multispectral data for deriving management decisions in precision farming. *Comput. Electron. Agric.* **2013**, *94*, 20–28. [CrossRef]
18. R Development Core Team. *R: A Language and Environment for Statistical Computing*; R Foundation for Statistical Computing: Vienna, Austria, 2008; ISBN 3-900051-07-0.
19. Thenkabail, P.S.; Lyon, J.G.; Huete, A. (Eds.) *Hyperspectral Remote Sensing of Vegetation*, 1st ed.; Crc Press Inc.: Boca Raton, FL, USA, 2012; doi:10.1201/b11222. [CrossRef]
20. Weis, M.; Andújar, D.; Peteinatos, G.G.; Gerhards, R. Improving the determination of plant characteristics by fusion of four different sensors. In *Precision Agriculture '13*; Stafford, J.V., Ed.; Wageningen Academic Publishers: Wageningen, The Netherlands, 2013; pp. 63–69.
21. Horler, D.N.H.; Dockray, M.; Barber, J. The red edge of plant leaf reflectance. *Int. J. Remote Sens.* **1983**, *4*, 273–288. [CrossRef]
22. Rouse, J.; Haas, R.; Schell, J.; Deering, D. Monitoring vegetation systems in the Great Plains with ERTS. In *Third ERTS Symposium*; NASA SP-351; Freden, S.C., Becker, M.A., Eds.; NASA: Washington, DC, USA, 1973; pp. 309–317.
23. Reusch, S.; Jasper, J.; Link, A. Estimating crop biomass and nitrogen uptake using Cropspec, a newly developed active crop-canopy reflectance sensor. In Proceedings of the 10th International Conference on Positron Annihilation (ICPA), Denver, CO, USA, 18–21 July 2010; p. 381.

24. Thenkabail, P.S.; Smith, R.B.; De Pauw, E. Hyperspectral vegetation indices and their relationships with agricultural crop characteristics. *Remote Sens. Environ.* **2000**, *71*, 158–182. [CrossRef]
25. Huete, A. A soil-adjusted vegetation index (SAVI). *Remote Sens. Environ.* **1988**, *25*, 295–309. 0034-4257(88)90106-X. [CrossRef]
26. Haboudane, D.; Miller, J.R.; Tremblay, N.; Zarco-Tejada, P.J.; Dextraze, L. Integrated narrow-band vegetation indices for prediction of crop chlorophyll content for application to precision agriculture. *Remote Sens. Environ.* **2002**, *81*, 416–426. [CrossRef]
27. Cerovic, Z.; Moise, N.; Agati, G.; Latouche, G.; Ben Ghozlen, N.; Meyer, S. New portable optical sensors for the assessment of winegrape phenolic maturity based on berry fluorescence. *J. Food Compos. Anal.* **2008**, *21*, 650–654. [CrossRef]
28. Ghozlen, N.B.; Cerovic, Z.G.; Germain, C.; Toutain, S.; Latouche, G. Non-destructive optical monitoring of grape maturation by proximal sensing. *Sensors* **2010**, *10*, 10040–10068, doi10.3390/s101110040. [CrossRef] [PubMed]
29. QGIS Development Team. *QGIS Geographic Information System*; Open Source Geospatial Foundation: Chicago, IL, USA, 2009.
30. Singh Brar, B.; Singh, J.; Singh, G.; Kaur, G. Effects of long term application of inorganic and organic fertilizers on soil organic carbon and physical properties in maize–wheat rotation. *Agronomy* **2015**, *5*, 220–238. [CrossRef]
31. Geipel, J.; Link, J.; Wirwahn, J.A.; Claupein, W. A Programmable aerial multispectral camera system for in-season crop biomass and nitrogen content estimation. *Agriculture* **2016**, *6*, doi10.3390/agriculture6010004. [CrossRef]
32. Geipel, J.; Jackenkroll, M.; Weis, M.; Claupein, W. A sensor web-enabled infrastructure for Precision Farming. *ISPRS Int. J. Geo-Inf.* **2015**, *4*, 385–399. [CrossRef]
33. Lichtenthaler, H.; Buschmann, C.; Rinderle, U.; Schmuck, G. Application of chlorophyll fluorescence in ecophysiology. *Radiat. Environ. Biophys.* **1986**, *25*, 297–308. [CrossRef] [PubMed]
34. Buschmann, C. Variability and application of the chlorophyll fluorescence emission ratio red/far-red of leaves. *Photosynth. Res.* **2007**, *92*, 261–271. [CrossRef] [PubMed]
35. Tremblay, N.; Wang, Z.; Cerovic, Z.G. Sensing crop nitrogen status with fluorescence indicators. A review. *Agron. Sustain. Dev.* **2011**, *32*, 451–464. [CrossRef]
36. Broge, N.; Mortensen, J. Deriving green crop area index and canopy chlorophyll density of winter wheat from spectral reflectance data. *Remote Sens. Environ.* **2002**, *81*, 45–57. [CrossRef]
37. Bill, R.; Nash, E.; Grenzdörffer, G. GIS in Agriculture. In *Springer Handbook of Geographic Information*; Danko, D.M., Kresse, W., Eds.; Springer: Berlin/Heidelberg, Germany, 2011; pp. 461–476.
38. Stone, M.L.; Raun, W.R. Sensing Technology for Precision Crop Farming. In *Precision Agriculture Technology for Crop Farming*; Zhang, Q., Ed.; CRC Press: Boca Raton, FL, USA, 2015; Chapter 2, pp. 21–54.
39. Sheffield, C. Selecting band combinations from multispectral data. *Photogramm. Eng. Remote Sens.* **1985**, *51*, 681–687.
40. Li, H.; Zhao, C.; Yang, G.; Feng, H. Variations in crop variables within wheat canopies and responses of canopy spectral characteristics and derived vegetation indices to different vertical leaf layers and spikes. *Remote Sens. Environ.* **2015**, *169*, 358–374. [CrossRef]
41. Peña, J.M.; Gutiérrez, P.A.; Hervás-Martínez, C.; Six, J.; Plant, R.E.; López-Granados, F. Object-based image classification of summer crops with machine learning methods. *Remote Sens.* **2014**, *6*, 5019–5041. [CrossRef]
42. Mortensen, A.K.; Gislum, R.; Larsen, R.; Jørgensen, R.N. Estimation of above-ground dry matter and nitrogen uptake in catch crops using images acquired from an octocopter. In *Precision Agriculture '15*; Wageningen Academic Publishers: Wageningen, The Netherlands, 2015; pp. 227–234.

A Comprehensive Study of the Potential Application of Flying Ethylene-Sensitive Sensors for Ripeness Detection in Apple Orchards

João Valente *[†], Rodrigo Almeida[†] and Lammert Kooistra

Laboratory of Geo-information Science and Remote Sensing, Wageningen University & Research,
6708 PB Wageningen, The Netherlands; rodrigo.almeida@wur.nl (R.A.); lammert.kooistra@wur.nl (L.K.)
* Correspondence: joao.valente@wur.nl
† These authors contributed equally to this work.

Abstract: The right moment to harvest apples in fruit orchards is still decided after persistent monitoring of the fruit orchards via local inspection and using manual instrumentation. However, this task is tedious, time consuming, and requires costly human effort because of the manual work that is necessary to sample large orchard parcels. The sensor miniaturization and the advances in gas detection technology have increased the usage of gas sensors and detectors in many industrial applications. This work explores the combination of small-sized sensors under Unmanned Aerial Vehicles (UAV) to understand its suitability for ethylene sensing in an apple orchard. To accomplish this goal, a simulated environment built from field data was used to understand the spatial distribution of ethylene when subject to the orchard environment and the wind of the UAV rotors. The simulation results indicate the main driving variables of the ethylene emission. Additionally, preliminary field tests are also reported. It was demonstrated that the minimum sensing wind speed cut-off is 2 ms^{-1} and that a small commercial UAV (like Phantom 3 Professional) can sense volatile ethylene at less than six meters from the ground with a detection probability of a maximum of 10%. This work is a step forward in the usage of aerial remote sensing technology to detect the optimal harvest time.

Keywords: apple orchards; modeling and simulation; unmanned aerial vehicles; fruit ripeness; ethylene gas detection

1. Introduction

Sustainable agriculture is a top priority for all the governments and nations worldwide. Our population is growing fast, and our resources are getting more scarce each day. By 2050, our population will reach nine billion, requiring crop production to double in order to meet food demands [1].

An efficient way to increase the upcoming demands is to avoid fruit spoiling during the harvesting. Immature fruits result in poor quality and are subject to mechanical damage, and overripe fruit results in a soft and flavorless quality, with a very short shelf-life. In general, if the harvesting is done too early or too late, physiological disorders in the fruits will be provoked with the consequence of a shorter shelf-life [2]. These issues becomes more relevant with international trade of fruit and vegetables is increasing, making the shelf-life become an important marketing tool [3]. Therefore, the Optimal Harvest Date (OHD) will dictate the resulting fruit yield.

The OHD is usually obtained from maturity indices that take into account fruit chemical composition, like total soluble solids or total acidity, fruit physical properties, like firmness or color, fruit physiological changes, like aroma and ethylene emission rate, and finally, chronological features, like the number of days after planting or blooming [4].

Fruits' and vegetables' lifespan can be broken down into three steps: maturation (i.e., increase in fruit size), ripening (i.e., increase in flavor), and senescence (i.e., tissue death) [3]. Fruits that ripen after harvesting are denoted as climacteric fruits [4]. For climacteric fruits, like apples, the optimal harvest date occurs when the pre-climacteric minimum happens, equivalent to the end of the maturation process or the beginning of the ripening process, as illustrated in Figure 1.

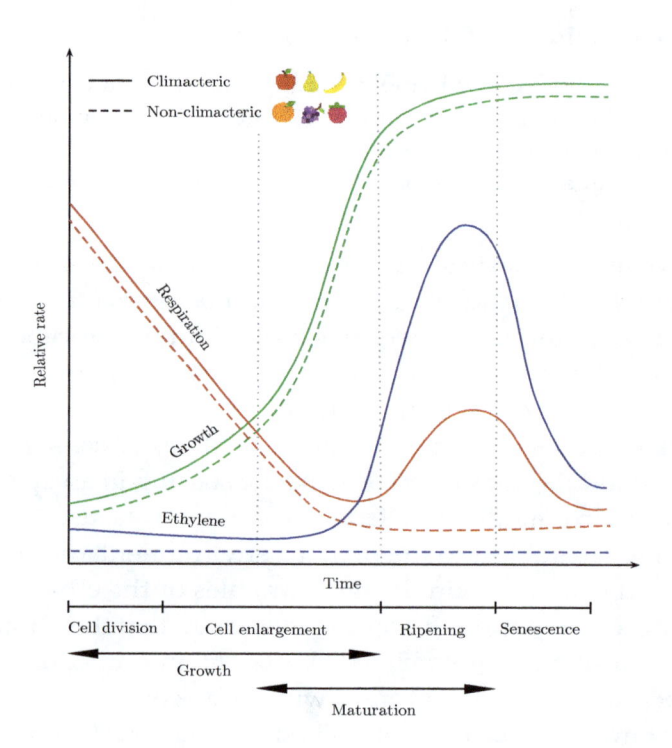

Figure 1. Relative rate of respiration, ethylene production, and growth in climacteric and non-climacteric fruits. Adapted from [5].

The fruits' distinctive aromas are characterized by a wide variety of Volatile Organic Compounds (VOCs) that are released during their maturation process [6]. The VOCs can be detected using a single-gas sensor or an array of gas sensors (also known as an electronic nose) [7]. An important VOC that is associated with fruit ripening is ethylene [8].

Ethylene (C_2H_4) is a gaseous phytohormone that regulates several growth and development processes in plants. In climacteric fruits, ethylene production regulates processes like flesh softening, color changes, and aroma emissions during ripening [9]. Ethylene can be measured via gas chromatography techniques, electrochemical sensors, and optical sensors [10].

Most current destructive and non-destructive methods of assessing fruit maturity require the sampling of individual fruit in the field and in some cases a further assessment in the lab [8,11,12]. That process is both labor intensive, since it requires an operator to physically go to the field and sample fruits, and dependent on the individual fruits that are sampled. Using the electronic noses and gas measurements with the fruit in concentration chambers provides less noise and augments the ethylene signal substantially, but it requires time and manpower to harvest and analyze the fruit, and at the same time, it is a method that is highly reliable on the sampling scheme used for the fruit [13,14].

The increasing availability of UAVs is a potential solution to acquire remotely and quickly data on a plot of land without the manual labor that would be required traditionally. The land manager/owner does not have to survey the plot manually, but can deploy a UAV. There are several aerial remote sensing applications in agriculture that were successful reported as an important contribution and step forward in Precision Agriculture (PA) practices [15].

Using the combination of airborne and electronic nose technology to map ethylene concentration in the orchard might give important information regarding fruit maturity in a fast and more representative way, without the need for additional labor. To the author's knowledge, no studies have been made so far regarding the potential limitations of this mapping application, but one could hypothesize that the sensitivity of the sensor and the atmospheric conditions during the measurements (i.e., wind speed and direction) are decisive.

Although plenty of research has been developed linking ethylene emission or VOC emission in apples to their maturity [8,16–18] and, in some literature, there are indications towards measuring ethylene in the field [11,19], to the authors' knowledge, no work of this sort has been carried out. This work should, because of this, be considered as a first attempt at understanding the potential and the limitations of such measurements, creating with it a theoretical framework from which further work can be developed.

On the other hand, in air quality monitoring systems, some development has occurred considering mobile measurement platforms such as a UAV, especially when it comes to gas source localization and adaptive path planning for gas plume estimations [20,21]. Additionally, the optimal position of a gas sensor in the UAV has been studied using a simulation approach by [22]. Although some successful gas sensing experiments have been reported with the sensor pointing down [23,24], no literature was found regarding the challenges of measuring in an orchard environment, especially when it comes to the dispersion dynamics and its effect on the measurement process. Additionally, most of these works were performed using artificial gas sources that are easily modeled and do not take into account the complexities of a natural emission source such as apples.

The main goal of this work is to evaluate if ethylene produced by apple orchards can be sensed using an electrochemical sensor mounted on a UAV. The evaluation is made using a model-based approach to identify the most influential factors for detection, after which the model results are compared to measurements from a UAV-mounted electrochemical sensor flown over an experimental apple orchard.

2. Materials and Methods

2.1. Study Area

The study area in which this research is based on is located in the Wageningen Plant Research for Flower bulbs, Nursery stock and Fruits in Randwijk, The Netherlands (see Figure 2a). A test plot of 0.17 ha of apple trees was selected (Study Area A). It had a length of 5 m in between tree rows and 1.1 m between trees in the row (5 × 1.1), which results in 14 lines of about 300 trees in the plot. Two apple (*Malus domestica*) cultivars are shown in Figure 2c: Junami and Golden Delicious (on the headers of each line for pollination purposes). Additionally, one other test plot was selected: Study Area B, which is a traditional apple orchard with 5 m between rows and 1 m between apple trees. The variety in this plot is Natyra. Only two lines were selected in Study Area B shown in Figure 2d.

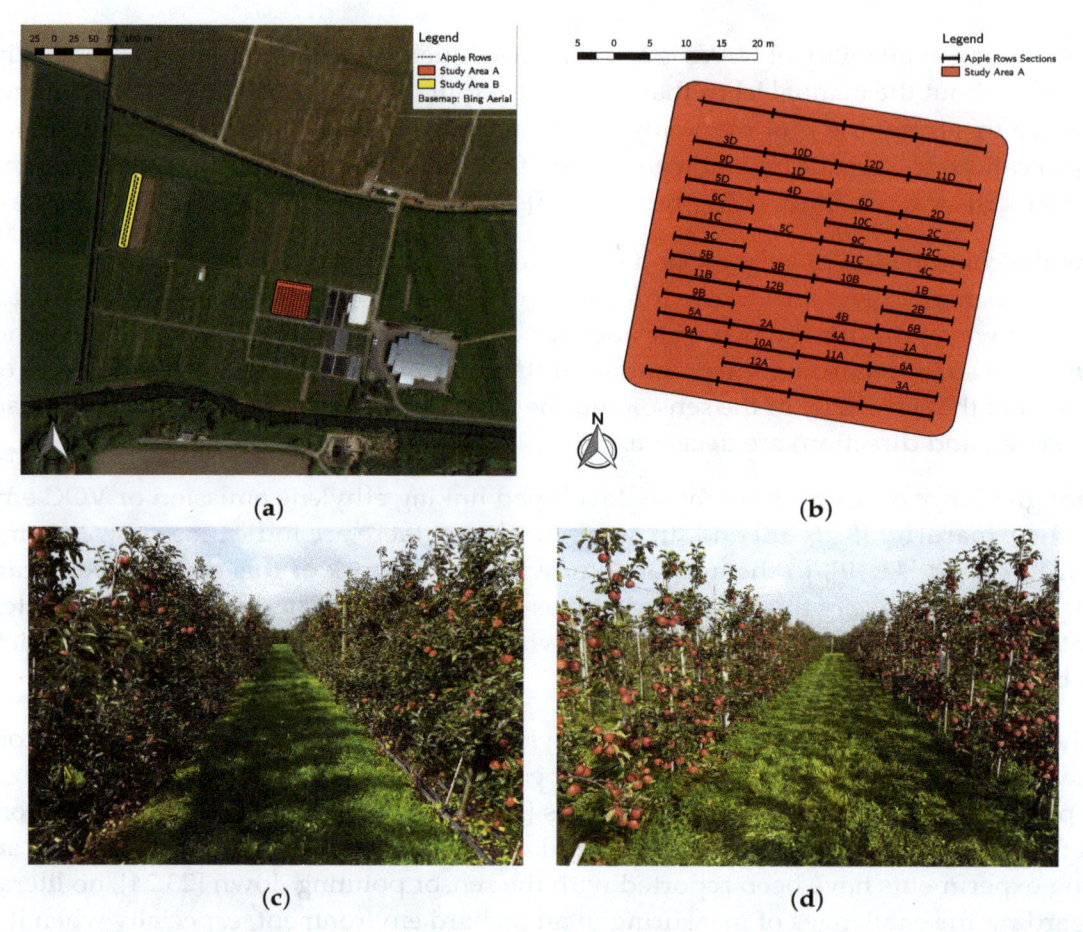

Figure 2. General and detailed map of the study area. (**a**) Map of the selected study areas in Randwijk (A and B). (**b**) Sections of apple lines used for fruit load assessment in Study Area A. Some lines show discontinuities since trees were removed in that section. (**c**) Junami and Golden Delicious cultivar. (**d**) Natyra cultivar.

2.2. Ethylene Flying Detector

The selected ethylene sensor was the Winsen ME4-C2H4, an electrochemical gas sensor, also referred to as a Taguchi gas sensor (TGS). According to the sensor specification sheet, it has a sensing range of 0–100 ppm of C_2H_4 and a response and recovery time of 100 s. Furthermore, the manufacturer indicates that the sensor has less than 10% of error.

In order to test both the ethylene sensor and the entire prototype, several preliminary experiments were conducted. With this, response times (amount of time it takes the sensor to detect the presence of ethylene) and recovery times (amount of time until the sensor signal returns to null after the ethylene source is removed) were tested. Figure 3 illustrates the results from the experiment in a controlled environment during four hours. The ethylene-sensitive sensor was placed inside a sealed plastic container of 40 cm × 50 cm × 40 cm with four *Junami* apples. After 3.5 h, the box was open, and after that, the sensor was placed outside the box.

The UAV-based measurements were conducted with the Phantom 3 Professional. This is a quadcopter drone weighing 1280 g with approximately 23 min of maximum flight time designed primarily for photo and video capture applications. The default payload (an HD camera) was removed and replaced with the ethylene sensor, as illustrated in Figure 4. The maximum payload of the UAV is 300 g, and the total payload was 218 g (very similar to the default payload).

Additionally, the sensor prototype is equipped with a memory card when the device is on, with a configurable measurement frequency, were measurements are recorded with the respective time-stamp and output signal from the sensor. In these experiments, one measurement per second (1 Hz) was determined as the measurement frequency.

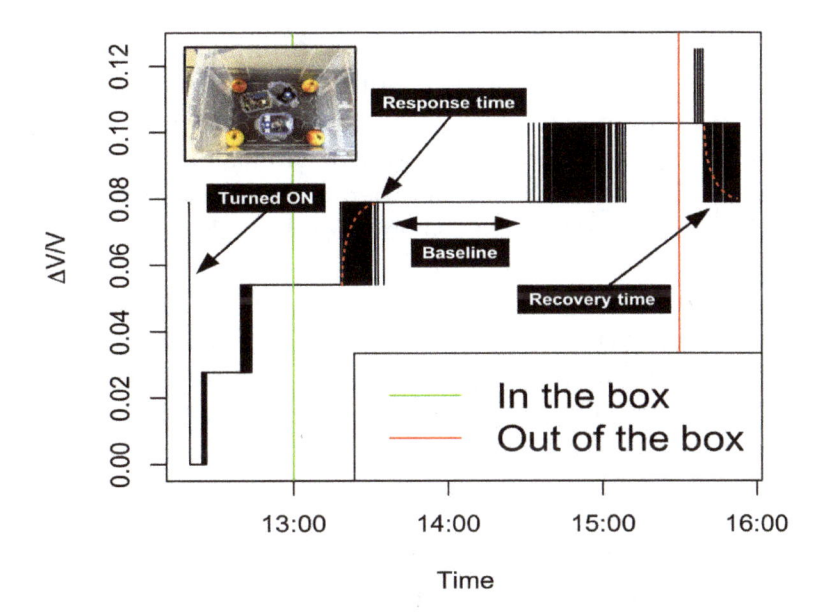

Figure 3. Tests conducted indoors in a sealed environment with an ethylene emission source (apples) that was placed in the box at the green line and removed at the red line.

The complete remote sensing system design for detecting and measuring ethylene is illustrated in Figure 4, and it has three main components: electrochemical sensor, Arduino board, and battery. The system was composed of commercially-available materials and open source tools. Finally, it can be easily acquired with a cost of less than 1000 Euros.

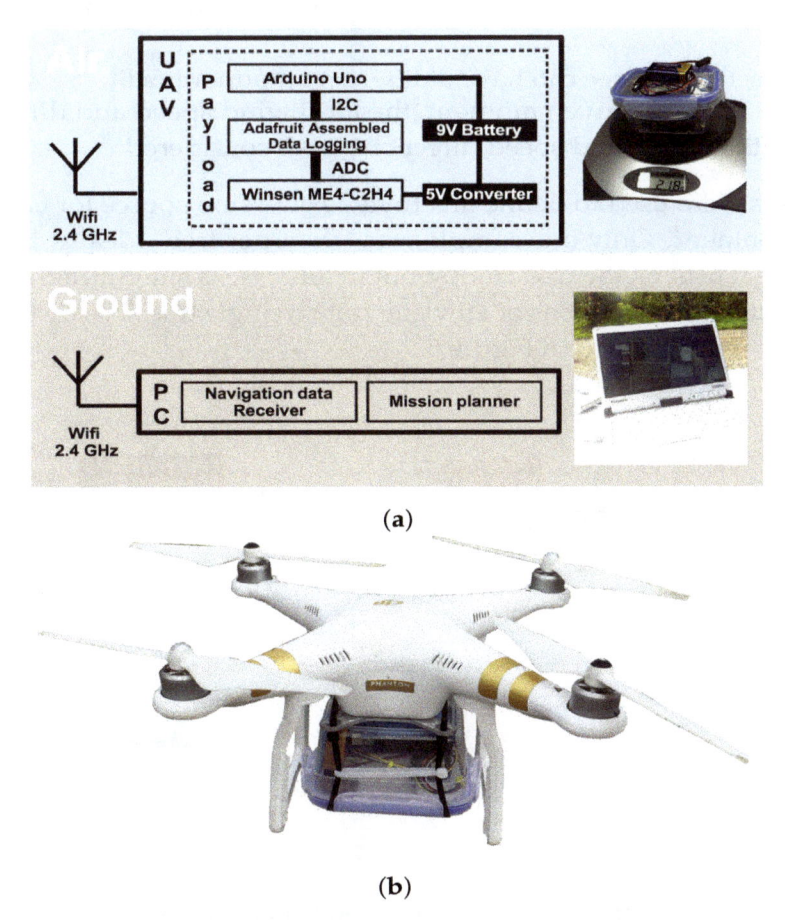

Figure 4. The ethylene flying-detector system: (**a**) air-ground system architecture and (**b**) Phantom 3 Professional (UAV) with the sensor prototype attached.

3. Determining the UAV Hovering Height

Understanding how the ethylene emission distributes above the orchard canopy is very important in order to define a starting sampling strategy. Determining the height above the orchard canopy where ethylene presents a higher concentration is not a trivial task mainly because there are several biophysical parameters such as the wind speed, temperature, and humidity. In this study, the wind speed (environment) and wind flow (UAV rotors) effect on the ethylene distribution was observed, while the temperature and humidity were omitted.

In order to determine the ideal sensing position, a modeling and simulation framework was developed to decrease the system deployment and testing times. Moreover, it allows a more reliable data acquisition by restricting the aerial sampling to areas within the orchard where a minimum ethylene concentration is expected. The modeling and simulation framework used GADEN, a gas dispersion simulation framework developed by [25], which is compatible with the ROS (Robot Operating System) [26].

3.1. Environment Wind Speed Modeling

Several parameters had to be obtained from the orchard field manager and from the research center where the experimental field is located in order to build the simulation workspace in GADEN. The parameters taken into account to simulate the ethylene distribution within the orchard when subject to wind were:

1. Ethylene emission (μ Lh^{-1} kg^{-1}). Each apple in the tree can in principle be at a different maturation stage and, therefore, have a different ethylene emission rate corresponding with three maturity stages.
2. Fruit position (height (m), direction (°)). Each apple in the tree can be at a different position in the canopy.
3. Fruit load (kg). Each apple tree can have a different amount of fruit.
4. Wind speed (ms^{-1}). In any given moment, the local wind speed and direction might vary. For this initial evaluation, two wind speed directions were considered.

These parameters were used to define an ethylene emission source for each tree represented in the simulation environment. Only one sample was taken per tree. Ideally, it would be possible to simulate each individual fruit on the tree canopy, but in this case, a simplification was performed using an artificial center for the total emissions of ethylene from a single tree. The distribution of the samples used in the simulations is illustrated in Figure 5.

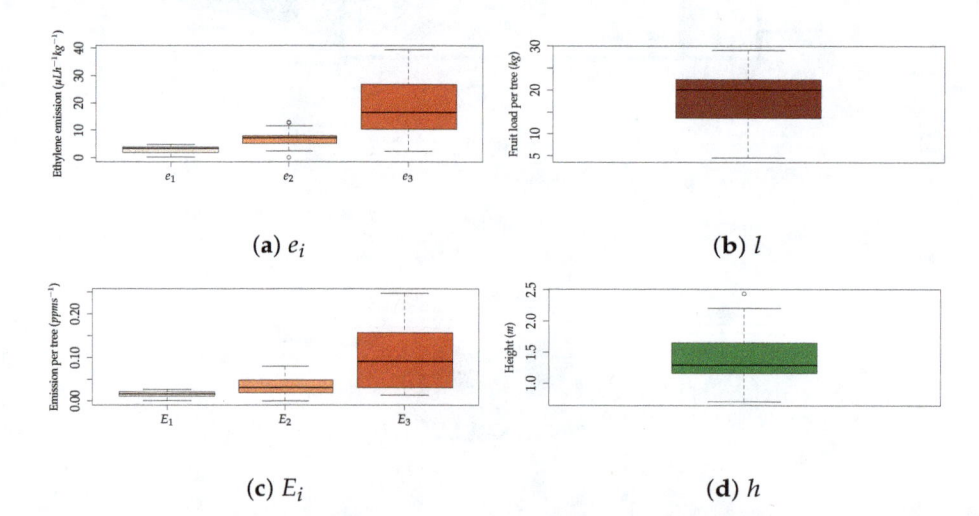

(a) e_i

(b) l

(c) E_i

(d) h

Figure 5. Distribution of the parameters used in the simulations: $\{e, E\}_1$, $\{e, E\}_2$, and $\{e, E\}_3$ stand for pre-climacteric, entering climacteric, and climacteric stages, respectively. Moreover, l stands for fruit load per tree and h for height.

This artificial center (P) can be described as the average position of the emission sources of the tree and is defined by a height (h) and direction in relation to the main stem (dir). This dir parameter in relation to the stem is defined in order to make the distribution of this parameter uniform, and therefore, the number of directions must be divisible by the amount of trees. In this case, six directions were defined, and each one was the sixth part of a circle, equivalent to 60°.

This simplification was applied mainly due to computational constraints. The simulator creates for each emission source a separate process, and for each process and time-step, the output is a simulation file of 90 MB. One can imagine that if each apple were simulated individually, the local memory of a standard computer would be very quickly surpassed. At the same time, according to our observations in the field, apples are usually clumped together in a branch, which means that the average distance between apples is in general small. Several branches can be further apart, but usually occupy one zone of the canopy. Figure 6 shows the orchard CAD model and the assumptions previously explained and used in the simulation process.

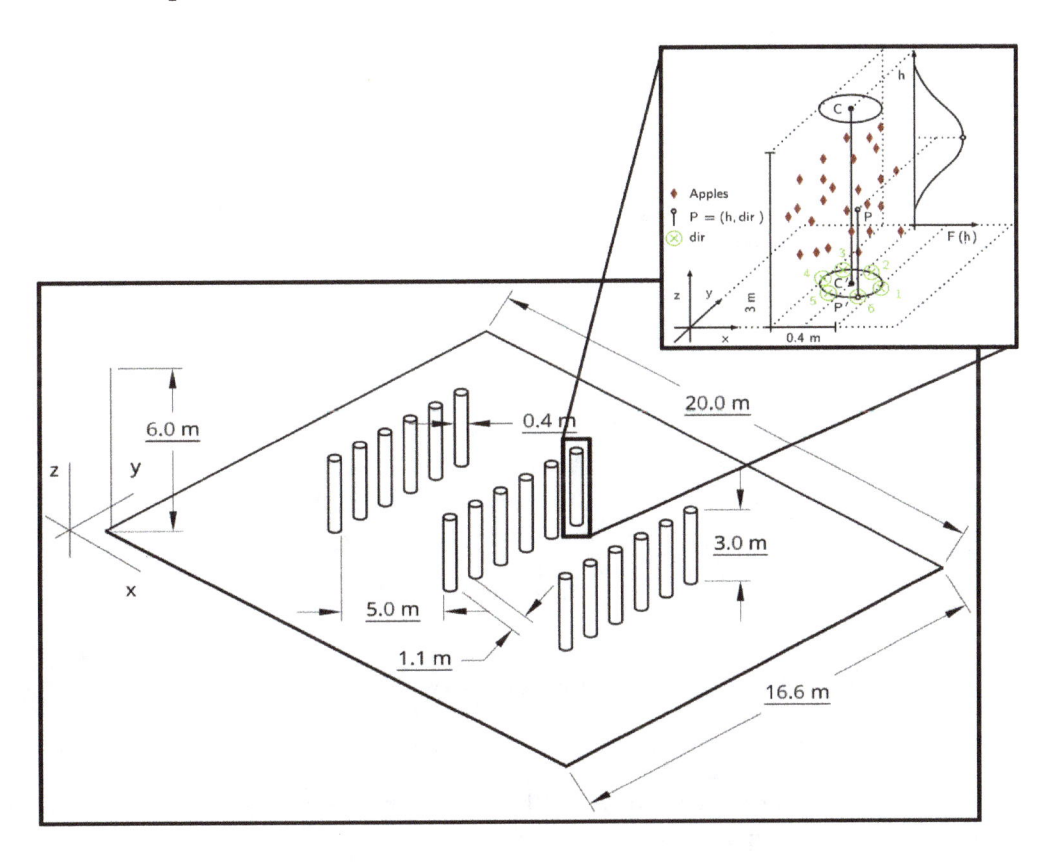

Figure 6. CAD model and respective parameters set in GADEN.

In the designed workspace, two environment inlets were set, the x-plane = 0 and the y-plane = 0. One of these inlets was chosen in order to simulate wind flow in a given direction: x for $\vec{x}$ and y for $\vec{y}$ (see Figure 7). The corresponding wind speed was assigned to this inlet, while the exact opposite plane (at the end of the environment) was set as a pressure outlet. All the other boundaries in the environment were set as walls with a slip setting. The computational fluid dynamics simulations were developed in SimScale, an online CFD software, with the recommended settings given in [25].

The number of ethylene-occupied cells in the environment is another important metric since it provides information on the probability of randomly finding an ethylene-filled cell. To get an understanding about which height is the most suitable to fly above the orchard, we must first look into the percentage of occupied cells with ethylene concentration above the canopy, as represented in Figure 8.

(a) $\vec{x} = 2$ ms^{-1}

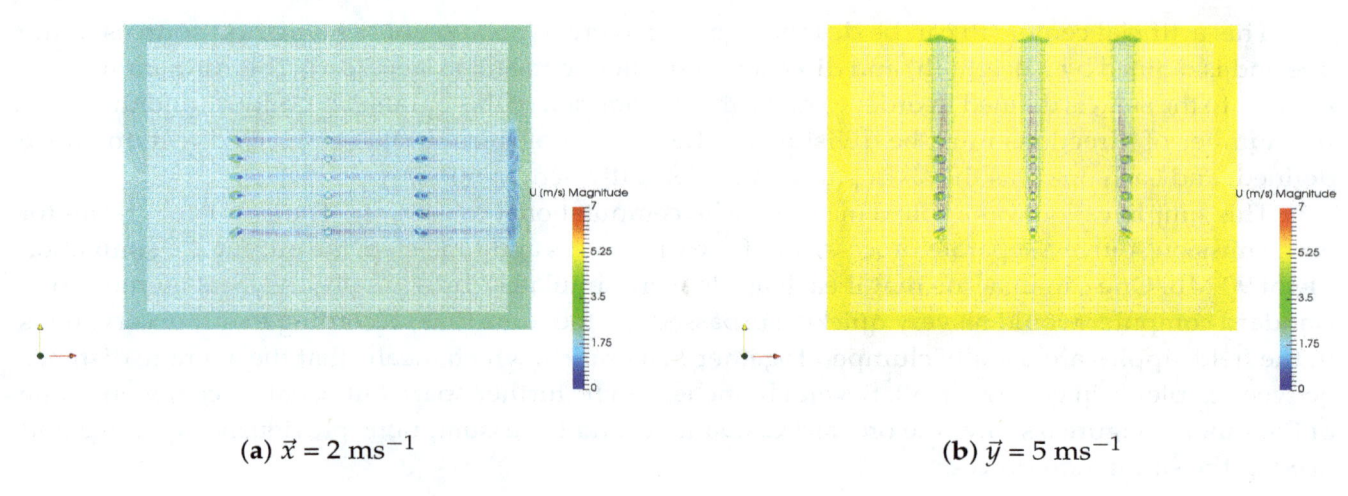

(b) $\vec{y} = 5$ ms^{-1}

Figure 7. Wind flow simulations used as input for GADEN without considering the rotors' airflow.

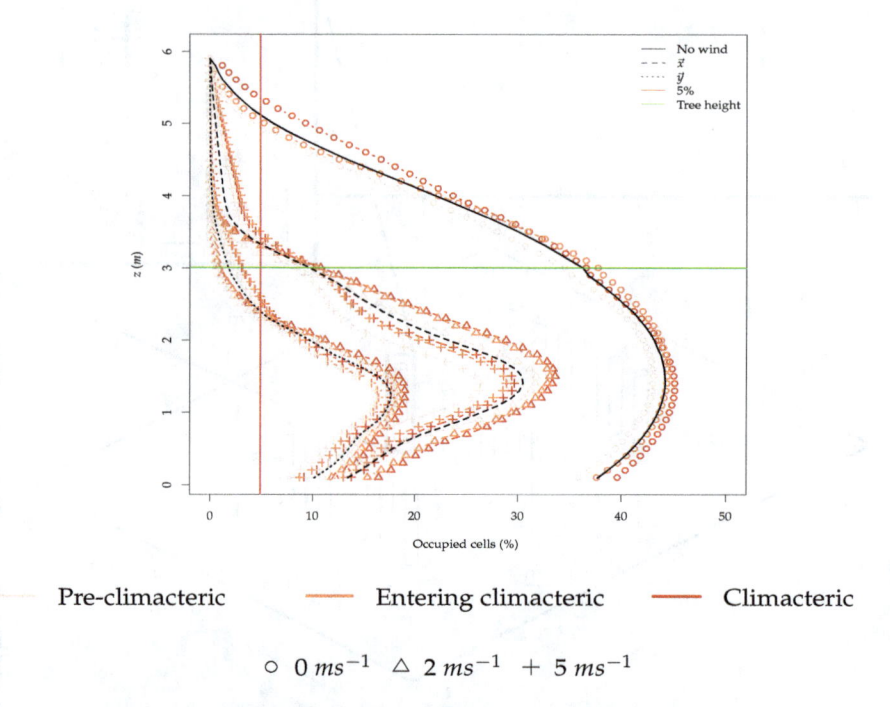

Figure 8. Percentage of occupied cells (cells with ethylene concentration higher than zero) in the environment across all time steps and simulations for the z-plane.

In almost all the simulations, less than 5% of the cells above the tree height were filled with ethylene. When wind speed was zero, there were more ethylene-filled cells above the tree height, but the majority of ethylene-filled cells can still be found under the tree height. It is also clear that the ethylene filled cells above the tree height had much lower ethylene concentration than the cells lower than the tree height. Therefore, the more likely place in the z axis to find ethylene-filled cells is between 1 and 2 m, where all simulations showed the biggest percentage of occupied cells.

To evaluate the impact of wind speed on the average ethylene concentration, Figure 9 was constructed. When looking at the environment, one can conclude that on average, a 1-ms^{-1} increase in wind speed results in a 30% decrease in average ethylene concentration. In the rows, the zone with higher average ethylene concentration, this decrease was 440%, while for in-between rows, this was only 110%. This difference is also accompanied by a very large difference in absolute ethylene concentration. This gives us an indication that choosing to sample in the rows might yield a higher concentration, but this measurement is very sensitive to the wind conditions.

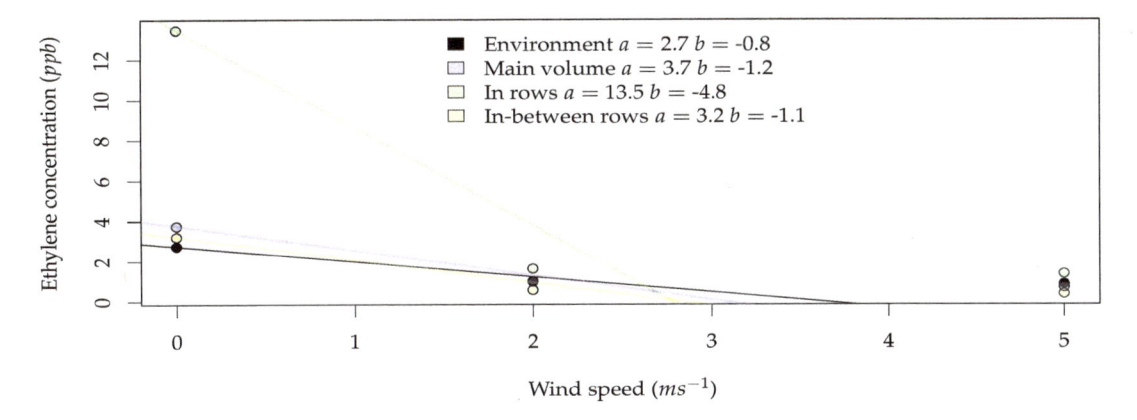

Figure 9. Relation between wind speed and average ethylene concentration in the four different zones. The colored lines represent the trend line for each zone, as given by the equation $y = a + bx$, where b is the decrease in average ethylene concentration (ppb) per additional unit of wind speed (ms^{-1}).

From Figures 8 and 9, it can be inferred that higher concentration levels of ethylene can be found below the trees and that the wind speed cut-off for the best practice is 2 ms^{-1}. In the next section, the rotors' airflow affect will be added to the environment to corroborate the results previously obtained omitting the rotors' airflow.

3.2. Rotors' Airflow Modeling

In order to simulate the effect of a UAV flying in the orchard, two different drone positions were considered: over the row (Position 1) and in between rows (Position 2). The drone over the row was positioned at 4 m, while the drone in between rows was positioned at 2 m. Only one wind scenario was considered for these simulations, $\vec{x} = 2$ ms^{-1}, in line with the results obtained in the previous section. This results in a total of six drone simulations, as exemplified in Table 1.

Table 1. Summary of drone simulator runs. The simulation number (#) will be used as a reference for naming each of these scenarios in the following sections.

#	Ethylene Emission	Wind Direction	Wind Speed (ms^{-1})	Drone Position
1.1	Pre-climacteric	$\vec{x}$	2	(10, 7.8, 4)
1.2	Entering climacteric			
1.3	Climacteric			
2.1	Pre-climacteric	$\vec{x}$	2	(12.5, 7.8, 2)
2.2	Entering climacteric			
2.3	Climacteric			

The wind flow caused by the rotors of the drone was modeled as four square air inlets with a given wind speed in the negative $\vec{z}$ direction. The squares had a width of 0.1 m, which is approximately the diameter of a single rotor in the Phantom 3 Professional. A relationship exists between the rotation speed of the propeller and the resulting wind speed generated, or thrust [27]. Taking the example of the Phantom 3 Professional in hovering flight in normal conditions, the rotors spin at around 8000 rpm [28], which results in an airflow of about 18 ms^{-1}. This wind speed was assigned to the velocity inlets mentioned above. The resulting wind flow simulations used as input for the GADEN simulations are displayed in Figure 10.

(a) $\vec{x} = 2$ ms^{-1}, Position 1.

(b) $\vec{x} = 2$ ms^{-1}, Position 2.

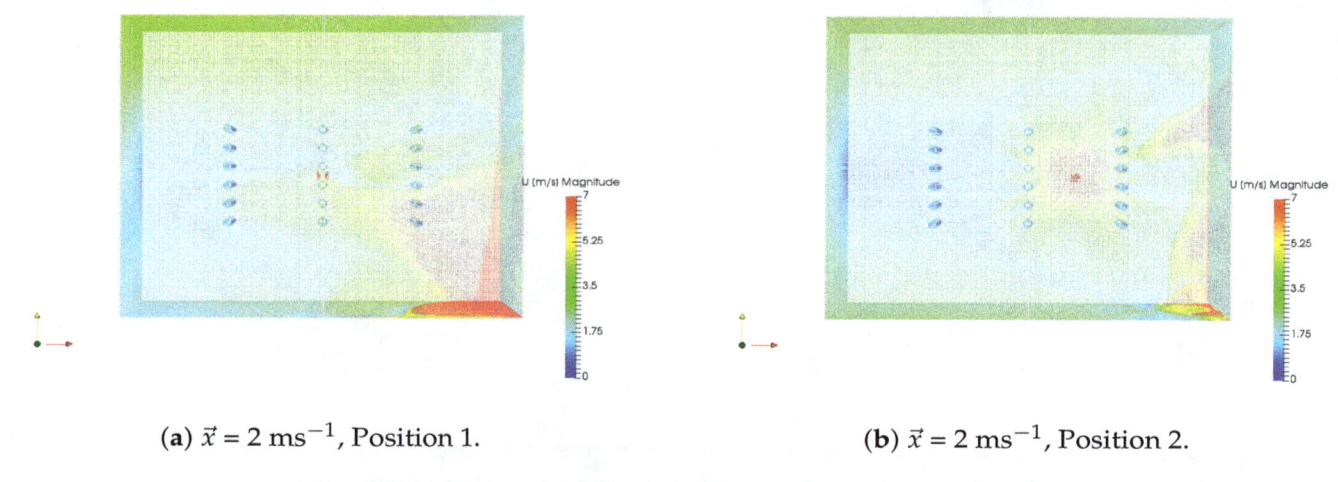

Figure 10. Drone wind flow simulations used as input for GADEN.

The biggest difference between the ethylene concentration distribution with and without a drone appears to be the range of values that are present. When looking at the climacteric simulations with the drone in both positions, the maximum concentration was about 250 ppb, while without the drone, the same conditions yielded a maximum of 300 ppb. This range also decreased substantially with the height of the drone (Position 2 to 1), from 250 to 150 ppb, as Figure 11 clearly shows. There was also a gas concentration effect right under the drone position where it appeared that the wind displacement of the gas decreased.

(a)

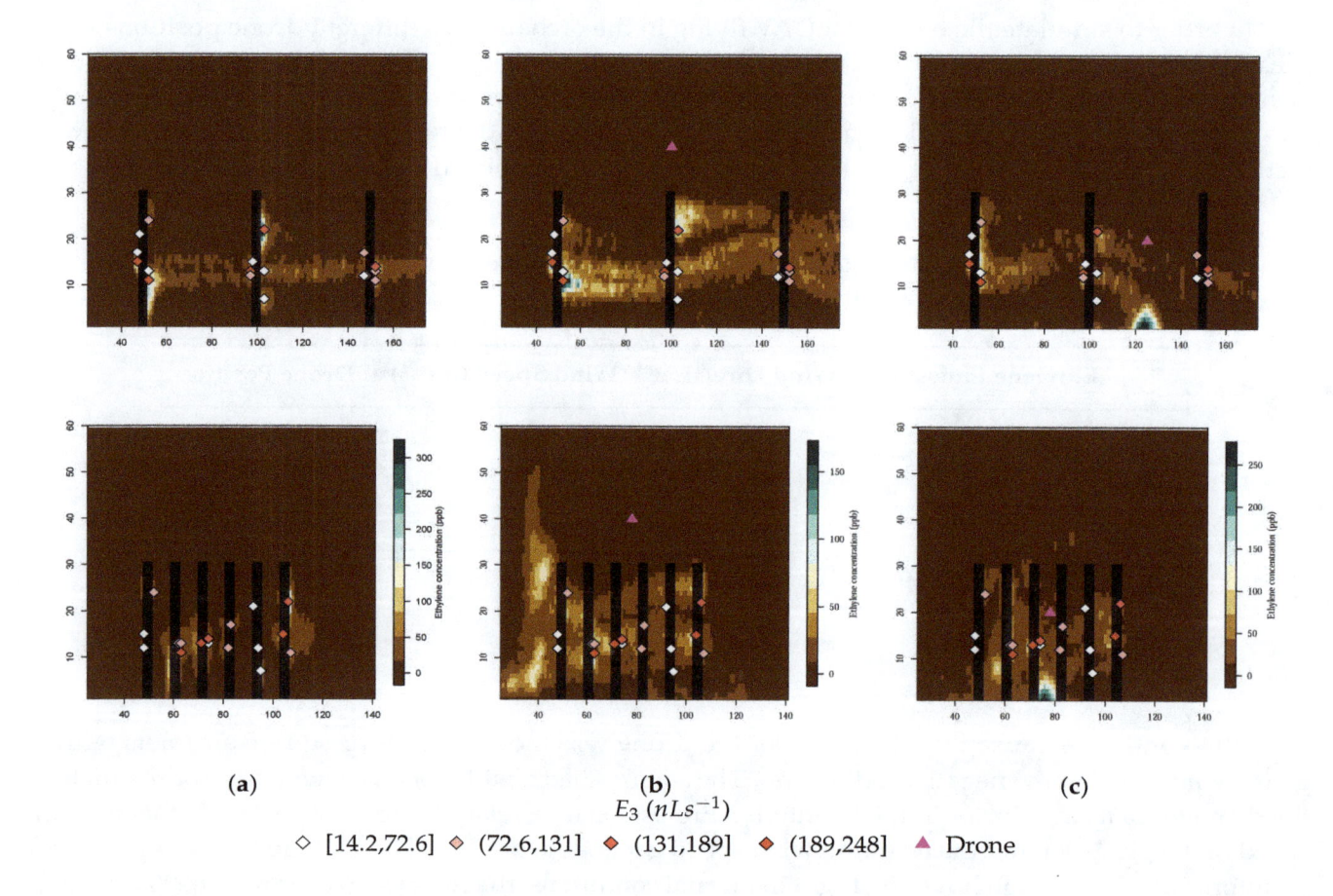

(b)
E_3 (nLs^{-1})
◇ [14.2,72.6] ◆ (72.6,131] ◆ (131,189] ◆ (189,248] ▲ Drone

(c)

Figure 11. Maximum ethylene concentration in the xz-plane (top plots) and yz-plane (bottom plots) for the drone simulations in the climacteric stage: (a) omitting rotor wind flow; (b) Drone Position 1; and (c) Drone Position 2. The ethylene sources' position and emission rate are also provided at the bottom.

When looking at the immediate vicinity of the position of the drone, a clear difference was detected between Position 1 and 2, as Figure 12 illustrates. While no ethylene was detected around Position 1, at Position 2, in every simulated time step, ethylene was present. This is a consequence of the concentration effect mentioned above.

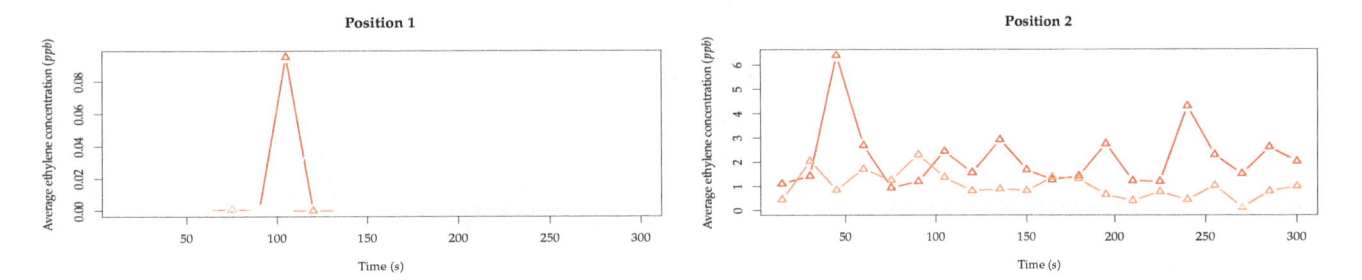

Figure 12. Average ethylene concentration across time in the vicinity of the drone position (± 0.2 m in xyz) for Positions 1 and 2.

The distribution of the occupied cells in the environment was also very different, as Figure 13 illustrates. The percentage of occupied cells was in general lower due to the increase in average wind speed in the environment, and especially on the z axis, a compression of the occupied cells closer to $z = 0$ was visible, depending on the height of the drone, which further confirms the concentration of ethylene effect described previously. This compression results in a higher percentage of occupied cells closer to the ground.

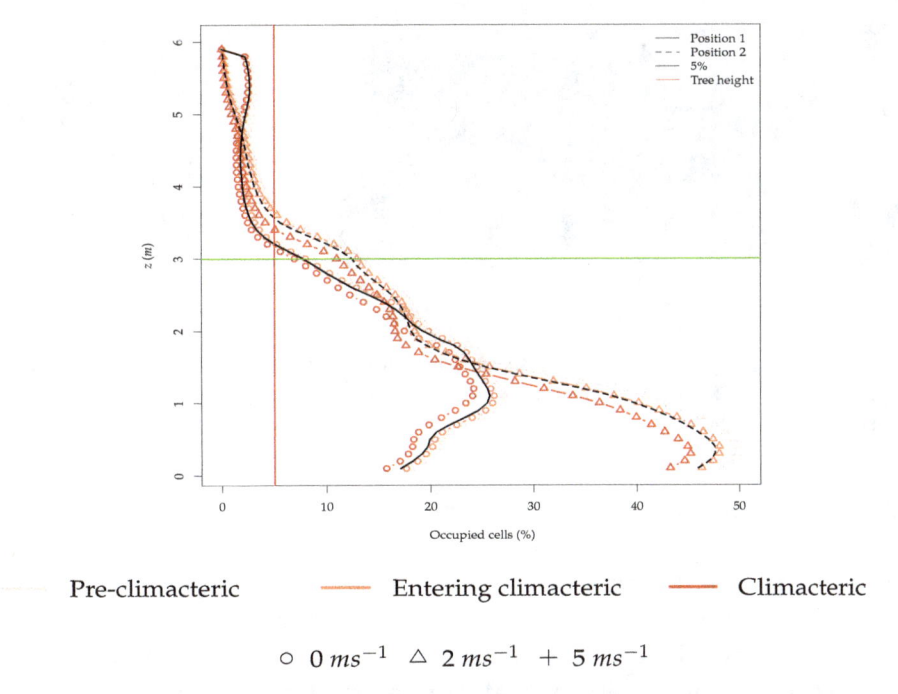

Figure 13. Percentage of occupied cells (cells with ethylene concentration higher than zero) in the environment across all time steps and simulations for the z-plane.

In general, we can say that the drone flying overhead had two main effects: a decrease in average ethylene concentration in the orchard, directly correlated with the height of the drone (4 m caused more gas dispersion than 2 m) and a concentration of gas directly under the drone, close to the ground (an effect that was more discernible at a 2-m height). In general, the drone flying overhead at 4 m caused a decrease in average ethylene concentration of 95%, while at 2 m, a decrease in 90%.

4. Field Tests on the Orchard

The results obtained in Figure 3 provide evidence that both the wind speed and rotor wind flow played an important role in ethylene sensing. The behavior observed in the simulation is used now as the reference to define boundaries in the field tests, observe other measuring heights, and analyze the feasibility of this practice in the real orchard environment from an aerial mission perspective.

4.1. Sampling Scheme

In order to analyze the sensor functioning and detect ethylene in the selected plot on the ground and using a UAV, both a spatial and temporal sampling scheme was defined. The measurements with the UAV were conducted also per measurement point: at 6 m and 12 m during 120 s. This difference in sensing time has to do with practical constraints related to the battery life of the UAV. Please refer to Figure 14.

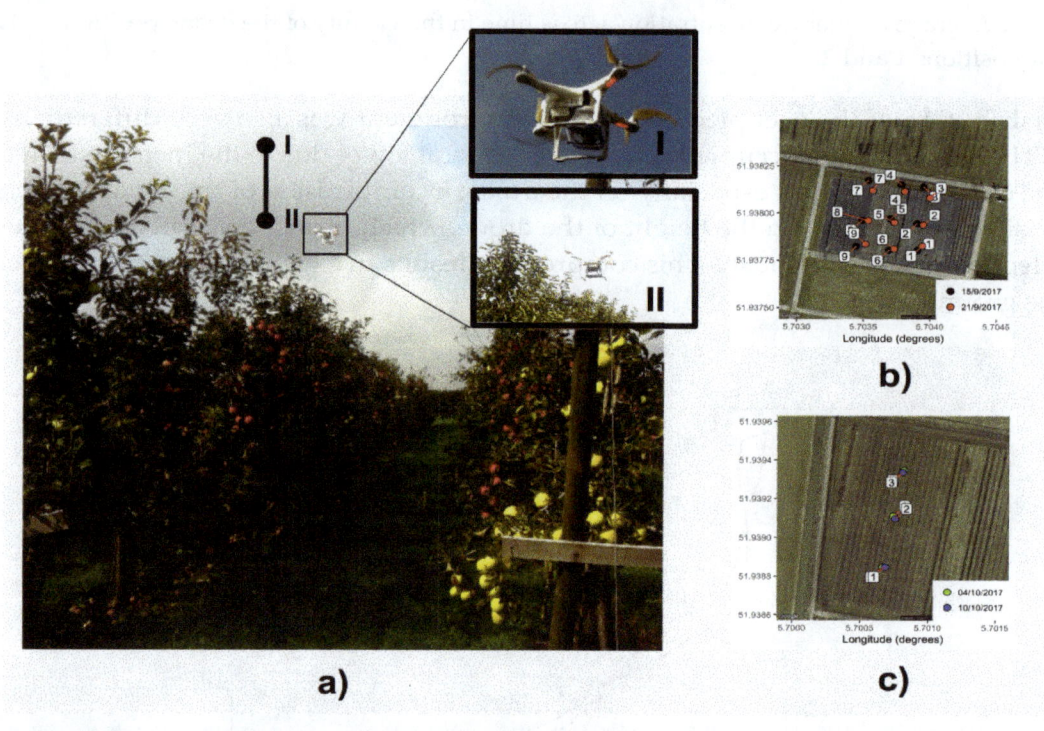

Figure 14. Hovering and sampling at two different position per experiment: (**a**) low and high heights within the orchard; (**b**) samples on Study Field A over the two days; and (**c**) samples on Study Field B over the two days.

For Study Area A (Figure 2a,c), nine measurement points were selected for UAV-based measurements on two different days: 15 and 21 September 2017. These points were selected using a three by three grid in the plot and placing the point roughly in the center of each grid. Additionally, two UAV-based measurements were conducted also per measurement point: at 6 m and 12 m above the ground (3 m and 9 m above the canopy), with a sensing time of less than 3 min. This difference in sensing time has to do with practical constraints related to the battery life of the UAV employed. With a battery charge from the Phantom 3 Professional, Study Area A was sampled a maximum of nine times.

For Study Area B (Figure 2a,d), the same spatial sampling approach as Study Area A was used, but in this case, only two lines were taken into account. Three points were selected in the middle of these two apple lines, where measurements were taken as described before, with a different height in the UAV measurements: 4 and 6 m. These measurements were also performed on two different dates: 4 and 10 October 2017.

4.2. System Deployment and Sensitivity Tests

A weather station adjacent to the plot was used to provide real-time atmospheric measurements of wind speed and direction, air temperature, and air moisture during the sampling period on the different dates. The average wind speed on 15 September was about 3 ms^{-1}, while on 21 September, it was about 3.2 ms^{-1}. The average wind speed on 4 October was about 5 ms^{-1}, while on 10 October, it was about 3.9 ms^{-1}. Furthermore, the first harvesting day for Study Areas A and B was, respectively, 28 September and 10 October.

The average measurements obtained per day are shown in Figure 15. Looking at the UAV-based measurements, no output was measured above 10% of the reference signal, thus achieving a maximum of 0.5 ppm (500 ppb). We can notice that there was no variation in the first two days of measurements (Study Field A). Nevertheless, there was an increase from the 4–10 October, which in this case suggested that there was more ethylene concentration on the second date (which was expected). The decrease in wind speed might also explain some of this variation from one date to the other.

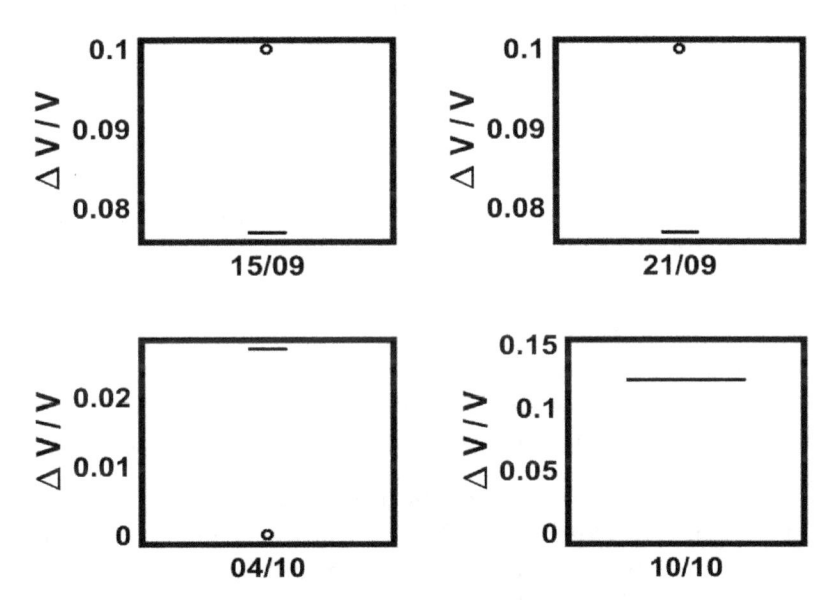

Figure 15. Sensor voltage output for aerial measurements on different days.

When flying at different heights (see Figure 16), an interesting behavior was observed: the measurements performed at a higher altitude (12 m) showed no variation from the baseline; at 6 m, there were more outliers that indicated more detection peaks, and at only 4 m, some variation was actually detected.

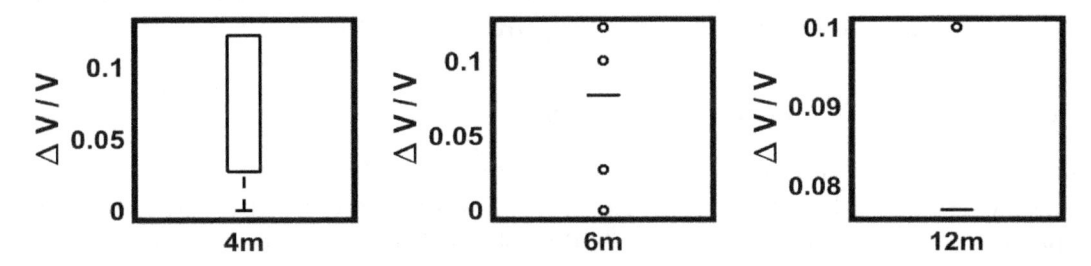

Figure 16. Sensor voltage output for aerial measurements at different heights.

5. Discussion

Although plenty of research has been developed linking ethylene emission or VOC emission in apples to their maturity [8,16–18] and in some literature there are indications towards measuring ethylene in the field [11,19], to the authors' knowledge, no work of this sort has been carried out, and there is to date no investigation addressing flying ethylene-sensitive sensor systems.

The modeling part of this study has shown that several factors influence the ethylene emission from the apple trees: these will be discussed in more detail below. We included the main influencing factors in the model, but additional ones, like time of the day, might be important, as well. These should be evaluated in follow-up studies.

In this section, the lessons learned from this study and major outcomes will be summed up.

5.1. Wind Speed and Rotors' Effect

The hypothesis that wind speed decreases the probability of ethylene detection was verified through simulation. Although, this supposition was expected, in the literature, there was not any discussion of the the maximum wind speed to ensure the minimum sensing. It was shown through simulation that the wind speed cut-off was 2 ms^{-1} (Figure 9).

Moreover, several authors discussed the rotor effect when using this sensor technology mounted in multi-rotor UAVs. It was stated that this is true, and there is a small margin left for detection that may be improved at a determinate hovering height (Figure 8).

5.2. Theoretical versus Practical Optimal Sampling Height

Simulation and practical results agreed that for wind speeds higher than 2 ms^{-1}, there would be very few or almost no detections. The minimum wind speed recorded during the field campaign was 3 $m\,s^{-1}$, and indeed, the sensor variation was very low.

The most important analytical outcome reinforced by the field tests was that flying lower would increase ethylene detection. Furthermore, flying close to or under the tree canopy gave a better result, although the margin for detection was limited, as shown in Figure 13. For the sake of the security and safety of the platform and the flying crew, the drone did not fly under the tree canopy, but it was stated that as the height decreased, more variation was observed (see Figure 16).

This study suggests that the UAV overflight should be performed at the lowest possible height to decrease the impact of the wind flow generated by the rotors on the ethylene distribution. However, further flying maneuvers should be explored when flying within or close to the orchard.

5.3. Discrete versus Continuous Sampling

It is important to note that the UAV measurements will only be considered during hovering flights over the determined measurement point. With that, the data acquired during the path of the UAV in the study area was not determinant, but it is an important point for further research considering moving and continuous measurements. However, the increased response time from the sensors (>90 s) should be taken into consideration in the sampling strategy to adopt.

5.4. Ethylene Detection over the Season and Inferring the OHD

The simulations showed that the range of ethylene concentration in an orchard was in the ppb range, and wind speed had a very big impact on this ethylene concentration. According to [29], the usual measuring range for VOCs starts at 100 ppb, and not very many sensors offer a sub-ppb range. If the state-of-the art of the technology does not provide a sensor with such characteristics, this might decrease the feasibility of this remote sensing strategy.

The observations when entering the climacteric stage were expected to be stronger than the ones observed in Figure 12. In order to determine the OHD, an ethylene increase during this stage was expected. Therefore, this strengths even more the idea that a sensor with more sensibility must be considered in future experiments.

5.5. Feasibility of Using Flying Ethylene-Sensitive Sensors

The challenges of measuring ethylene with a UAV in an orchard environment especially when it comes to the dispersion dynamics and its effect on the measurement process was not found in the literature.

On the other hand, in air quality monitoring systems, some development has happened considering mobile measurement platforms such as a UAV, especially when it comes to gas source localization and adaptive path planning for gas plume tracking [20,21]. The benefits of using the mobile platform for air quality monitoring and also for the purpose of this research are similar: they can offer high resolution sampling both at a spatial and temporal level at a low cost [23]. However, most of these works are performed using artificial gas sources that are easily modeled, and none takes into account the complexities of a natural emission source such as apples trees.

The optimal position of the gas sensor also has been discussed [22,23] and could have been a valuable reference in this study. However, in one work, the authors performed experiments indoors, inside a garage, and in the other work, the outcomes provided were very limited. Both authors suggested different sensors placements: pointing down separated from the main frame and on the top of the platform. In this study, the sensor was used pointing down because the frame of the UAV did not allow other configurations. Further, studies are needed to explore the position suggestion from the previous authors. Nevertheless, the simulations provided in this study reveal that higher concentrations values will be found mostly below the platform (see Figure 13).

5.6. UAV vs. UGV

The discussion of which mobile vehicle will perform better in a determinate agricultural management task is not new. In general, UAVs have more of a sensing role, like aerial surveying, where there is the need to increase spatial resolution; while Unmanned Ground Vehicles (UGV) have more of an actuation role, where there an action should be performed, such as mechanical weeding [30].

In this study, we were interested in a versatile platform that could carry different instrumentation and sample the orchard on different 2D and 3D positions. Moreover, while this could be achieved at different heights with the UAV, it would be limited to a static height with the UGV. Moreover, UAVs are considerable better than UGVs, as regards the price, maintenance, and portability. Summing up, they can offer high resolution sampling both at a spatial and temporal level at a low cost [23].

6. Conclusions

This is the first study to investigate the feasibility of using a flying ethylene-sensitive sensor systems in a fruit orchard only some days before being harvested. A simulated environment built from field data was used to understand the spatial distribution of ethylene within the apple orchard, to define the field sampling boundaries, and to evaluate how this influences the detection from a miniaturized sensor on a UAV. Finally, some preliminary tests in the orchard field were carry out to elucidate the sensor;s sensitivity and to contrast with the theoretical study.

The drone flight effect on the ethylene distribution was tested, and we concluded that flying at a higher altitude will cause more disturbance and lower the average ethylene concentration than flying lower. At the same time, at higher altitude, almost no ethylene is present in the vicinity of the drone. In general, the drone flying overhead at 4 m causes a decrease in average ethylene concentration of 95%, while at 2 m, a decrease of 90%. The detection margin is short and not sufficient to infer the fruit maturity, where increased variability over the season is expected. With these results, the issue of the measurement system sensitivity is further confirmed: a requirement for a sub-ppb ethylene sensor is clearly supported.

The use of a UAV to perform ethylene measurements in an uncontrolled environment such as an apple orchard still needs to be further explored, but it is suggested that future practices using this system are imminent with further research. The effect different UAV propeller spans on the intensity of dispersion of the gas and also detailed response models of different sensor models are, among others, pressing issues to be considered in the future.

Author Contributions: Conceptualization, J.V. and R.A.; methodology, J.V. and R.A.; software, R.A.; validation, R.A.; formal analysis, J.V. and R.A.; investigation, J.V. and R.A.; resources, J.V. and L.K.; writing, original draft preparation, all; writing, review and editing, all; supervision, J.V. and L.K.; project administration, J.V.; funding acquisition, L.K.

Acknowledgments: The authors would also like to thank Pieter van Dalfsen, whom is with Wageningen Plant Research, for providing the field test and data that helped us to understand more about the apple orchard lifecycle.

References

1. Alexandratos, N.; Bruinsma, J. *World Agriculture Towards 2030/2050: The 2012 Revision*; ESA Working Paper No. 12-03; FAO, Agricultural Development Economics Division: Roma, Italy, 2012.
2. Kader, A.A. Fruit maturity, ripening, and quality relationships. *Int. Symp. Effect Pre- Postharvest Factors Fruit Storage* **1997**, *485*, 203–208. [CrossRef]
3. Barbosa-Cánovas, G.V. *Handling and Preservation of Fruits and Vegetables by Combined Methods for Rural Areas: Technical Manual*; Number 149 in 1; Food & Agriculture Organization: Roma, Italy, 2003.
4. Knee, M. *Fruit Quality and Its Biological Basis*; CRC Press: Boca Raton, FL, USA, 2002; Volume 9.
5. Paul, V.; Pandey, R.; Srivastava, G.C. The fading distinctions between classical patterns of ripening in climacteric and non-climacteric fruit and the ubiquity of ethylene—An overview. *J. Food Sci. Technol.* **2011**, *49*, 1–21. [CrossRef] [PubMed]
6. Baietto, M.; Wilson, A. Electronic-Nose Applications for Fruit Identification, Ripeness and Quality Grading. *Sensors* **2015**, *15*, 899–931. [CrossRef] [PubMed]
7. Arshak, K.; Moore, E.; Lyons, G.; Harris, J.; Clifford, S. A review of gas sensors employed in electronic nose applications. *Sens. Rev.* **2004**, *24*, 181–198. [CrossRef]
8. Ma, L.; Wang, L.; Chen, R.; Chang, K.; Wang, S.; Hu, X.; Sun, X.; Lu, Z.; Sun, H.; Guo, Q.; et al. A Low Cost Compact Measurement System Constructed Using a Smart Electrochemical Sensor for the Real-Time Discrimination of Fruit Ripening. *Sensors* **2016**, *16*, 501. [CrossRef] [PubMed]
9. Génard, M.; Gouble, B. ETHY. A theory of fruit climacteric ethylene emission. *Plant Physiol.* **2005**, *139*, 531–545. [CrossRef] [PubMed]
10. Cristescu, S.M.; Mandon, J.; Arslanov, D.; Pessemier, J.D.; Hermans, C.; Harren, F.J.M. Current methods for detecting ethylene in plants. *Ann. Bot.* **2012**, *111*, 347–360. [CrossRef] [PubMed]
11. Łysiak, G. Measurement of ethylene production as a method for determining the optimum harvest date of "Jonagored" apples. *Folia Hortic.* **2014**, *26*. [CrossRef]
12. Gómez, A.H.; Wang, J.; Hu, G.; Pereira, A.G. Electronic nose technique potential monitoring mandarin maturity. *Sens. Actuators B Chem.* **2006**, *113*, 347–353. [CrossRef]
13. Kathirvelan, J.; Vijayaraghavan, R. An infrared based sensor system for the detection of ethylene for the discrimination of fruit ripening. *Infrared Phys. Technol.* **2017**, *85*, 403–409. [CrossRef]
14. Lihuan, S.; Liu, W.; Xiaohong, Z.; Guohua, H.; Zhidong, Z. Fabrication of electronic nose system and exploration on its applications in mango fruit (M. indica cv. Datainong) quality rapid determination. *J. Food Meas. Charact.* **2017**, *11*, 1969–1977. [CrossRef]
15. Zhang, C.; Kovacs, J.M. The application of small unmanned aerial systems for precision agriculture: A review. *Precis. Agric.* **2012**, *13*, 693–712. [CrossRef]
16. Brezmes, J.; Llobet, E.; Vilanova, X.; Saiz, G.; Correig, X. Fruit ripeness monitoring using an electronic nose. *Sens. Actuators B Chem.* **2000**, *69*, 223–229. [CrossRef]
17. Saevels, S.; Lammertyn, J.; Berna, A.Z.; Veraverbeke, E.A.; Di Natale, C.; Nicolai, B.M. Electronic nose as a non-destructive tool to evaluate the optimal harvest date of apples. *Postharvest. Biol. Technol.* **2003**, *30*, 3–14. [CrossRef]
18. Young, H.; Rossiter, K.; Wang, M.; Miller, M. Characterization of Royal Gala Apple Aroma Using Electronic Nose TechnologyPotential Maturity Indicator. *J. Agric. Food Chem.* **1999**, *47*, 5173–5177. [CrossRef] [PubMed]

19. Pathange, L.P.; Mallikarjunan, P.; Marini, R.P.; O'Keefe, S.; Vaughan, D. Non-destructive evaluation of apple maturity using an electronic nose system. *J. Food Eng.* **2006**, *77*, 1018–1023. [CrossRef]
20. Neumann, P.P.; Bennetts, V.H.; Lilienthal, A.J.; Bartholmai, M.; Schiller, J.H. Gas source localization with a micro-drone using bio-inspired and particle filter-based algorithms. *Adv. Robot.* **2013**, *27*, 725–738. [CrossRef]
21. Rossi, M.; Brunelli, D.; Adami, A.; Lorenzelli, L.; Menna, F.; Remondino, F. Gas-Drone: Portable gas sensing system on UAVs for gas leakage localization. In Proceedings of the IEEE SENSORS 2014 Proceedings, Valencia, Spain, 2–5 November 2014. [CrossRef]
22. Roldán, J.; Joossen, G.; Sanz, D.; del Cerro, J.; Barrientos, A. Mini-UAV Based Sensory System for Measuring Environmental Variables in Greenhouses. *Sensors* **2015**, *15*, 3334–3350. [CrossRef] [PubMed]
23. Villa, T.F.; Gonzalez, F.; Miljievic, B.; Ristovski, Z.D.; Morawska, L. An overview of small unmanned aerial vehicles for air quality measurements: Present applications and future prospectives. *Sensors* **2016**, *16*, 1072. [CrossRef] [PubMed]
24. Neumann, P.P. Gas Source Localization and Gas Distribution Mapping with a Micro-Drone. Ph.D. Thesis, Bundesanstalt für Materialforschung und-prüfung (BAM), Berlin, Germany, 2013.
25. Monroy, J.; Hernandez-Bennets, V.; Fan, H.; Lilienthal, A.; Gonzalez-Jimenez, J. GADEN: A 3D Gas Dispersion Simulator for Mobile Robot Olfaction in Realistic Environments. *Sensors* **2017**, *17*, 1479. [CrossRef] [PubMed]
26. Quigley, M.; Conley, K.; Gerkey, B.P.; Faust, J.; Foote, T.; Leibs, J.; Wheeler, R.; Ng, A.Y. ROS: An open-source Robot Operating System. *ICRA Workshop Open Source Softw.* **2009**, *3*, 5.
27. Allain, R. Modeling the Thrust from a Quadcopter. Available online: https://www.wired.com/2014/05/modeling-the-thrust-from-a-quadcopter/ (accessed on 1 April 2018).
28. DJI. *Phantom 3 Professional User Manual*; DJI: Shenzhen, China, 2017.
29. Spinelle, L.; Gerboles, M.; Kok, G.; Persijn, S.; Sauerwald, T. Review of Portable and Low-Cost Sensors for the Ambient Air Monitoring of Benzene and Other Volatile Organic Compounds. *Sensors* **2017**, *17*, 1520. [CrossRef] [PubMed]
30. Conesa-Muñoz, J.; Valente, J.; del Cerro, J.; Barrientos, A.; Ribeiro, A. A Multi-Robot Sense-Act Approach to Lead to a Proper Acting in Environmental Incidents. *Sensors* **2016**, *16*, 1269. [CrossRef] [PubMed]

Acquisition of Sorption and Drying Data with Embedded Devices: Improving Standard Models for High Oleic Sunflower Seeds by Continuous Measurements in Dynamic Systems

Simon Munder *, Dimitrios Argyropoulos and Joachim Müller

Institute of Agricultural Engineering, Universität Hohenheim, Garbenstrasse 9, 70599 Stuttgart, Germany; dimitrios.argyropoulos@uni-hohenheim.de (D.A.); joachim.mueller@uni-hohenheim.de (J.M.)
* Correspondence: info440e@uni-hohenheim.de

Abstract: Innovative methods were used to determine both sorption and drying data at temperatures typically found in the handling of agricultural products. A robust sorption measurement system using multiple microbalances and a high precision through flow laboratory dryer, both with continuous data acquisition, were employed as the basis for a water vapor deficit based approach in modeling the sorption and drying behavior of high oleic sunflower seeds. A coherent set of data for sorption (Temperature T = 25–50 °C, water activity a_w = 0.10–0.95) and for drying (T = 30–90 °C, humidity of the drying air x = 0.010–0.020 kg·kg^{-1}) was recorded for freshly harvested material. A generalized single-layer drying model was developed and validated (R^2 = 0.99, MAPE = 8.3%). An analytical solution for predicting effective diffusion coefficients was also generated (R^2 = 0.976, MAPE of 6.33%). The water vapor pressure deficit-based approach allows for an easy integration of meaningful parameters recorded during drying while maintaining low complexity of the underlying equations in order for embedded microcontrollers with limited processing power to be integrated in current agro-industrial applications.

Keywords: dynamic vapor sorption; high-precision dryer; modeling; water vapor pressure deficit

---

1. Introduction

Sunflower (*Helianthus annuus* L.) is one of the major oilseeds produced in the world. It is cultivated in different climatic zones with varying grain moisture content during the harvesting period. Drying is typically required in order to achieve an optimum final moisture content for safe storage. Excessive moisture levels may lead to a generally increased activity of microorganisms, heating of the product, dry matter losses, and high levels of free fatty acids in the extracted oil [1,2]. Several studies point out that only the non-fat components of the seeds are the critical parts for stability considerations in moisture-dependent storage [2–4]. Water activity a_w holds information on the availability of water for the growth of microorganisms and thus allows inference on threshold levels, above which spoilage is unlikely to occur [5]. It is defined as the partial vapor pressure of water in the measured food, divided by the partial vapor pressure of pure water [6]. This is equal to the equilibrium relative humidity, at which the measured food is in equilibrium with the surrounding atmosphere and does not adsorb nor desorb water. Sorption isotherms describe the relationship between the equilibrium moisture content (MC_e), formed at a given temperature and at the relative humidity, if the food is in equilibrium with the atmospheric surroundings. In general, water activity increases at higher moisture content and, consequently, microorganisms, such as molds, yeasts, and bacteria increasingly grow at a_w > 0.70, while enzymatic activity is also promoted by high values of a_w [5]. The commonly applied threshold value for safe farm level storage of agricultural products is found at water activities between $0.6 \leq a_w \leq 0.7$ [3,4,7].

In practice, vapor pressure manometers, capacitance hygrometers, and chilled mirror dew point hygrometers represent fast and robust techniques for the indirect measurement of water activity from a common set of partially dried samples [8]. However, these sorption techniques can not generate kinetic data. For the gravimetric measurement of moisture sorption, the static gravimetric method is considered a standard technique. Climatic test chambers have been used before. However, the sorption experiments were realized with discontinuous weight measurements on external balances or with balances inside the test chamber [9,10]. To minimize the negative effects associated with the discontinuous weight measurements, instruments using controlled atmosphere microbalances such as a Dynamic Vapor Sorption apparatus (DVS) have been employed for the automated moisture sorption analysis of food ingredients and other homogenous materials [11–17]. The DVS method is used to measure the equilibrium moisture content of a material at any desired relative humidity and selected temperatures in a short period of time. However, as DVS is designed for extremely small sample mass, bias in sorption measurements may occur when dealing with agricultural products and only a small part of a heterogeneous organ is being measured [7]. Based on the dynamic vapor sorption principle, an innovative experimental system for determining moisture sorption properties of heterogeneous agricultural products is needed. In addition, the system should enable monitoring of moisture sorption by measuring the weight gain or loss at regular time intervals and the automatic acquisition of mass data in more than one high precision ultra-microbalance.

The experimental determination of sorption characteristics allows for the description of moisture diffusivity to be a function of moisture content, partial vapor pressure, and temperature without the requirement to pre-define the mechanisms controlling diffusion [6]. Thus, these isotherms constitute a suitable tool in describing and modeling drying processes for agricultural products. Remarkably, the occurrence of hysteresis between adsorption and desorption is directly affected by the oil content while no significant difference can be observed [4] at values above 48.6%. Most studies on the sorption and drying of sunflower seeds have in common data from different varieties and harvest years, which are combined to provide an empirical basis for model development [3,18,19]. Additionally, seeds are usually remoistened to a desired moisture content, which potentially leads to experimental errors and an alteration of the drying behavior [19]. It is commonly agreed that moisture movement at the surface is negligible, compared to internal resistance, and, thus, the influence of air velocity becomes insignificant after a threshold of approximately 0.1 m·s^{-1} [19–21]. Sunflower seeds are comparable to multi-domain composite foods, consisting of a fibrous outer shell, and an oily kernel. Both hulls and kernels show significantly different sorption behavior [2]. In addition, whole seeds and kernels are significantly different in most physical properties, such as volume and equivalent diameter [19,22]. Sunflower kernels show significantly slower moisture diffusivity compared to hulls and, thus, are the limiting factor in drying [3,19].

However, a literature research revealed remarkable differences between the seed/hull-ratio of examined traditional oil-seed varieties on which most sorption and drying studies are based. Given these distinct variations in physical properties, it is obvious that sunflower sorption and drying models should be updated and validated for modern high oleic varieties.

The aim of the present work is, therefore, to develop a robust semi-empirical drying model using a coherent set of experimental sorption and drying data for high oleic sunflower seeds (*Helianthus annuus* L.). The objectives are: (i) to experimentally determine a broad set of equilibrium moisture content data by an automatic, gravimetric analyzer, (ii) to obtain single-layer drying kinetic data in a high precision laboratory dryer at different temperature and absolute humidity of the drying air, and (iii) to establish a generalized single-layer drying model in which its parameters are a function of air conditions. In addition, these related datasets are used to analytically determine moisture effective diffusion coefficients.

2. Materials and Methods

2.1. Plant Material

High oleic sunflower seeds (*Helianthus annuus* L.), F1 of hybrid cultivar 'PR65H22' were harvested mid October 2014 from a farm 30 km east of Würzburg (Germany). The original bulk of approximately 500 kg was reduced to a representative sample of 110 kg. To evaluate the moisture content at harvest (0.317 ± 0.008 kg·kg^{-1}), 10 samples of 3 g each were used.

2.2. Moisture Content Determination

The moisture content *MC* of seeds and hulls was measured from samples of 3 g by a standard thermogravimetric analysis in a convection oven at 103 ± 2 °C, according to ISO 665:2000 [23] and expressed in kg water per kg dry matter (kg·kg^{-1}). All analyses were performed in triplicates, which is commonly applied in other studies [18,22].

2.3. Determination of Dynamic Vapor Sorption Isotherms

The adsorption isotherms of seeds and hulls were separately determined by using an automated system designed at the Institute of Agricultural Engineering, University of Hohenheim (Stuttgart, Germany) (Figure 1). The system consists of a climatic test chamber (C + 10/600, CTS GmbH, Hechingen, Germany), which regulates air temperature between 10 °C and 95 °C while maintaining a relative humidity between 10% and 98% and ensuring air circulation. A weight measuring system, consisting of five high precision load cells (WZA1203-N, Sartorius AG, Goettingen, Germany) with an accuracy of 1 mg was mounted on top of the chamber to record the change in mass, caused either by adsorption or desorption of water vapor. Each load cell carries a perforated sample holder, which is suspended through the chamber ceiling. Load cell and climate chamber control is realized remotely. Direct control via an attached computer is also possible. Relative humidity is varied gradually at a pre-set temperature, which is held ceteris paribus during an experiment.

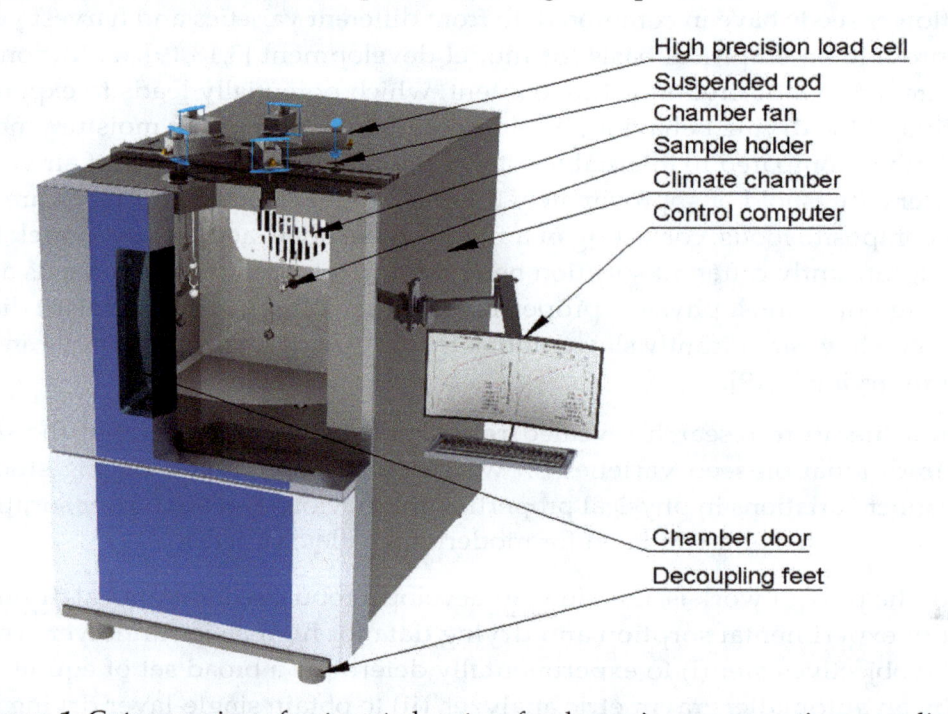

Figure 1. Cutaway view of automated system for dynamic vapor sorption recording.

The test material was dried at 60 °C to a moisture content of 0.006 ± 0.004 kg·kg^{-1} and manually cleaned from impurities. The seeds were ground to a particle size of approximately 5 mm. To obtain hulls, seeds were manually hulled. About 12 g of dried seeds or hulls were used per sorption experiment. The samples were loaded into the sample holder and the adsorption isotherms were measured at 25 °C and 50 °C, which increased the relative humidity gradually from 10% to 85% with increments of 10% and a final step of 5%. Mass, temperature, and humidity data were recorded in 20 min intervals. The equilibrium was considered to have been reached when observing a change in weight of less than 5% of the initial sample weight during 10 consecutive measurements. Three repetitions per temperature and material were performed, which resulted in a total of 160 individually determined equilibrium moisture content data points.

2.4. Sorption Isotherm Models

Five commonly applied, three parameter moisture sorption models were tested for their accuracy to describe the experimental sorption data [3,5,6,24]. The models are presented in terms of equilibrium moisture content MC_e (kg·kg^{-1}), water activity a_w ($p_{vs} \cdot p_{sat}^{-1}$), and a, b, c as model constants. The applied model equations and their ranges of validity are shown in Table 1.

Table 1. Models for sorption isotherms. MCe = equilibrium moisture content. T = temperature in °C. a_w = water activity. a, b, c = model constants [7].

Model	Original Plant Material	Validity (a_W)
Modified Chung-Pfost $MC_e = \frac{-1}{a} \ln\left(-\frac{(T+b)}{c} \ln(a_W)\right)$	Maize and maize components	0.1–0.9
Modified Oswin $MC_e = (a + b \times T)\left(\frac{a_W}{1-a_W}\right)^{\frac{1}{c}}$	Various	0.3–0.5
Modified Halsey $MC_e = \left(\frac{-exp(a+b \times T)}{\ln(a_W)}\right)^{\frac{1}{c}}$	Maize, wheat flour, laurel, nutmeg	0.1–0.8
Modified Henderson $MC_e = \left(-\frac{\ln(1-a_W)}{a(T+b)}\right)^{\frac{1}{c}}$	Maize	-
Modified G.A.B. $MC_e = \frac{ab\left(\frac{c}{T}\right)a_W}{(a-ba_W)\left(1-ba_W+\left(\frac{c}{T}\right)ba_W\right)}$	Various	<0.94

2.5. Thin-Layer Drying Experiments

Thin-layer drying experiments were conducted using a high precision hot-air laboratory dryer designed at the Institute of Agricultural Engineering, University of Hohenheim (Stuttgart, Germany), which allowed the control of the desired drying conditions over a wide range of operating parameters. For the drying experiments, 500 kg of freshly harvested seeds were manually cleaned and 70 subsamples of 1.5 kg each were randomly taken, vacuum-sealed in PEHD bags and stored at 4 °C for no longer than four weeks. The experimental system has been previously described in detail by Argyropoulos et al. [25]. In total, 63 individual drying experiments were conducted at temperatures T between 30 and 90 ± 0.1 °C in steps of 10 °C and an absolute humidity x of 0.010, 0.015, and 0.020 kg water per kg of dry air. During the individual drying experiments, temperature and absolute humidity were kept constant and a uniform air flow through the sample was maintained at 0.6 ± 0.05 m·s^{-1}. The initial seed moisture during all drying experiments was 0.317 ± 0.008 kg·kg^{-1}. An initial mass of 0.400 ± 0.001 kg was evenly spread on a perforated drying tray, which resulted in a layer depth of 15 mm. The tray was supported on PC6 load cells (Flintec GmbH, Meckesheim, Germany). The weight was measured every 10 min. Meanwhile, a bypass valve was opened to prevent floating of the tray. The seeds were dried until a constant weight was achieved, once the total weight change for three consecutive measurements was below 0.5 g. The drying experiments were repeated at least three times for each drying condition. The moisture content of seeds before and after drying was determined as described above (Section 2.2). In total, 63 experiments with 1291 single measurements were conducted.

2.6. Empirical Drying Model

By including equilibrium moisture information from the previous experiments, the normalized moisture ratio (MR) was computed. MR (t) is the average moisture ratio at time t (minutes), MC_t the moisture content at time t, MC_e the equilibrium moisture content, and MC_0 the initial moisture content (kg·kg^{-1}). The moisture ratio over time can be well depicted by the semi-empirical page equation (Equation (2)). This approach offers a compromise between inclusion of the physical theory and ease of use, which results in a semi-empirical model [6]. Fickian moisture migration, constant moisture diffusion coefficients isothermal conditions, and negligible shrinkage are the basics of this approach [26]. Two coefficients have to be fitted including k as the rate constant (min^{-1}) and n as the dimensionless coefficient to improve the fit [19].

To provide a generalized, semi-empirical model capable of representing different temperatures and humidity of the drying medium, k and n were transformed to a function of drying conditions at time t. It is important to highlight that the actually recorded conditions from the drying chamber were employed in the model instead of the set-point conditions.

Two different approaches, based on (i) temperature T (°C) and absolute humidity x (kg·kg^{-1}) (Equation (3)) and water vapor pressure deficit ΔP (Pa) (Equation (4)), both described by an Arrhenius-type equation, were followed [27]. ΔP (Pa) was computed as the vapor pressure deficit between the drying air and saturated air under the same temperature conditions with P_{sat} (Pa) derived from the Magnus-equation, T (°C), and rh as the relative humidity (%) (Equation (1)).

$$\Delta P = P_{sat} \times \left(1 - \frac{rh}{100}\right) = 611.2 \times \exp\left(\frac{17.62T}{243.12 + T}\right) \times \left(1 - \frac{rh}{100}\right) \tag{1}$$

$$MR(t) = \frac{MC_t - MC_e}{MC_0 - MC_e} = \exp(-kt^n) \tag{2}$$

with the following conditions: $k_{T1} < k_{T2}$, $n_{T1} > n_{T2}$ for $T_1 < T_2$ and $x_1 = x_2$, $k_{x1} > k_{x2}$, $n_{x1} < n_{x2}$ for $T_1 = T_2$ and $x_1 < x_2$

$$f(k, n) = d \times \exp\left(\frac{e}{T}\right) + \frac{f}{x} \tag{3}$$

$$MR(t) = \exp\left(-\left(d \times \exp\left(\frac{e}{T}\right) + \frac{f}{x}\right)t^{d \times \exp\left(\frac{e}{T}\right) + \frac{f}{x}}\right) \tag{2a}$$

$$f(k, n) = g \times \log(\Delta P)^h \tag{4}$$

$$MR(t) = \exp\left(-g \times \log(\Delta P)^h t^{g \times \log(\Delta P)^h}\right) \tag{2b}$$

2.7. Analytical Estimation of Diffusion Coefficients

The high oleic sunflower seeds used in this study are shaped like compressed, oval bodies with a sphericity of 0.45 to 0.55 and an equivalent diameter of 5.48 ± 0.64 mm [22]. Average moisture diffusivity D was assumed to remain constant during the relevant drying period [28]. Under these conditions, the special "short time" solution to the differential diffusion equation proposed by Becker (1959) can be applied. The physical basis of this solution is the restriction of changes in moisture to the vicinity of the surface [29]. The validity is limited to the fast drying region of $0.2 < MR < 1$ including the name "short times solution". The mathematical derivation was intensively discussed by Giner and Mascheroni [29], and Becker's short time analytical solution was proven to be accurate, fast, and applicable in the practical drying range for agricultural products. For spheres, the thin layer equation takes the form of Equation (5) where a_v is the kernel's surface specific area in m^2·m^{-3}.

$$MR = \frac{MC_t - MC_e}{MC_0 - MC_e} = 1 - \frac{2}{\sqrt{\pi}} a_v \sqrt{Dt} + \frac{f''(0)}{2} a_v^2 Dt \tag{5}$$

with the following conditions: $MC_t = MC_0$ for $t = 0$, $MC_t = MC_e$ for $t \to \infty$

The term $f''(0)$ is a shape dependent factor, derived from the slope of Equation (6), expressed as a straight line [29].

$$Y = \frac{1 - MR}{a_v \sqrt{Dt}} = \frac{2}{\sqrt{\pi}} - \frac{f''(0)}{2} a_v \sqrt{Dt} \qquad (6)$$

D can be derived from a moisture independent, Arrhenius type relationship, proposed by Sun and Woods [30].

$$D = 1.126 \times R^2 \times exp\left(-\frac{2806.5}{T + 273.15}\right) \qquad (7)$$

where R is the seeds' equivalent radius. After correction of the shape dependent factor $f''(0)$, the individual diffusion coefficients D_s can be determined with Equation (5). The limits of validity for spherical bodies, where D_s is no longer essentially constant, are found at $MR = 0.2$ [28]. The same approximation was also adapted by Giner and Mascheroni [31] for wheat as well as by Santalla and Mascheroni [19] for sunflower seeds.

2.8. Statistical Analysis

All statistical analyses were conducted using the R Project for Statistical Computing [32]. The sorption models and individual Page equations were fitted to the experimental data using R's nls2 procedure [33]. The coefficient of determination R^2 and the mean absolute percentage error $MAPE$ with $MC_{e,exp}$ and MR_{exp} as the observed and $MC_{e,pre}$ and MR_{pre} as the predicted equilibrium moisture content and moisture ratio were taken as the main criteria for the goodness of fit.

$$R^2 = 1 - \frac{\sum (MR_{exp} - MR_{pre})^2}{\sum (MR_{exp} - \overline{MR_{exp}})^2} \qquad (8)$$

$$MAPE = \frac{100}{n} \sum \frac{|MR_{exp} - MR_{pre}|}{MR_{exp}} \qquad (9)$$

The effect of independent variables on the overall model constants was determined by regression analysis and analysis of variance (ANOVA). Asterisks mark significance at $p < 0.05$ (*), $p < 0.01$ (**), and $p < 0.001$ (***) level.

3. Results and Discussion

3.1. Analysis of Moisture Sorption Models

Figure 2a shows the experimentally derived sorption data for hulls and seeds at 25 °C and 50 °C and the curves as predicted by the selected models. At a constant temperature, the equilibrium moisture content increased with increasing water activity, which indicates an asymptotic convergence at $a_w = 1$. In general, at constant water activity, the equilibrium moisture content decreased with increasing temperature. Remarkably, for hulls, this trend was only observed for $a_w \leq 0.82$ and for seeds for $a_w \leq 0.72$. Above these threshold values, the equilibrium moisture content increased with increasing temperature, which is a phenomenon that usually can be observed in high sugar foods [34,35].

Those sorption models are not designed to depict this inverse phenomenon. Therefore, data at $a_w > 0.82$ for hulls and $a_w > 0.72$ for seeds were excluded from the non-linear fitting procedure. A random 60% of the remaining dataset was used for fitting while the entire remaining dataset was used for model performance evaluation in terms of R^2 and $MAPE$ value. The fitted curves are shown as solid lines and continued as dashed lines when the above-mentioned limits of a_w are exceeded, which shows what the corresponding models would predict. Figure 2b provides the predicted versus observed plots for employed sorption equations.

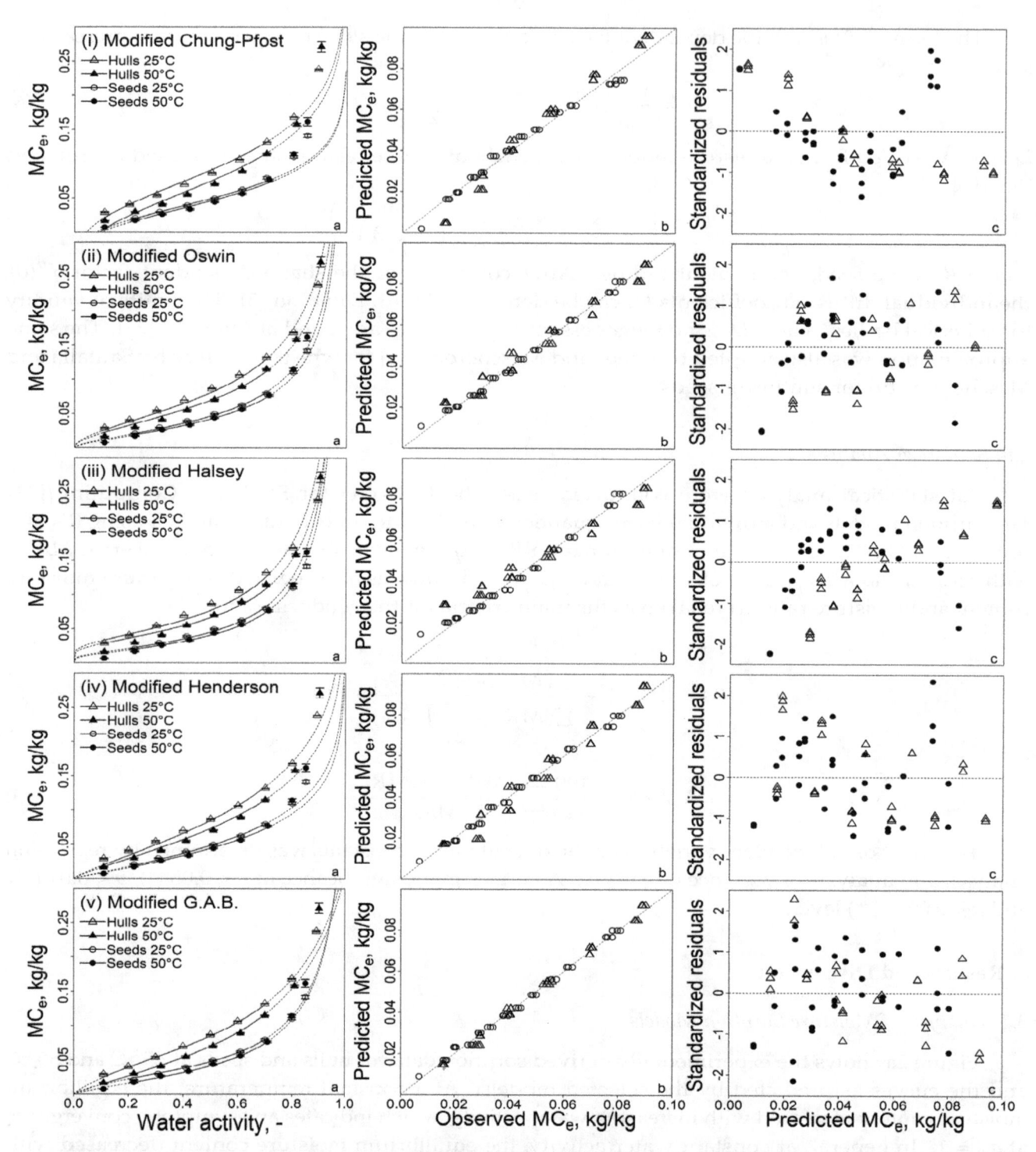

Figure 2. (a) Sorption isotherm for high oleic sunflower seeds (circles) and hulls (triangles) at 25 °C (hollow) and 50 °C (solid) predicted with different models as specified in the plot. Slashed lines are an extrapolation beyond the dataset used for fitting. (b) observed equilibrium moisture content vs. predicted equilibrium moisture content and (c) standardized residuals vs. predicted equilibrium moisture content.

In Table 2, the model coefficients determined by the nonlinear least squares procedure and the corresponding performance parameters are summarized. For seeds, the Mod. Oswin, Mod. Henderson and Mod. G.A.B. equation all achieved $R^2 > 0.99$ and $MAPE < 10\%$. However, the residual plots (Figure 2c) reveal a patterned shape for all but the Mod. Henderson equation. For hulls, only the modified Oswin and modified G.A.B. equation showed $R^2 > 0.99$. However, only the modified G.A.B. equation showed random distribution of residuals.

Table 2. Coefficients a, b, and c, coefficient of determination R^2 and MAPE for the (i) Mod. Chung-Pfost, (ii) Mod. Oswin, (iii) Mod. Halsey, (iv) Mod. Henderson, and (v) Mod. G.A.B. equation fitted for high oleic sunflower seeds in the range of 10–70% equilibrium relative humidity at 31 degrees of freedom and hulls in the range of 10–80% equilibrium relative humidity and 34 degrees of freedom.

Equation		a	b	c	R^2	MAPE, %
(i) Mod. Chung-Pfost		28.181 ***	208.987 ***	611.811 ***	0.988	7.900
(ii) Mod. Oswin		0.048 ***	-1.58×10^{-4} ***	1.607 ***	0.992	8.131
(iii) Mod. Halsey	Seeds	-3.895 ***	-4.60×10^{-3} ***	1.159 ***	0.978	13.884
(iv) Mod. Henderson		0.102 ***	207.939 ***	1.145 ***	0.995	5.408
(v) Mod. G.A.B.		2.88×10^{-2} ***	0.919 ***	167.831 ***	0.994	5.923
(i) Mod. Chung-Pfost		17.761 ***	47.204 ***	240.673 ***	0.985	9.173
(ii) Mod. Oswin		9.94×10^{-2} ***	-0.001 ***	1.776 ***	0.992	6.618
(iii) Mod. Halsey	Hulls	-3.255 ***	-9.87×10^{-3} ***	1.245 ***	0.982	9.964
(iv) Mod. Henderson		0.208 ***	54.690 ***	1.299 ***	0.991	6.112
(v) Mod. G.A.B.		0.070 ***	0.775 ***	122.616 ***	0.997	3.207

Significant contribution to the model on the $p \leq 0.05$, $p \leq 0.01$, and $p \leq 0.001$ level, respectively, are indicated by one, two, or three asterisks.

Furthermore, the experimental data for both seeds and hulls indicate the convex shape of a type III sorption isotherm, according to Brunauer's classification [6]. This type is relatively uncommon and not generally found [36]. Physically, the type III isotherm does not show monolayer adsorption [36]. It is well known that plant seed sorption behavior should rather be depicted with type II sorption isotherms, which is generally adopted in many other studies [7,37,38]. Type II isotherms are an extension of Langumir-like monolayer adsorption isotherms to include unrestricted mono and multilayer adsorption [36], which result in the characteristic sigmoidal shape with a nearly straight segment. It is generally assumed that the beginning of this straight line portion represents the most likely point, where a completely saturated monolayer occurs [39]. From Figure 2, however, no transition from monolayer to multilayer adsorption is derivable, which indicates that the surface area was already covered with a monolayer below the lowest measured a_w value. Sunflower seeds are very much comparable to complex multi-domain foods, with the hulls being a fibrous matrix, allowing two to four times faster diffusivity of moisture than the kernel [19]. In addition, it is clear that the oil content in kernels highly affects the equilibrium moisture content. Similar observations have been reported by other studies [2,40]. At a constant temperature, hulls usually reach a higher MC_e than seeds, which again show a higher MC_e than kernels [2,18,19]. The Mod. Henderson was the only model showing both satisfying fit and random residuals for seeds. Opposed to this, hulls alone were best modeled by the Mod. G.A.B. equation. Given the fact that safe moisture levels for oilseeds are found in the region of $a_w = 0.64$–0.70, the limits of model validity and the observed anomaly at high values of a_w were considered non-critical for the development of a drying model [1,3]. Based on these insights, the fitted Mod. Henderson equation and the coefficients found for seeds were integrated in Equation (2) in order to describe the drying process.

3.2. Modelling of Thin-Layer Drying Behavior

The individual drying data for T from 30 °C to 90 °C and x of 0.010, 0.015, and 0.020 kg·kg^{-1} were described well by the Page equation with high R^2 (>0.99) and low MAPE (<5%) values. The equilibrium moisture content MC_e (kg·kg^{-1}) for seeds was calculated with the Mod. Henderson equation with $a = 0.102$, $b = 207.939$, and $c = 1.145$ (Table 2). The kinetic parameter k increased with increasing T at constant x, while n decreased. With increasing absolute humidity at constant T, k decreased and n increased. Equation (10) and Equation (10a) best described model parameters k and n as a function of T and x. The resulting course is shown in Figure 3a. Inclusion in Equation (2a) resulted in a temperature and absolute humidity-based generalized model describing the experimental data with an R^2 value of 0.982 and $MAPE$ of 9.4%.

$$k = 1.982 \times \exp\left(\frac{-112.9}{T}\right) + \frac{8.384 \times 10^{-4}}{x}, R^2 = 0.980, MAPE = 6.13\% \tag{10}$$

$$n = 0.266 \times \exp\left(\frac{15.10}{T}\right) + \frac{-1.624 \times 10^{-4}}{x}, R^2 = 0.960, MAPE = 1.98\% \tag{10a}$$

The description of k and n as a function of the water vapor pressure deficit ΔP was best described by Equation (11) and Equation (11a). The course of k and n vs. ΔP is depicted in Figure 2b. By inclusion in Equation (2b), a ΔP based, generalized, drying model is achieved, which is capable of describing the experimentally derived drying data with an overall R^2 value of 0.988 and $MAPE$ of 8.3%. This is slightly better than the model based on T and x.

$$k = 1.126 \times 10^{-5} \cdot \log(\Delta P)^{4.540}; R^2 = 0.988, MAPE = 4.25\% \tag{11}$$

$$n = 2.494 \times \log(\Delta P)^{-0.880}; R^2 = 0.934, MAPE = 2.25 \tag{11a}$$

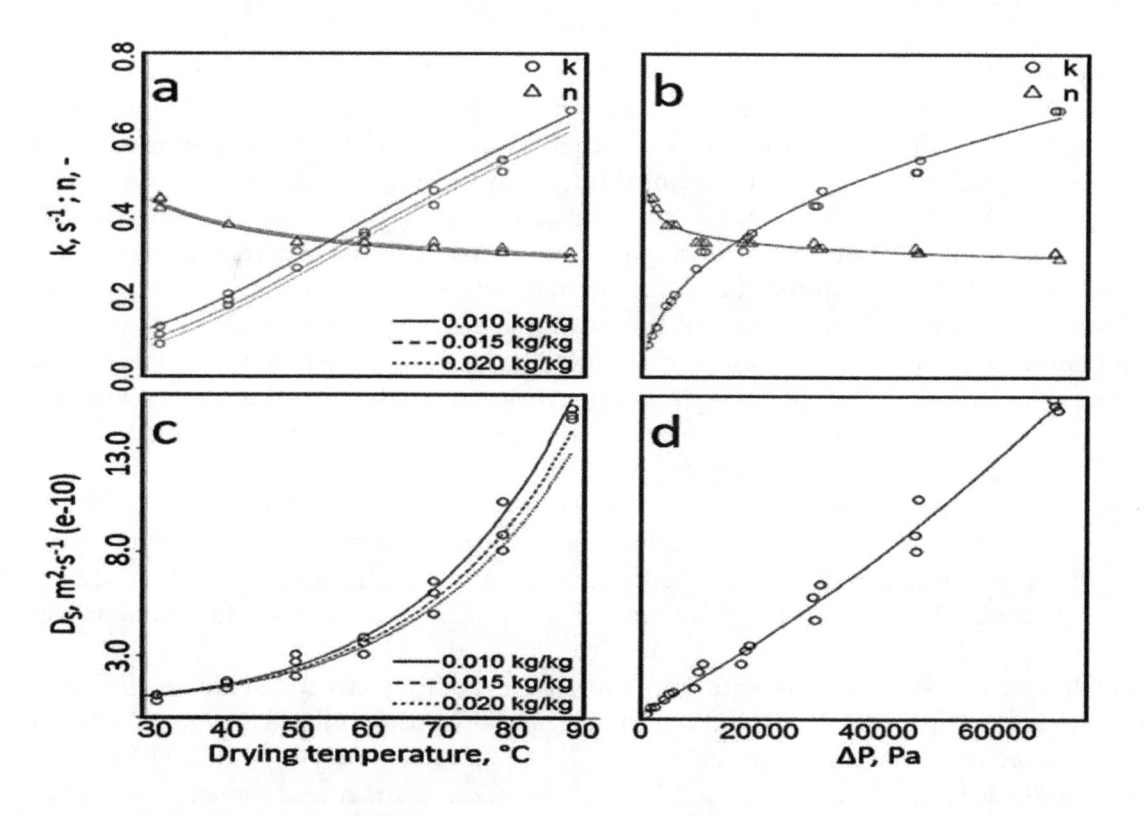

Figure 3. Parameters k, n, and the fitted models from (**a**) Equations (10) and (10a) for $k, n = f(T,x)$ in °C and kg·kg^{-1} and (**b**) modeled with Equation (11) and (11a) for $k, n = f(\Delta P)$ in Pa. (**c**) $D_S = f(T,x)$ for temperature from 30 °C to 90 °C and absolute humidity of the drying air of 0.010, 0.015, and 0.020 kg·kg^{-1} modeled with Equation (12) and (**d**) $D_S = f(\Delta P)$ for the same temperature and humidity range, modeled with Equation (13).

Since ΔP is mainly dependent on T and x, it is a promising physical parameter for the description of drying processes. Regarding the slightly superior fit of the generalized model described by ΔP and the requirement of only four model constants instead of six, this relationship was chosen to further describe the drying process. The predicted vs. observed plot in Figure 4a shows a nearly straight line, which indicates a satisfying performance of the generalized model. The residuals plot in Figure 4b shows a random distribution, supporting the validity of the derived model.

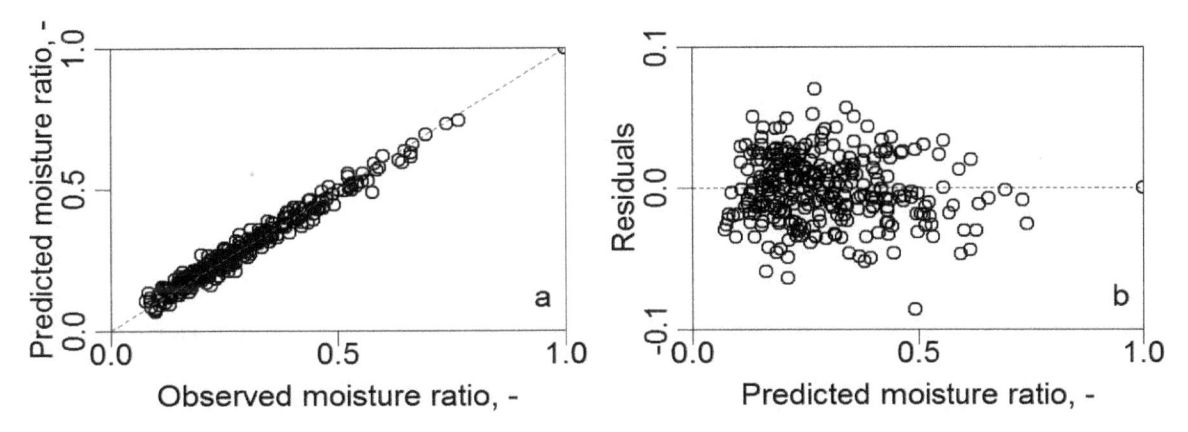

Figure 4. Model performance for MR expressed as a function of ΔP calculated with Equation (2b), **(a)** predicted MR versus observed MR values are concentrated closely to the perfect fit line of $x = y$, and **(b)** the residuals are equally distributed and do not show any trend.

Application of the short times equation indicated an increase of D_S with increasing temperature T. With increasing absolute humidity x, D_S showed a decreasing trend for temperatures up to 70 °C, from where no obvious trend was derivable. For T below 50 °C, the threshold of $MR < 0.2$ usually was not undershot. The shape dependent factor $f''(0)$ found to be 0.428 with a standard error of 1.3%. By inclusion in Equation (6), the diffusion coefficients for short times ($0.2 < MR < 1$) can be calculated (Table 3).

Table 3. Moisture diffusion coefficients D for short times ($MR \geq 0.2$), calculated with Equation (5) for different absolute humidity x and temperature T of the drying air, coefficient of determination R^2 and $MAPE$.

x	T, °C	D (m²·s⁻¹)·10⁻¹⁰		R^2	MAPE, %
0.010 kg·kg⁻¹	30	0.643	***	0.979	4.088
	40	1.317	***	0.985	6.935
	50	2.620	***	0.991	6.217
	60	3.467	***	0.970	10.903
	70	6.277	***	0.989	7.547
	80	10.190	***	0.996	5.059
	90	14.800	***	0.995	7.399
0.015 kg·kg⁻¹	30	0.593	***	0.984	4.303
	40	1.231	***	0.987	5.591
	50	2.258	***	0.974	11.780
	60	3.222	***	0.979	11.238
	70	4.627	***	0.969	11.187
	80	7.796	***	0.990	7.829
	90	14.310	**	0.997	5.611
0.020 kg·kg⁻¹	30	0.342	***	0.979	3.556
	40	0.978	***	0.980	5.790
	50	1.529	***	0.954	10.944
	60	2.634	***	0.988	5.715
	70	5.696	***	0.982	9.844
	80	8.559	***	0.993	5.203
	90	14.490	**	0.999	2.152

Significant contribution to the model on the $p \leq 0.05$, $p \leq 0.01$, and $p \leq 0.001$ level, respectively, are indicated by one, two, or three asterisks.

The mean coefficient of determination was 0.984 with a MAPE of 7.09% for short times. Visual analysis of Figure 3c, which compares the measured data to the fit of Equation (12) revealed an overestimation when the limit of validity is approached. The derived diffusion coefficients, however, are in the same range as the ones reported by Santalla and Mascheroni [19]. In addition to their findings, that initial moisture content did not affect diffusion coefficients. The results from this study indicate an effect of x only in the lower temperature regions up to 70 °C. In an analogy with the approach for fitting a generalized empirical drying model, the same procedure was applied for the analytical solution. A generalized analytical model was fitted for $D_S = f(T,x)$, which is well described by Equation (12).

$$D_S = 4.125 \times 10^{-13} \cdot \exp\left(T^{0.4851} \times x^{-0.033}\right); \ R^2 = 0.984, \ MAPE = 12.75\% \tag{12}$$

Integration of Equation (12) in Equation (5) yielded a generalized model with an overall R^2 of 0.966 and $MAPE$ = 7.37%. The description of $D_S = f(\Delta P)$ was described well with a similar function:

$$D_S = 1.296 \times 10^{-12} \times \exp\left(\Delta P^{0.1747}\right); \ R^2 = 0.986, \ MAPE = 10.03\% \tag{13}$$

Integration of Equation (13) in Equation (5) resulted in a generalized model based on ΔP, which was fitted with an overall R^2 of 0.976 and $MAPE$ = 6.33%. This is slightly better than the approach based on temperature and absolute humidity. The fit to the experimental data of both Equations (12) and (13) is given in Figure 3c,d.

An applicability to determine the effective diffusion coefficients in the initial, most relevant phase of drying is obviously given and commonly applied [19,29,31]. Especially, the calculation of the shape-depending factor f''(0), based on the physical properties of the investigated product, adds further meaning to the application of Becker's equation, and is expected to increase the applicability of the thin-layer equation for dryer control and simulation.

4. Conclusions

The embedded systems employed for data recording during the sorption and drying experiments in the current study did allow the acquisition of a large amount of experimental inline-data from one year's harvest. This information was used to fit semi-empirical and analytical sorption and drying models that are simple enough to be run on embedded microcontrollers with limited processing power. The sorption experiments revealed that, for high values of a_w, the equilibrium moisture content increased with increasing temperature. An anomaly, which is usually found in high sugar foods, was not yet reported for sunflower seeds. It is remarkable that most of the commonly applied equations to describe sorption isotherms showed patterned residuals, which restricted their applicability and validity. This study found the Page equation to satisfactorily describe the drying process of high oleic sunflower seeds for a wide range of drying air temperatures and humidity. It also proposes the application of water vapor pressure deficit ΔP to calculate the parameters of the Page equation including both the effect of temperature and the humidity of the drying air. This approach was found to be superior to a prediction based on temperature. Values of diffusivity ascertained with the Becker equation and a correction of the shape dependent factor are comparable to those reported in other studies, which corroborates the applicability and accuracy in thin layer drying of high oleic sunflower seeds.

Author Contributions: Literature review: S.M., Methodology: S.M., D.A., J.M.; Concept and design: S.M.; Investigation: S.M., D.A.; Formal analysis: S.M.; Writing—Original Draft Preparation: S.M.; Writing—Review and Editing: S.M., D.A., J.M.

Abbreviations

a_w	water activity
MC	moisture content
MC_e	equilibrium moisture content
MC_t	moisture content at time t
DVS	dynamic vapour sorption apparatus
kg	kilogram
g	gram
mg	milligram
m	meter
mm	millimeter
rh	relative humidity, %
min	minutes
P_{vs}	water vapor partial pressure, Pa
P_{sat}	saturation vapor pressure, Pa
T	temperature, °C
a, b, c, d, e, f, g	model constants
G.A.B.	Guggenheim, Anderson, DeBoer
x	absolute humidity, kg water per kg of dry air
s	second
MR	moisture ratio
t	time
k	rate constant, min^{-1}
n	dimensionless coefficient of page equation
P	pressure, Pa
D	moisture diffusivity, $m^2 \cdot s^{-1}$
a_v	kernel's surface specific area in $m^2 \cdot m^{-3}$
ANOVA	analysis of variance
MAPE	mean absolute perecentage error
R^2	coefficient of determination
p	probability level at which significance is assumed

References

1. Robertson, J.A.; Chapman, G.W.; Wilson, R.L. Effect of moisture content of oil type sunflower seed on fungal growth and seed quality during storage. *J. Am. Oil Chem. Soc.* **1984**, *61*, 768–771. [CrossRef]
2. Mazza, G.; Jayas, D.S.S. Equilibrium moisture characteristics of sunflower seeds, hulls, and kernels. *Trans. ASAE* **1991**, *34*, 534–538. [CrossRef]
3. Giner, S.A.; Gely, M.C. Sorptional parameters of sunflower seeds of use in drying and storage stability studies. *Biosyst. Eng.* **2005**, *92*, 217–227. [CrossRef]
4. Maciel, G.; De la Torre, D.; Izquierdo, N.; Cendoya, G.; Bartosik, R. Effect of oil content of sunflower seeds on the equilibrium moisture relationship and the safe storage condition. *Agric. Eng. Int.* **2015**, *17*, 248–258.
5. Mujumdar, A.S.; Beke, J. Grain drying: Basic principles. In *Handbook of Postharvest Technology: Cereals, Fruits, Vegetables, Tea, and Spices*; Chakraverty, A., Mujumdar, A.S., Raghavan, G.S.V., Ramaswamy, H.S., Eds.; Marcel Dekker: New York, NY, USA, 2003; pp. 119–138.
6. Guillard, V.; Bourlieu, C.; Gontard, N. Theoretical background. In *Food Structure and Moisture Transfer*; Springer: New York, NY, USA, 2013; pp. 3–33.
7. Argyropoulos, D.; Alex, R.; Kohler, R.; Müller, J. Moisture sorption isotherms and isosteric heat of sorption of leaves and stems of lemon balm (*Melissa officinalis* L.) established by dynamic vapor sorption. *LWT-Food Sci. Technol.* **2012**, *47*, 324–331. [CrossRef]
8. Argyropoulos, D.; Müller, J. Effect of convective-, vacuum- and freeze drying on sorption behaviour and bioactive compounds of lemon balm (*Melissa officinalis* L.). *J. Appl. Res. Med. Aromat. Plants* **2014**, *1*, 59–69. [CrossRef]

9. Arslan, N.; Toğrul, H. Moisture sorption isotherms for crushed chillies. *Biosyst. Eng.* **2005**, *90*, 47–61. [CrossRef]
10. Stubberud, L.; Arwidsson, H.G.; Graffner, C. Water-solid interactions: I. A technique for studying moisture sorption/desorption. *Int. J. Pharm.* **1995**, *114*, 55–64. [CrossRef]
11. Hill, C.A.S.; Norton, A.J.; Newman, G. The water vapour sorption properties of Sitka spruce determined using a dynamic vapour sorption apparatus. *Wood Sci. Technol.* **2010**, *44*, 497–514. [CrossRef]
12. Kachrimanis, K.; Noisternig, M.F.; Griesser, U.J.; Malamataris, S. Dynamic moisture sorption and desorption of standard and silicified microcrystalline cellulose. *Eur. J. Pharm. Biopharm.* **2006**, *64*, 307–315. [CrossRef] [PubMed]
13. Kohler, R.; Dück, R.; Ausperger, B.; Alex, R. A numeric model for the kinetics of water vapor sorption on cellulosic reinforcement fibers. *Compos. Interfaces* **2003**, *10*, 255–276. [CrossRef]
14. Liu, H.; Zhang, L.; Han, Z.; Xie, B.; Wu, S. The effects of leaching methods on the combustion characteristics of rice straw. *Biomass Bioenergy* **2013**, *49*, 22–27. [CrossRef]
15. Vollenbroek, J.; Hebbink, G.A.; Ziffels, S.; Steckel, H. Determination of low levels of amorphous content in inhalation grade lactose by moisture sorption isotherms. *Int. J. Pharm.* **2010**, *395*, 62–70. [CrossRef] [PubMed]
16. Xie, Y.; Hill, C.A.S.; Jalaludin, Z.; Curling, S.F.; Anandjiwala, R.D.; Norton, A.J.; Newman, G. The dynamic water vapour sorption behaviour of natural fibres and kinetic analysis using the parallel exponential kinetics model. *J. Mater. Sci.* **2011**, *46*, 479–489. [CrossRef]
17. Nurhadi, B.; Roos, Y.H. Dynamic water sorption for the study of amorphous content of vacuum-dried honey powder. *Powder Technol.* **2016**, *301*, 981–988. [CrossRef]
18. Santalla, E.M.; Mascheroni, R.H. Equilibrium moisture characteristics of high oleic sunflower seeds and kernels. *Dry. Technol.* **2003**, *21*, 147–163. [CrossRef]
19. Santalla, E.M.; Mascheroni, R.H. Moisture diffusivity in high oleic sunflower seeds and kernels. *Int. J. Food Prop.* **2010**, *13*, 464–474. [CrossRef]
20. Henderson, S.M.; Pabis, S. Grain drying theory IV, The effect of airflow rate on the drying index. *J. Agric. Eng. Res.* **1962**, *7*, 85–89.
21. Hutchinson, D.; Otten, L. Thin-layer air drying of soybeans and white beans. *Int. J. Food Sci. Technol.* **1983**, *18*, 507–522. [CrossRef]
22. Munder, S.; Argyropoulos, D.; Müller, J. Class-based physical properties of air-classified sunflower seeds and kernels. *Biosyst. Eng.* **2017**, *164*, 124–134. [CrossRef]
23. *ISO 665:2000 Oilseeds—Determination of Moisture and Volatile Matter Content*; International Organization for Standardization (ISO): Geneva, Switzerland, 2000.
24. Al-Muhtaseb, A.H.; McMinn, W.A.M.; Magee, T.R.A. Moisture sorption isotherm characteristics of food products: A review. *Food Bioprod. Process.* **2002**, *80*, 118–128. [CrossRef]
25. Argyropoulos, D.; Heindl, A.; Müller, J. Assessment of convection, hot-air combined with microwave-vacuum and freeze-drying methods for mushrooms with regard to product quality. *Int. J. Food Sci. Technol.* **2011**, *46*, 333–342. [CrossRef]
26. Shahari, N.A. Mathematical Modelling of Drying Food Products: Application to Tropical Fruits. Ph.D. Thesis, University of Nottingham, Nottingham, UK, 2012.
27. Udomkun, P.; Argyropoulos, D.; Nagle, M.; Mahayothee, B.; Janjai, S.; Müller, J. Single layer drying kinetics of papaya amidst vertical and horizontal airflow. *LWT-Food Sci. Technol.* **2015**, *64*, 67–73. [CrossRef]
28. Becker, H.A. A study of diffusion in solids of arbitrary shape, with application to the drying of the wheat kernel. *J. Appl. Polym. Sci.* **1959**, *1*, 212–226. [CrossRef]
29. Giner, S.A.; Mascheroni, R.H. PH—Postharvest technology. *J. Agric. Eng. Res.* **2001**, *80*, 351–364. [CrossRef]
30. Sun, D.W.; Woods, J.L. Low temperature moisture transfer characteristics of wheat in thin layers. *Trans. ASAE* **1994**, *37*, 1919–1926. [CrossRef]
31. Giner, S.A.; Mascheroni, R.H. PH—Postharvest technology. *Biosyst. Eng.* **2002**, *81*, 85–97. [CrossRef]
32. R Core Team. R: A Language and Environment for Statistical Computing 2015. Available online: https://www.r-project.org (accessed on 17 September 2015).
33. Grothendieck, G. nls2: Non-Linear Regression with Brute Force 2013. Available online: https://cran.r-project.org/web/packages/nls2/nls2.pdf (accessed on 17 September 2015).
34. Djendoubi Mrad, N.; Bonazzi, C.; Boudhrioua, N.; Kechaou, N.; Courtois, F. Influence of sugar composition on water sorption isotherms and on glass transition in apricots. *J. Food Eng.* **2012**, *111*, 403–411. [CrossRef]

35. Saravacos, G.D.; Tsiourvas, D.A.; Tsami, E. Effect of temperature on the water adsorption isotherms of sultana raisins. *Dry. Technol.* **1986**, *4*, 633–649. [CrossRef]

36. Ling, M. Manometrische Bestimmung der NO2-Sorptionsisothermen von Superberliner Blau—Derivaten und Charakterisierung der inneren Oberflächen mittels der BET—Methode. Ph.D. Thesis, Universität Hamburg, Hamburg, Germany, 2001.

37. Simha, H.V.V.; Pushpadass, H.A.; Franklin, M.E.E.; Kumar, P.A.; Manimala, K. Soft computing modelling of moisture sorption isotherms of milk-foxtail millet powder and determination of thermodynamic properties. *J. Food Sci. Technol.* **2016**, *53*, 2705–2714. [CrossRef]

38. Mathlouthi, M.; Rogé, B. Water vapour sorption isotherms and the caking of food powders. *Food Chem.* **2003**, *82*, 61–71. [CrossRef]

39. Brunauer, S. *The Adsorption Of Gases And Vapors Vol I*; Oxford University Press: London, UK, 1943.

40. Pixton, S.W.; Warburton, S. Moisture content relative humidity equilibrium, at different temperatures, of some oilseeds of economic importance. *J. Stored Prod. Res.* **1971**, *7*, 261–269. [CrossRef]

UAV-Borne Dual-Band Sensor Method for Monitoring Physiological Crop Status

Lili Yao, Qing Wang, Jinbo Yang, Yu Zhang, Yan Zhu, Weixing Cao * and Jun Ni *

National Engineering and Technology Center for Information Agriculture, Key Laboratory for Crop System Analysis and Decision Making, Ministry of Agriculture, Jiangsu Key Laboratory for Information Agriculture, Nanjing Agricultural University, Nanjing 210095, Jiangsu, China; 2017201083@njau.edu.cn (L.Y.); 2016101037@njau.edu.cn (Q.W.); 2018101166@njau.edu.cn (J.Y.); zhangyu@njau.edu.cn (Y.Z.); yanzhu@njau.edu.cn (Y.Z.)

* Correspondence: caow@njau.edu.cn (W.C.); nijun@njau.edu.cn (J.N.).

Abstract: Unmanned aerial vehicles (UAVs) equipped with dual-band crop-growth sensors can achieve high-throughput acquisition of crop-growth information. However, the downwash airflow field of the UAV disturbs the crop canopy during sensor measurements. To resolve this issue, we used computational fluid dynamics (CFD), numerical simulation, and three-dimensional airflow field testers to study the UAV-borne multispectral-sensor method for monitoring crop growth. The results show that when the flying height of the UAV is 1 m from the crop canopy, the generated airflow field on the surface of the crop canopy is elliptical, with a long semiaxis length of about 0.45 m and a short semiaxis of about 0.4 m. The flow-field distribution results, combined with the sensor's field of view, indicated that the support length of the UAV-borne multispectral sensor should be 0.6 m. Wheat test results showed that the ratio vegetation index (RVI) output of the UAV-borne spectral sensor had a linear fit coefficient of determination (R^2) of 0.81, and a root mean square error (RMSE) of 0.38 compared with the ASD Fieldspec2 spectrometer. Our method improves the accuracy and stability of measurement results of the UAV-borne dual-band crop-growth sensor. Rice test results showed that the RVI value measured by the UAV-borne multispectral sensor had good linearity with leaf nitrogen accumulation (LNA), leaf area index (LAI), and leaf dry weight (LDW); R^2 was 0.62, 0.76, and 0.60, and RMSE was 2.28, 1.03, and 10.73, respectively. Our monitoring method could be well-applied to UAV-borne dual-band crop growth sensors.

Keywords: CFD; airflow field test; monitoring method; spectral sensor; crop growth

1. Introduction

Accurate management of crop water and fertilizer in crop fields is an important prerequisite for ensuring high yield and quality of crops, sustainable use of cultivated land, and healthy development of the environment [1]. High-throughput, accurate, and real-time acquisition of crop-growth information is an important basis for the accurate management of crop water and fertilizer [2]. Monitoring technology based on the characteristics of the reflection spectrum has the advantages of being nondestructive, providing real-time information, and delivering high-efficiency analysis. Thus, it is widely used in crop-growth parameter acquisition. Various research institutions have developed spectral sensors to monitor crop growth, providing effective technical support for field-crop production management [3–5].

In 2004, Moya et al. [6] designed a chlorophyll-fluorescence test device, which uses sunlight as a light source. During the test, the leaf blade was required to be in a relatively static state, and the reflection=spectrum information of chlorophyll fluorescence in the 510 and 570 nm bands could be obtained from a short distance. Quantitative inversion of chlorophyll fluorescence could be achieved by the physiological reflectance index (PRI) calculated from the test results. In 2010, Ryu et al. [7] developed a normalized vegetation index spectroscopy sensor to achieve the inversion of vegetation-growth indicators. It has its own LED light source for illumination. When the test height was less than 3 m, the best test results could be achieved. Several commercial instruments are currently available for crop-growth monitoring. For example, the Greenseeker spectral sensor designed by Trimble USA can obtain the spectral information of reflection characteristics in crop canopy red and near-infrared bands and calculate the relevant vegetation index. For this device, the test distance should be kept at a height of 60–180 cm from the canopy [8,9]. The ASD FieldSpec4 spectrometer developed by American ASD Company can realize reflection-spectrum acquisition of the 350–2500 nm crop-canopy band, the data information is rich, and accuracy is high. Relying on sunlight as a light source, the test needs to be carried out at noon without wind or clouds, the test height should be kept between 30–120 cm, and the crop canopy needs to remain relatively static [10–12]. Holland Scientific designed and developed active light-source spectral sensors for monitoring crop growth, such as Crop Circle and Rapidscan. These instruments can emit light and receive reflection-spectrum information of a crop canopy in real time through their own light-source system. The test height should be kept within 3 m from the canopy, and the canopy structure needs to maintain a steady state [13–17]. These spectral crop-growth monitoring devices are simple in operation, easy to carry, high in test accuracy, intuitive in results, and can provide nondestructive access to crop-growth information, but they also have shortcomings, such as a small monitoring range, high labor intensity, and discontinuous monitoring, which cannot provide high-throughput information and real-time decision-making for large-scale crop-production management in the field.

Unmanned-aerial-vehicle (UAV) operation is highly efficient, flexible, easy, and has strong terrain applicability. Thus, UAVs are widely used in agricultural information-acquisition platforms [12], but so far there are few studies on the use of UAV-borne spectral sensors for monitoring crop growth. Krienke et al. attempted to measure the normalized vegetation index of lawn using a MikroKopterOktoKopter XL UAV equipped with a RapidScan CS-45 spectroscopy sensor. Flight height was maintained at 0.5–1.5 m above the lawn. However, test results were poor because the disturbance of the turf canopy from the downwash airflow field was ignored [18]. Shafian et al. mounted an image sensor on a fixed-wing UAV and collected the image information of a sorghum planting area at an altitude of 120 m. Pix4D software was used to splice, correct, and extract vegetation indices from each acquired image. The leaf area index (LAI) value of sorghum was simultaneously sampled and tested. The results show that the Normalized Difference Vegetation Index (NDVI) value extracted from the acquired image information had a higher linear fit with the sorghum LAI value obtained with the sampling test [19]. Schirrmann et al. used a UAV to work at an altitude of 50 m and obtain an RGB image of the wheat growth period. The acquired image was calibrated by Agisoft lens software, the distortion correction was modeled by Brown distortion model, and the final image mosaic and surface model generation results were improved by radiation pretreatment, from which the information such as crop coverage and plant height were extracted [20]. Zheng et al. used an OktoXL UAV equipped with a Cuubert UHD 185 hyperspectral camera to acquire hyperspectral images of a rice canopy at an altitude of 50 m. Images in the acquisition results were corrected using ENVI software. By comparing the synchronous test results of a ground-object spectrometer and the agronomic parameters obtained from laboratory chemical analysis, they proved that the image information acquired by the UAV platform equipped with imaging instruments could be used for the quantitative inversion of agronomic crop parameters [21]. Stroppianaet al. acquired a large number of multispectral images at a height of 70 m from the ground by using a 3DRobotics SOLO quadrotor UAV equipped with a Parrot Sequoia multispectral camera. By screening the acquired images, and then correcting and extracting the vegetation index, they proposed an automatic classification method for weeds and crops, which can be used for the classification and management of specific weeds in the field [22]. However, these studies simply installed image sensors on UAVs to obtain crop images from a high altitude. Although the influence of the UAV downwash airflow field is small, captured images can only be stored in the memory. Scientific-research personnel are required to use special software for image correction, cropping, splicing, enhancement, and other offline processing, and then analyze the relationship between the images and crop-growth parameters. The process is complex, requires specialized knowledge, and cannot acquire information in real time, which is not conducive to popularization [23,24].

In this paper, we studied the dual-band crop-growth sensor independently developed by Nanjing Agricultural University [25]. First, we investigated the monitoring method of the UAV-borne dual-band crop-growth sensor based on its spectral-monitoring mechanism and structural-design features. Then, we analyzed the spatial-distribution characteristics of the airflow field under the low-altitude hovering operation of the UAV. Finally, we built a UAV-borne crop-growth monitoring system to achieve high throughput and real-time access to rice- and wheat-growth information.

2. Materials and Methods

2.1. Test Equipment

2.1.1. Dual-Band Crop-Growth Sensor

There is a certain relationship between crop growth and the spectral reflectance of the crop canopy. As shown in Figure 1, the reflectivity of a wheat canopy is relatively low when the band is under 710 nm. Reflectivity linearly rises in the 710–760 nm band, and reaches a comparatively high level in the 760–1210 nm band. Wheat-canopy reflectivity shows significant differences between different nitrogen levels for the 460–730 and 760–1210 nm bands. In the 460–730 nm band, spectral reflectivity is negatively correlated with the amount of nitrogen fertilizer, with reflectivity being the lowest at N5 and highest at N1. In the 760–1210 nm band, spectral reflectivity is positively correlated with the amount of nitrogen fertilizer, with reflectivity being the highest at N5 and lowest at N1. The 710–760 nm band is an apparent transition zone. The spectral reflectivity of the wheat canopy at above 1150 nm is not very susceptible to the amount of nitrogen fertilizer, and there is little difference between spectral reflectivity at N2–N5. According to currently available research, there is a good linear relationship between leaf nitrogen accumulation (LNA) and spectral-reflectivity changes near 550, and 600–700 and 720 nm; there is a good linear relationship between leaf dry weight (LDW), and spectral-reflectivity changes at 580–700 and 770–900 nm; the spectral-reflectivity change at 460–680 nm and near 810 nm is closely correlated the LAI. Considering the sensitive bands of the three agronomic parameters, crop growth can be well-inverted with the ratio vegetation index (RVI) index constructed using the 730 and 810 nm bands [26–28].

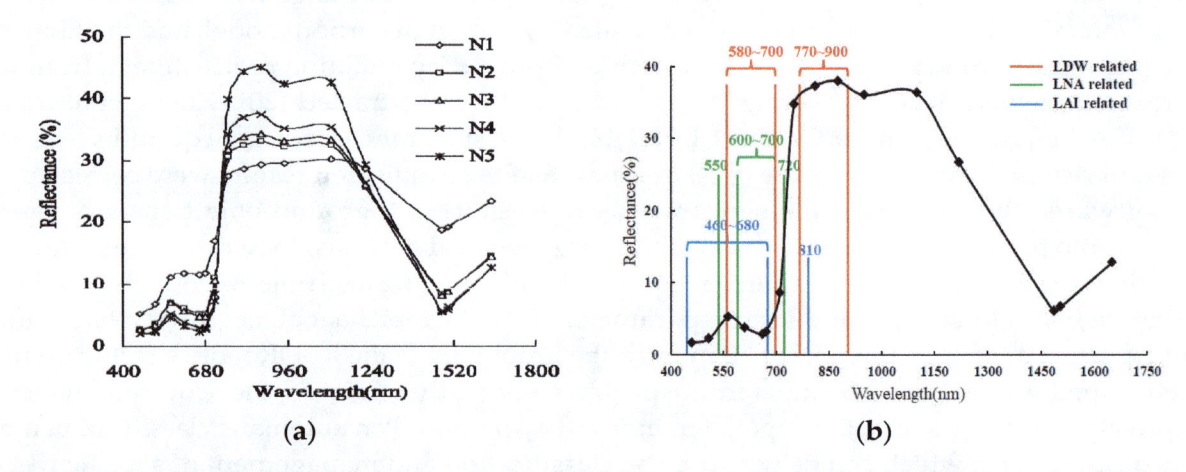

Figure 1. Characteristic curve of the spectral reflectivity of wheat canopy. (**a**) Multispectral reflectivity of wheat canopy under application of different amounts of nitrogen, and (**b**) characteristic spectral bands of agronomic wheat parameters.

The dual-band crop-growth sensor was designed by Nanjing Agricultural University, it is equipped with a dual-band detection lens, and its structure can be divided into an upward light sensor and a downward light sensor, as shown in Figure 2. The upward light sensor is used to acquire sunlight-radiation information at 730 and 815 nm wavelengths, and the downward light sensor was configured to receive crop-canopy-reflected light-radiation information of a corresponding wavelength. The structure is shown in Figure 2. It is packaged in a nylon case and weighs 11.34 g, with a test field of view of 27°. The crop-canopy RVI can be output in real time, and wireless transmission can achieve long-distance transmission and analysis.

With sunlight as the light source, dual-band crop-growth sensors require the testing object (i.e., wheat canopy) to remain relatively static so that the canopy presents the Lambertian reflection characteristics and the field of view of the sensor points vertically downward. During measurement, optical-radiation energy is converted to electric-energy signals by the sensor. Therefore, to ensure high sensitivity, measuring height should be maintained at 1.0–1.5 m above the canopy. The principle is shown in Figure 2.

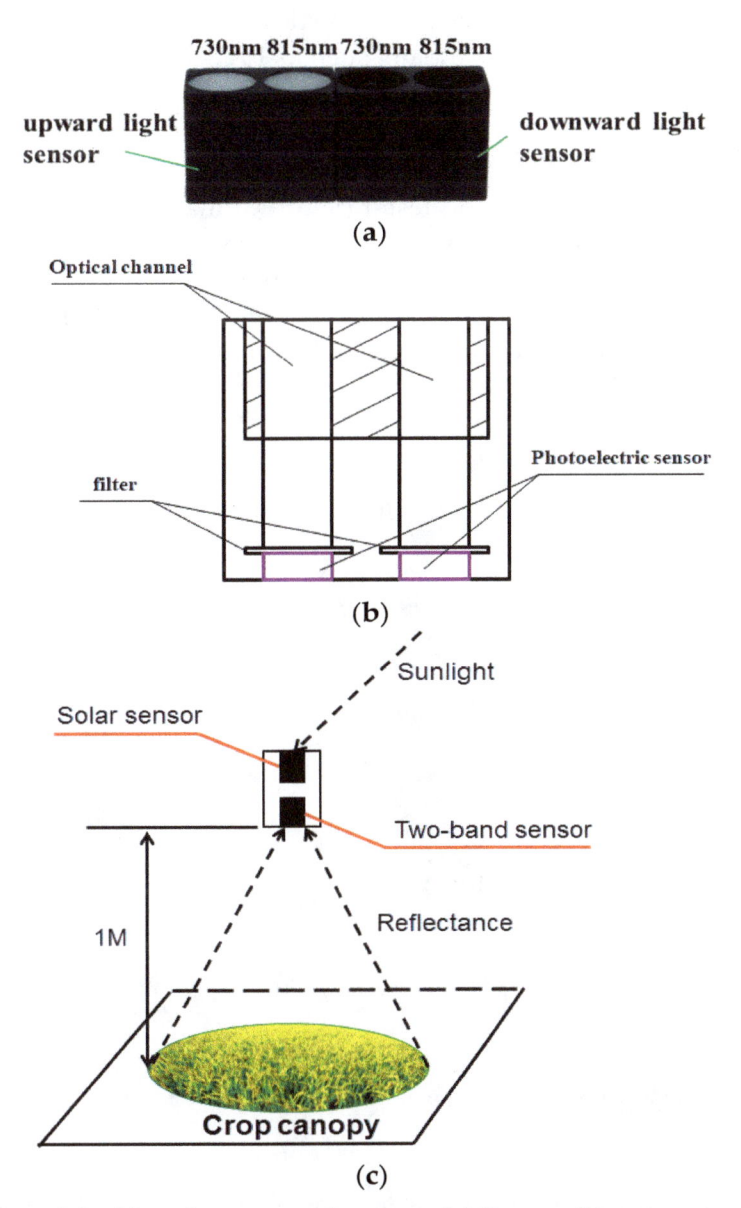

Figure 2. Introduction of dual-band crop-growth sensor. (**a**) Sensor; (**b**) sensor structure; (**c**) schematic diagram of dual-band crop-growth sensor test.

2.1.2. ASD FieldSpec HandHeld 2 Handheld Spectrometer

Developed by American Analytical Spectral Devices (AS, made by Advanced Systems Development, Inc., Alexandria, VA, America), the ASD FieldSpec HandHeld 2 handheld spectrometer can be used for the reflection-spectrum acquisition of different objects such as crops, marine organisms, and minerals. The device has the advantages of portability, simple operation, and accurate results. Test wavelength range is 325–1075 nm, wavelength accuracy is ±1 nm, spectral resolution is less than 3 nm, and test field of view is 25° [29,30].

2.1.3. LAI-2200C Vegetation Canopy Analyzer

The LAI-2200C (Made by LI-COR, Lincoln, NE, USA) is a vegetation-canopy analyzer manufactured by LI-COR, United States. The analyzer is light in weight, consumes little power, and is very suitable for outdoor measurements. In addition, the analyzer can work independently, perform unattended long-term continuous measurement, and automatically record data. The measurement principle is to measure the transmitted light at five angles above and below the vegetation canopy by using the "fisheye" optical sensor (which has a 148° vertical field of view and a 360° horizontal field of view), and calculate canopy-structure parameters such as LAI, average leaf dip angle, void ratio, and aggregation index by using the radiation-transfer model of a vegetation canopy [31–33].

2.1.4. Three-Dimensional Airflow Field Tester

The three-dimensional (3D) airflow field tester (South China Agricultural University, Guangzhou, China) uses three wind-speed sensors to test X-, Y-, and Z-axial airflow velocity. The test results are transmitted to a computer through the Zigbee module in real time. The power consumption of the sensor is little, and it can maintain continuous operation for a long time. The center point of the three test axes was fixed to a height of 60 cm from the ground, and the distance from each wind speed sensor to the center point was 15 cm. The three test axes were kept perpendicular to each other. The structure is shown in Figure 3.

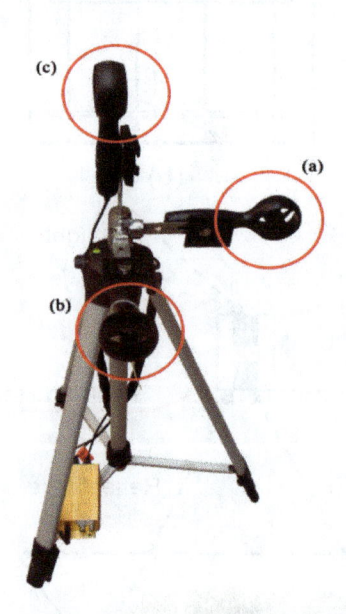

Figure 3. Structure of three-dimensional airflow field tester. (**a**) X-axial velocity test axis; (**b**) Y-axial velocity test axis; (**c**) Z-axial velocity test axis.

2.2. Research Methods

When the dual-band crop-growth sensor is measuring, test height is required to maintain a distance of 1–1.5 m from the crop canopy, and the crop canopy must remain relatively static, exhibiting Lambertian reflection characteristics. However, while the UAV is hovering at a low altitude, rotors rotate at a high speed, which causes the surrounding airflow to shrink and contract, forming an airflow field; this airflow field acts on the crop canopy, causing disturbance to it, destroying the Lambertian reflection characteristics of the crop canopy, affecting the test results, and even preventing completion of the test. Therefore, when a UAV is equipped with a dual-band crop-growth sensor for crop-growth monitoring, it is necessary to consider the disturbance effect of the downwash airflow field on the crop canopy.

To solve this problem, we analyzed the spatial distribution of the UAV rotor downwash airflow field by using computational fluid dynamics (CFD) and 3D airflow field testers. We then determined an acceptable deployment location for the dual-band crop-growth sensor based on the surface velocity and distribution range of the airflow field.

2.2.1. Airflow-Field Numerical Simulation

CFD is a branch of fluid mechanics. It uses computers as tools and applies various discrete mathematical methods to conduct numerical experiments, computer simulations, and analytical studies on various fluid-mechanics problems. The advantage of CFD lies in its ability to simulate the experimental process from basic physics theorems instead of expensive fluid-dynamics experiment equipment. According to the specific process of CFD numerical simulation analysis [25], we obtained 3D-sized data of the DJI phantom drone (a type of UAV) with a 3D scanning system, converted the 3D information of the UAV into a digital signal that could be processed by the computer, and used CFD to perform UAV mesh generation and numerical solution. The second-order upwind style was selected for calculation to improve accuracy [34–36]. The 3D scanning technology was only used to measure the profile data of the UAV fuselage and rotor, and construct a 3D UAV model. This is not a 3D model of the crop-canopy structure. The specific calculation process is shown in Figure 4.

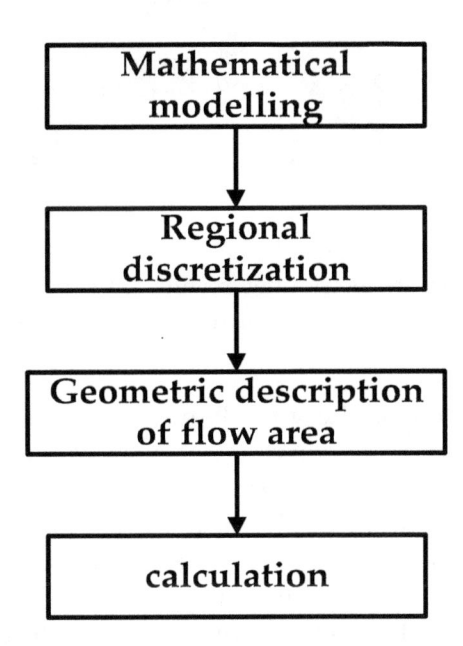

Figure 4. Computational fluid dynamics (CFD) mesh generation and numerical solution process.

Physical parameters such as UAV flight area, and rotor-rotation speed and direction, were set by CFD software, and UAV hovering-operation-state simulation analysis was carried out. In the grid simulation area with a diameter of 1.2 m and a height of 1.85 m, the velocity of each grid node is composed of the X-axle, Y-axle, and Z-axle velocity components. The flying height of the UAV was set to 1 m from the ground, and the ground-velocity nephogram distribution result was displayed by CFX's own post processing module, as shown in Figure 5.

(a)   (b)

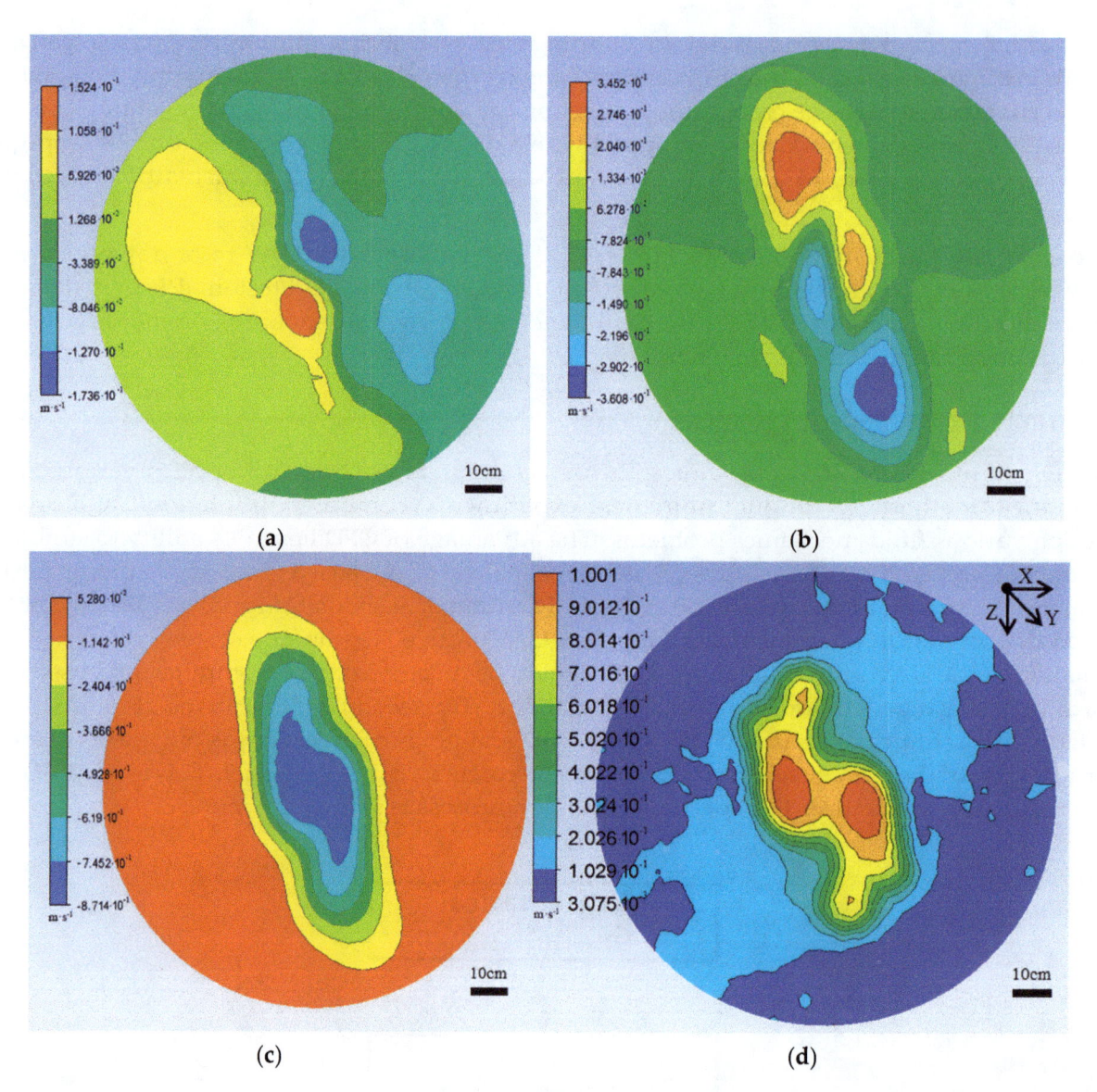

(c)   (d)

Figure 5. Ground-velocity nephogram distribution results. (**a**) X-axial velocity distribution nephogram; (**b**) Y-axial velocity distribution nephogram; (**c**) Z-axial velocity distribution nephogram; (**d**) combined velocity distribution nephogram.

Velocity intensity was analyzed, as shown in Figure 5. When the UAV was hovering at a height of 1 m from the ground, the Z-axial velocity component was much larger than the X-axial and Y-axial velocity components. The combined velocity depends mainly on the axial velocity component. The size of the axial velocity nephogram was elliptical, the central region had the highest wind speed, and it gradually decreased toward the periphery. The results of the combined velocity nephogram distribution show that, due to the opposite rotation of the two pairs of rotors of the quadrotor UAV, forward-rotating rotors generate a downwash flow, so the region with the highest wind speed is distributed directly below the two forward-rotating rotors, wind speed gradually decreases toward the periphery, and is also distributed in an elliptical shape. The long semiaxis of the region is about 0.35 m, and the short semiaxis is about 0.3 m.

2.2.2. Actual Test of Airflow Field

In order to verify the numerical simulation analysis results of the airflow field, the 3D airflow field testers were used to carry out a real-world test of the downwash airflow field under the hovering operation state of the UAV. Nine 3D airflow field testers were used to test the X-axial, Y-axial, and

Z-axial velocity components of the downwash airflow field. The testers were arranged in an array structure of three rows and three columns at equal intervals. The UAV was hovering and flying at a vertical position above the center of the array. Flight height was 1 m from the array plane. When the distance between adjacent testers was 0.6 m, the wind-speed data collected by the testers at the edge of the array were 0 m/s. Therefore, tester spacing was adjusted to 0.5 m, so testers at the edge of the array collected nonzero data, which met the test requirements. In the test, after the flight attitude of the UAV was stabilized, each three-dimensional airflow field tester stopped the test after collecting 100 datasets. The test process is shown in Figure 6.

Figure 6. Actual test of downwash flow field.

To reduce the error, the 10 maximum and 10 minimum values were removed from the 100 collected datasets before calculating the average value from the remaining data. The obtained result was solved by interpolation using the four-point spline-interpolation (V4) algorithm. When the adjacent tester spacing was 0.6 m, the edge-tester test result was 0 m/s. Therefore, the interpolation boundary was set to 0.6 m away from the center. The V4 interpolation algorithm is also called the interpolation algorithm based on biharmonic Green's functions. The difference surface is a linear combination of Green's functions centered on each sample. The surface is passed through various points by adjusting the weight of each point. Green's function of the spline satisfies the following biharmonic equation:

$$\frac{d^4\phi}{dx^4} = 6\delta(x) \qquad (1)$$

The specific solution of Equation (1) is

$$\phi(x) = |x|^3 \qquad (2)$$

When Green's function is used to interpolate N data points, the problem of w_i interpolation at x_i is

$$\frac{d^4 w}{d^4} = \sum_{j=1}^{N} 6\alpha_j \delta(x - x_j) \qquad (3)$$

$$w(x_i) = w_i \qquad (4)$$

The specific solution to Equations (3) and (4) is that Green's function is linearly combined around each data point, eliminating the need for a uniform solution.

$$w(x) = \sum_{j=1}^{N} \alpha_j |x - x_j|^3 \tag{5}$$

Green's function of the plane space is shown in Table 1:

Table 1. Function in flat space.

Dimension m	Green's Function $\phi_m(x)$
1	$|x|^3$
2	$|x|^2(\ln|x|-1)$
3	$|x|$

Weight value α_j is obtained by using the x values and $w(x)$ of N points, and the interpolation results are obtained by substituting the weight value into Equation (5).

The data of each 3D airflow field tester were sorted, invalid data were eliminated, valid data were retained, and data results were analyzed and processed. According to the V4 interpolation principle and calculated by MATLAB software, the interpolation results of each 3D airflow field tester node data are displayed in Figure 7.

From the results of Figure 7, we can see that, when the UAV hovering flight height was 1 m from the test plane of the 3D airflow field testers, the Z-axial velocity component of the tester was much larger than the X-axial and Y-axial velocity components. The combined velocity mainly depends on the axial velocity component. In the distribution range, the axial velocity component is elliptical, and the center velocity is the largest and gradually decreases toward the periphery. The combined-velocity distribution results show that the influence range of the UAV downwash airflow field was also elliptical, the center-point velocity was the largest, and peripheral speed was gradually reduced. The long semiaxis of the affected area was about 0.45 m, and the short semiaxis was about 0.4 m. Test results are consistent with the CFD numerical simulation results.

(a)

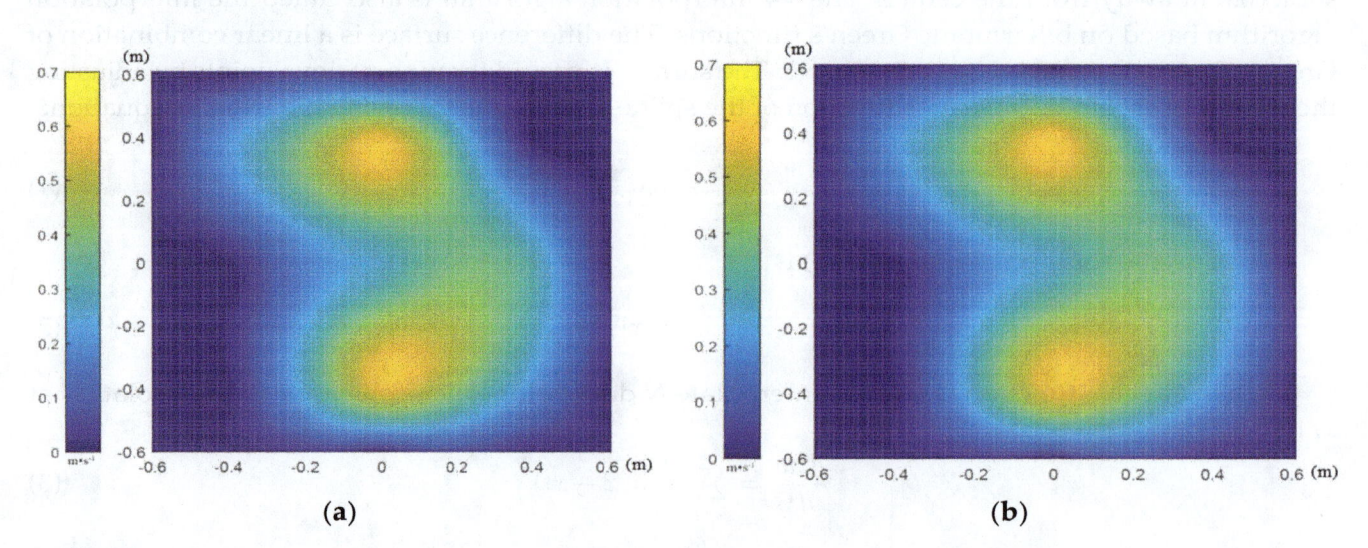

(b)

Figure 7. Cont.

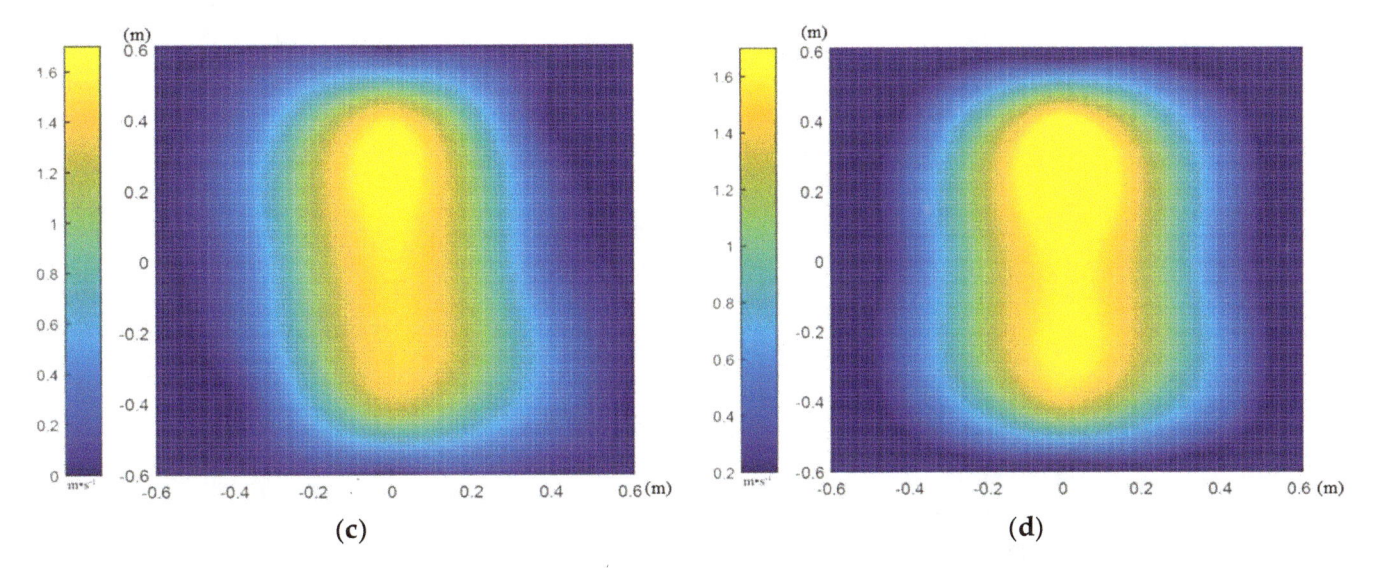

Figure 7. Three-dimensional tester interpolation results when the DJI phantom drone (unmanned aerial vehicle, UAV) hover-flight height is 1 m from 3D airflow field testers. (**a**) X-axial velocity profile; (**b**) Y-axial velocity profile; (**c**) Z-axial velocity profile; (**d**) combined velocity profile.

2.2.3. Dual-Band Crop-Growth-Sensor Deployment Method

According to CFD numerical simulation analysis and the actual test results of the 3D airflow field, combined with the test field of view of the dual-band crop-growth sensor, when the dual-band crop-growth sensor was deployed 60 cm away from the UAV rotors, the test area of the crop canopy retained Lambert characteristics, and measurement results were not affected by the airflow field, enabling normal testing. Therefore, we here designed a carbon-fiber sensor support with a length of 60 cm. One end of the support was fixed under the UAV spiral wing, and the other end extends outward along the UAV arm. The sensor was fixed to the rear end of the support through a mechanical structure. The support is connected to the UAV by a cantilever beam structure. We also designed supports of other lengths for experimental comparison.

In order to avoid the vibration impact of rotor high-speed rotation on the dual-band crop-growth sensor and the support during the flight, a damping rod was designed for shock absorption in order to maintain the stable state of the support and the dual-band crop-growth sensor. The support and damping rod were fixed with a triangular structure to improve the overall stability and shock resistance, as shown in Figure 8. The angle between support and damping rod is very important for the stability of the structure and the balance performance of the aircraft. Therefore, the optimal value of the angle was calculated by static equation analysis.

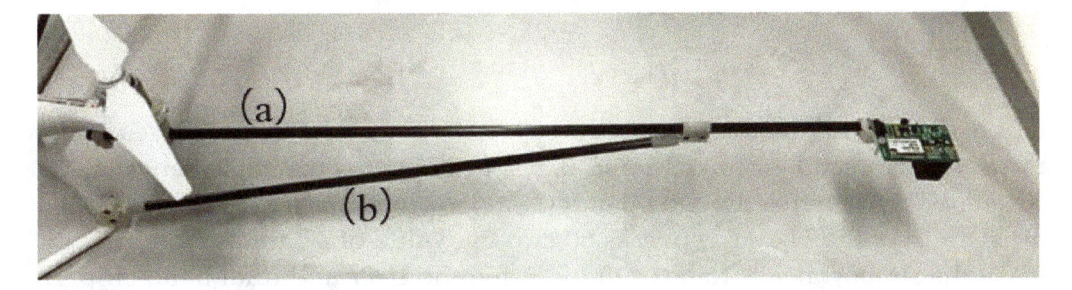

Figure 8. Design of UAV support and damping rod. (**a**) Support; (**b**) damping rod.

Taking the sensor support as the research object, the analysis diagram of the force sustained by the support is shown in Figure 9.

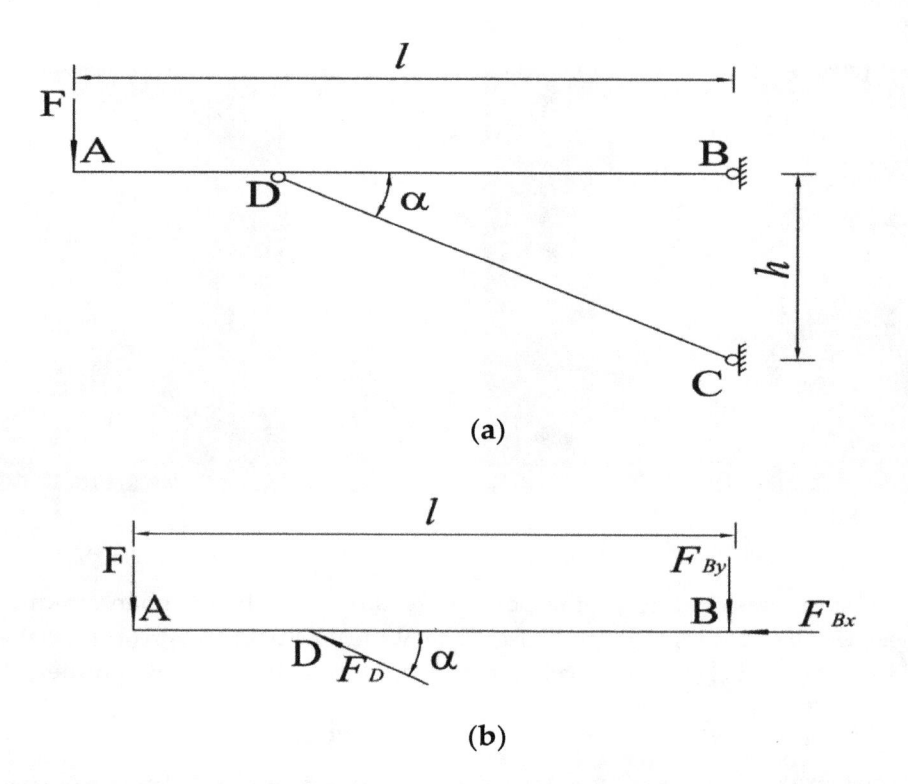

Figure 9. Analysis of the force sustained by UAV sensor support. (**a**) Schematic diagram of the structure of the support and damping rod; (**b**) analysis of the force sustained by each part of the support.

In Figure 9, AB is the sensor support, CD is the support damping rod, points C and D are the fixed points of the sensor support and the damping rod at the end of the UAV, and the force sustained by the sensor support is analyzed by the following static equations:

$$F = -F_{By} + F_D \sin \alpha \tag{6}$$

$$F_D \cos \alpha = -F_{Bx} \tag{7}$$

$$Fl = F_D \sin \alpha \cdot \frac{h}{\tan \alpha} \tag{8}$$

Simultaneous calculations were performed on Equations (6)–(8) to obtain the following:

$$F_D = \frac{Fl}{h \cos \alpha} \tag{9}$$

$$F_{Bx} = -\frac{Fl}{h} \tag{10}$$

$$F_{By} = F\left(\frac{l}{h}\tan\alpha - 1\right) \tag{11}$$

In Equations (6)–(11), F_D is the internal force of CD, F_{Bx} and F_{By} are the reaction forces in the horizontal and vertical directions of fixed-point B, respectively, α is the angle between AB and CD, h is the height difference between fixed points B and C, and l is the length of AB.

When the length of the UAV support is 60 cm, the value of α ranges from 9.5° to 90°. Since a certain length of the support needs to be used for fixing the UAV and dual-band crop-growth sensor, the actual value range of α is 13°–90°. From the derivation results of Equations (9)–(11), it can be seen that F_D and F_{By} decrease with the decrease of α, and the smaller the values of F_D and F_{By} are, the more stable the force sustained by the support structure is. Therefore, the optimal angle between sensor support and damping rod is 13°.

2.3. Field Trial

2.3.1. Test Design

Field Trial 1 was conducted at the Baipu Town (32°14′58.88 N 120°45′44.26 E) test base in Rugao City, Jiangsu Province, China, from February to May 2016. The test varieties were Ningmai 13 (V1) and Huaimai 33 (V2), three different gradients of nitrogen fertilizer treatment were set up, which were N_0 (0 kg/hm^2), N_1 (180 kg/hm^2), and N_2 (270 kg/hm^2), and each variety was repeated three times. Each planting area was 30 m^2 (5 × 6 m). In addition, the application rate of phosphate fertilizer was 120 kg/hm^2, and the application rate of potassium fertilizer was 135 kg/hm^2, which was applied once in the base fertilizer. Other cultivation-management measures were the same as those in general high-yield fields.

Field Trial 2 was conducted at the Lingqiao Township (33°35′53.27 N 118°51′11.01 E) Test Base in Huai'an City, Jiangsu Province, China, from July to October 2016. The test varieties were Nanjing 9108 (V1) and Lianjing 10 (V2), four different gradients of nitrogen fertilizer were applied, which were N_0 (0 kg/hm^2), N_1 (120 kg/hm^2), N_2 (240 kg/hm^2), and N_3 (360 kg/hm^2), each variety was repeated three times, and each planting area was 30 m^2 (5 × 6 m). In addition, the application rate of phosphate fertilizer was 105 kg/hm^2 and was applied once in the base fertilizer, the potassium fertilizer was 135 kg/hm^2, the base fertilizer was applied 50%, and, at the early boot stage, application was 50%. The other cultivation-management measures were the same as those in general high-yield fields.

2.3.2. Test Method

Field Trial 1 was used to test whether the proposed monitoring method can effectively avoid the disturbance range of the UAV downwash airflow field when acquiring data from the crop canopy. Additionally, Field Trial 1 was used to verify the accuracy and stability of the dual-band crop-growth sensor test results. The experiment was carried out in the middle of the wheat jointing stage. The test was carried out on a clear, windless, and cloudless day. Test time was between 10:00 and 14:00. The UAV was flown 1 m above the wheat canopy, and, as shown in Figure 10, the dual-band crop-growth sensor was deployed in three different horizontal distances from the UAV rotors: 0 (i.e., directly below the UAV), 30, and 60 cm. The sensors determined the RVI value of the wheat canopy by measuring three random points in each subarea and repeating the measurement of each point three times to obtain an average value. The ASD FieldSpec HandHeld 2 was used to measure the RVI value of the wheat canopy at the same time.

Figure 10. Comparison of field flights with the conditions of three support lengths.

Field Trial 2 was used to evaluate the applicability and accuracy of UAV-borne spectral sensors for crop-growth parameters. The test was carried out in the tillering, jointing, booting, and heading stages of the rice, test weather was sunny and windless, and test time was between 10:00 and 14:00. In the test, the UAV was made to hover at a height of 1 m above the rice canopy in different test plots, and the dual-band crop-growth sensor was deployed at a horizontal distance of 60 cm from the UAV rotors to obtain the RVI value at the 730 and 815 nm bands of the rice canopy. Three points were randomly measured in each subarea, and the measurement of each point was repeated three times to obtain an average value. The FieldSpec HandHeld 2 and LAI2200 testers were synchronously used to obtain the RVI and LAI values of the rice leaf layer. At the same time, in parallel with the test, the rice sample was destructively sampled, and the sample was placed at 105 °C for 30 min for fixing, then baked at 80 °C to constant weight, and weighed to obtain the LDW. After the sample was pulverized, the LNA was determined by the Kjeldahl method.

2.3.3. Analysis of Field-Test Data

The field-test datasets were statistically analyzed with Microsoft Excel 2010 software; the correlation of the model was evaluated by the coefficient of determination (R^2) and root mean square error (RMSE).

3. Results and Discussion

3.1. Field-Test Results

Figure 11 shows the test results of Field Trial 1, in which the dual-band crop-growth sensor was located at different positions relative to the UAV in the middle of the wheat-jointing stage. Parts a, b, and c of Figure 11 show the simple linear fitting results of the RVI test value of the UAV-borne spectral sensor when the length of the support was 0, 30, and 60 cm, respectively, and the RVI value of the handheld ASD FieldSpecHandHeld 2 spectrometer in the corresponding growth period. When the length of the support was 0 and 30 cm, the results were close, and R^2 values were 0.63 and 0.66, respectively. When the length of the support was extended to 60 cm, the curve fitting degree was obviously improved; R^2 was 0.81, and RMSE was 0.38.

(a)

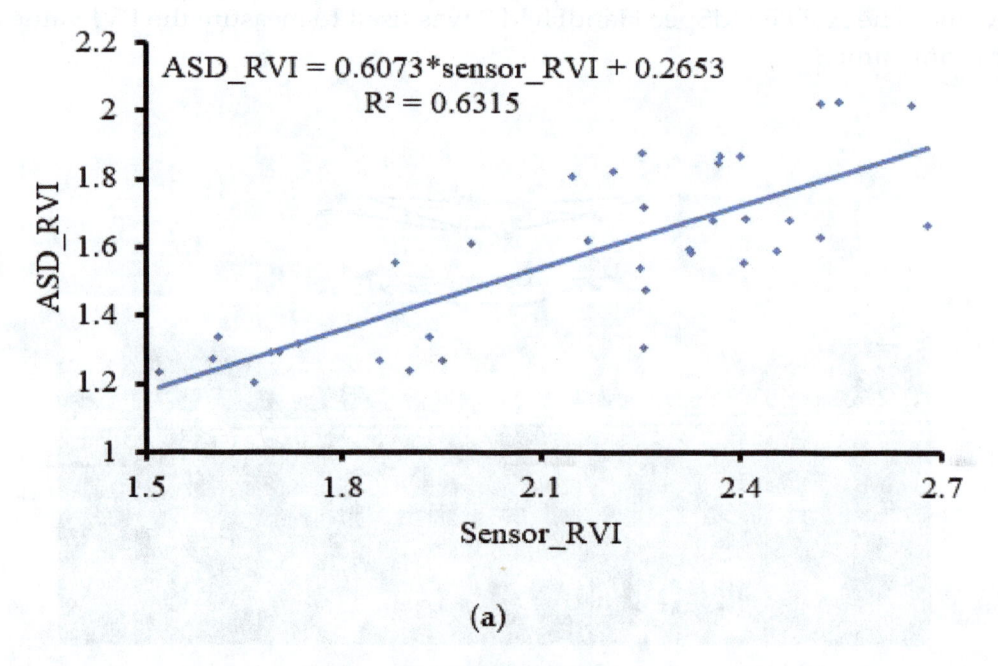

Figure 11. *Cont.*

(b)

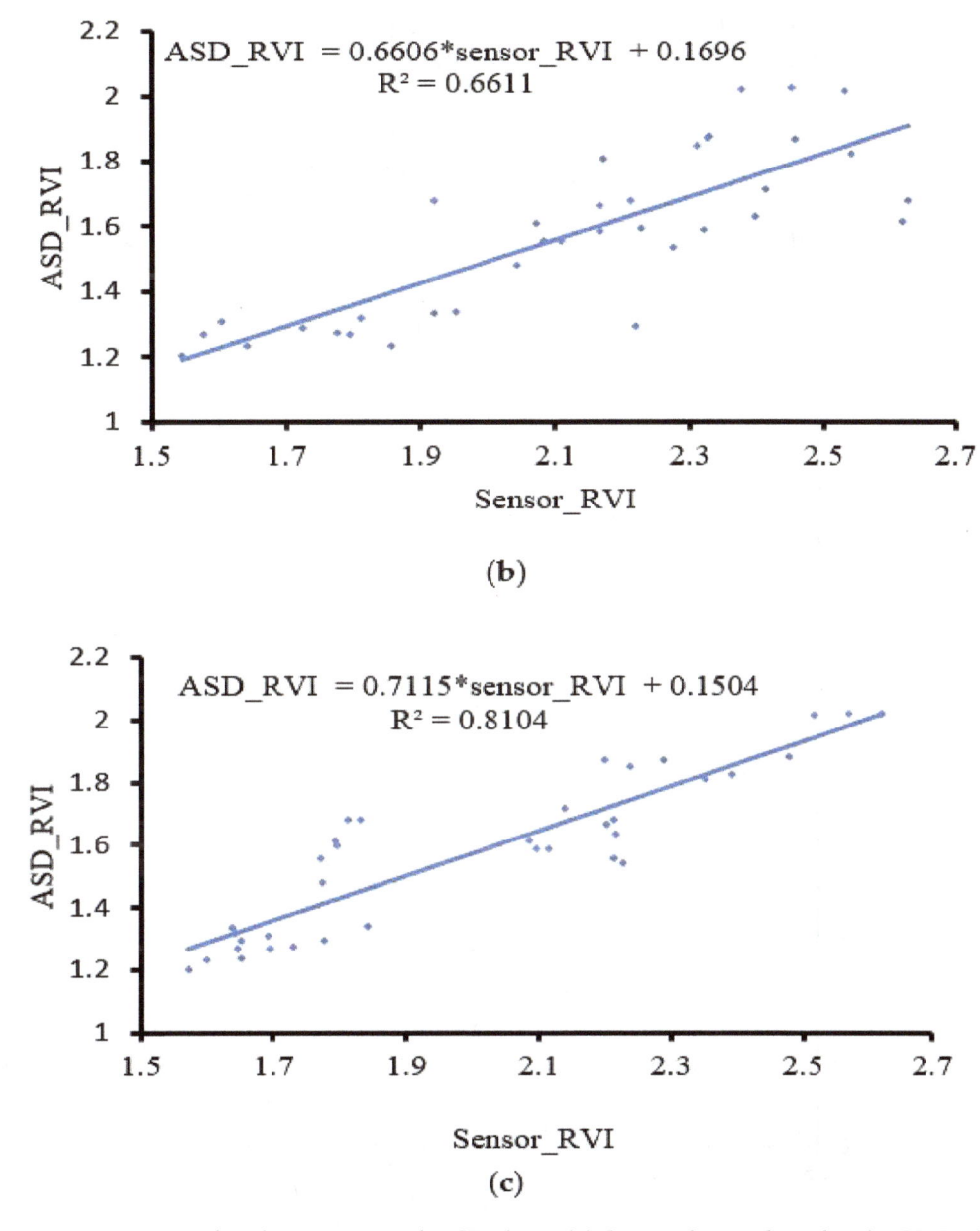

$$ASD\_RVI = 0.6606*sensor\_RVI + 0.1696$$
$$R^2 = 0.6611$$

(c)

$$ASD\_RVI = 0.7115*sensor\_RVI + 0.1504$$
$$R^2 = 0.8104$$

Figure 11. Fitting curves for the sensor and ASD data. (**a**) Sensor located under the UAV; (**b**) sensor located 30 cm from the rotor; (**c**) sensor located 60 cm from the rotor.

It can be seen from the above test results that the farther the dual-band crop growth sensor was deployed from the UAV rotors, the better the correlation between the test data and the ASD test results. According to CFD numerical simulation results and 3D airflow field-test results, when the dual-band crop-growth sensor was deployed directly below the UAV, the sensor-test field of view was completely within the disturbance range of UAV rotor downwash airflow field. When the dual-band crop-growth sensor was deployed 30 cm away from the UAV rotors, the sensor-test field of view included both the disturbance and nondisturbance zone of the rotor downwash airflow field. The correlation of the data results improved slightly, but it was still not ideal. When the dual-band crop-growth sensor was deployed 60 cm away from the rotors of the UAV, the sensor-test field was completely in the nondisturbance zone, and the correlation of the data results was significantly improved. In summary, the downwash airflow field generated by the rotation of the UAV rotors has a certain influence on the results of the dual-band crop-growth sensor. The proposed UAV-borne spectral sensor crop-growth monitoring method can effectively target areas of the crop canopy outside the disturbance range of the UAV downwash airflow field.

In Field Trial 2, the flying height of the UAV was 1 m from the rice canopy, the dual-band crop-growth sensor was deployed 60 cm away from the UAV rotors, and measurements were taken throughout the entire growth period of rice. Figure 12 shows the linear fitting results of the RVI values obtained by our method, and the LNA, LAI, and LDW obtained from the field test and the indoor chemical analysis test. R^2 values were 0.62, 0.76, and 0.60, and RMSE values were 2.28, 1.03, and 10.73, respectively. Using the proposed UAV-borne spectral sensor crop-growth monitoring method, rice-growth parameters in the entire growth period could accurately be obtained.

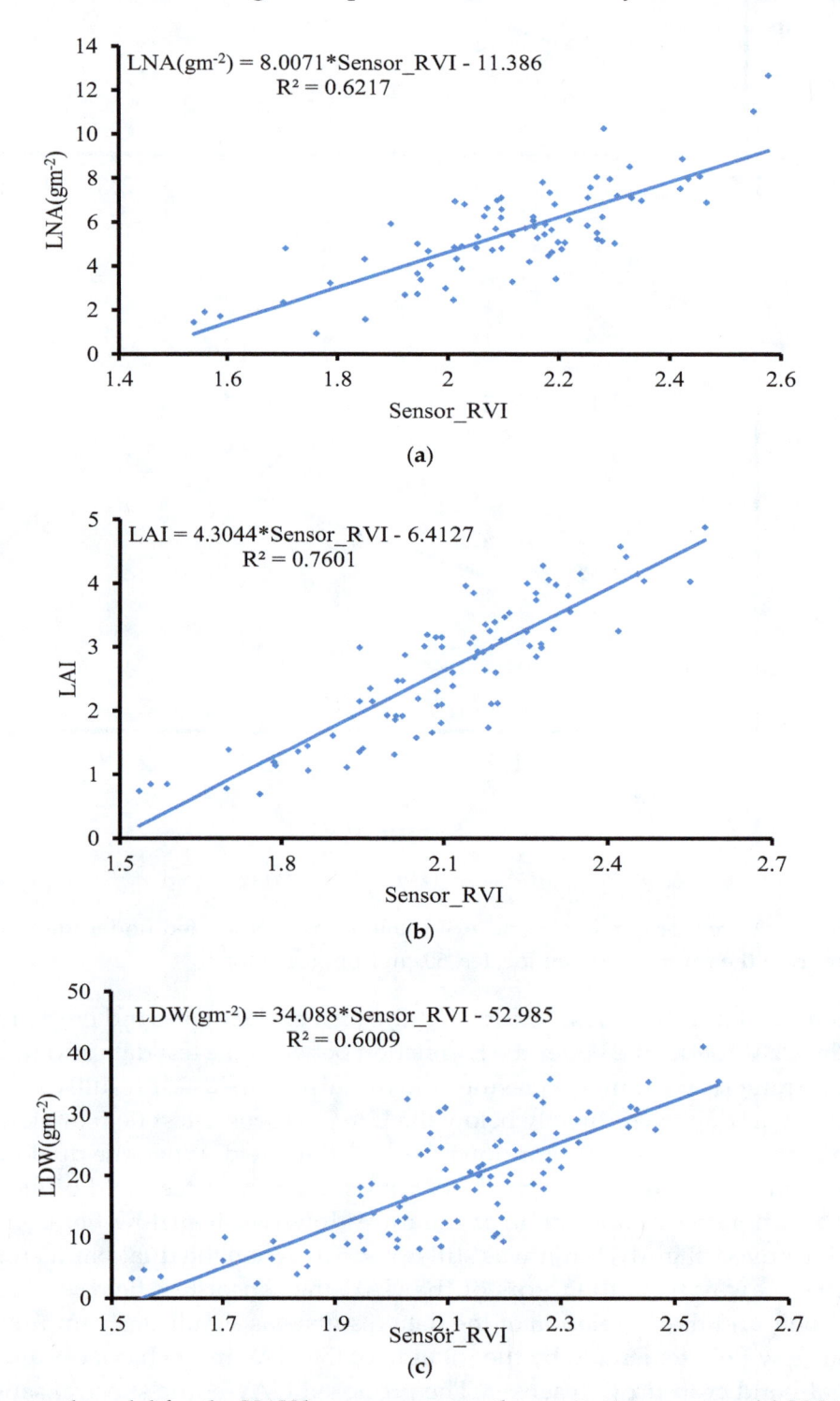

Figure 12. Spectral model for the UAV-borne crop-growth monitoring system. (**a**) LNA–RVI fitting curve; (**b**) LAI–RVI fitting curve; (**c**) LDW–RVI fitting curve.

3.2. Discussion

UAVs have the characteristics of simple operation and high efficiency. At present, there is little targeted research on monitoring crop growth by UAV-borne dual-band crop-growth sensors. Under a low-altitude hovering operation, the downwash airflow field generates strong disturbance on the crop canopy that disrupts the Lambertian reflection characteristics of the crop canopy and thus has a serious impact on the accuracy of the test, even causing the test to fail. Therefore, we used CFD numerical simulation and real-world 3D airflow field testing to analyze the spatial distribution of the downwash airflow field when the UAV is hovering at a height of 1 m above the crop canopy and determine the disturbance range on the crop canopy. Most of the current studies simply mounted the energy-type spectrum sensor on UAVs, lacking consideration for the disturbance influence of the crop canopy by the UAV downwash airflow field during the test [37]. Our study filled the gap in these studies. Our test showed that the influence range of the downwash airflow field of the UAV is elliptical, wind speed at the center point is the largest, and gradually decreases toward the periphery. The long semiaxis was about 0.45 m, and the short semiaxis was about 0.4 m. According to the distribution range of the downwash airflow field, we designed a support with a length of 0.6 m to deploy a dual-band crop-growth sensor so that the test field of view is extended beyond the distribution range of the downwash airflow field, and the disturbance effect of the downwash airflow field on the crop canopy was avoided. When the flight height of the UAV was kept at 1 m from the wheat canopy and the dual-band crop-growth sensor was deployed 0.6 m from the rotors, the linear fit R^2 of the test output RVI value and the handheld ASD Fieldspec2 spectrometer, the test RVI value was 0.81, and RMSE was 0.38. Therefore, the UAV-borne spectral sensor crop-growth monitoring method can effectively target areas of the crop canopy outside the disturbance range of the UAV downwash airflow field.

In the test process of the 3D airflow field testers, the wind-speed results of different dimensions measured by the testers were larger than the CFD numerical-simulation results. The main reason was that the test plane of the 3D airflow field testers was 60 cm away from the ground, and the arrangement was lattice. When the downwash airflow diffused downward through the lattice plane, the direction of the airflow changed. In the CFD numerical simulation analysis, the downwash airflow directly reached the ground plane, so the distribution states of the real-world test and numerical simulation differed. Although the analyzed UAV in this paper is a DJI Phantom drone, the proposed UAV-borne spectral sensor crop-growth monitoring method can be applied to various types of multirotor UAV structures.

4. Conclusions

1. We identified the UAV-borne spectral sensor crop-growth monitoring method, used CFD numerical simulation and an actual test of 3D airflow field to determine the distribution range of the UAV downwash airflow field above the surface of the crop canopy, and designed sensor supports to target areas of the crop canopy outside the disturbance range of the UAV downwash airflow field.

2. When the flying height of the UAV was 1 m from the crop canopy, the influence range of the downwash airflow field of the UAV was elliptical, central wind speed was the largest and gradually decreased toward the periphery, the long semiaxis was about 0.45 m, and the short semiaxis was about 0.4 m. When the designed sensor support length was 60 cm, the sensor-test field of view was completely outside the disturbance range of the UAV downwash airflow field.

3. The wheat test showed that, when the dual-band crop-growth sensor was deployed at 0, 30, and 60 cm from the UAV, the linear fit R^2 of the RVI value obtained by our method, and the RVI value measured by the ASD FieldSpec HandHeld 2 spectrometer, was 0.63, 0.66, and 0.81, respectively. When the length of the support was 60 cm, the fitting degree was obviously improved. The UAV-borne spectroscopy sensor crop-growth monitoring method can effectively avoid the disturbance range of the UAV downwash airflow field on the crop canopy.

4. The rice experiment showed that the RVI value measured by the UAV-borne spectral sensor had a good linear fitting relationship with the LNA, LAI, and LDW obtained from the field test and the indoor chemical analysis test. R^2 values were 0.62, 0.76, and 0.60, respectively, and RMSE values

were 2.28, 1.03, and 10.73, respectively. Using the UAV-borne spectral sensor crop-growth monitoring method, rice-growth parameters during the entire growth period could be accurately obtained.

Author Contributions: J.N., Y.Z. (Yan Zhu) and W.C. designed the research, J.N., Q.W., Y.Z. (Yu Zhang) and L.Y. performed the research, L.Y., J.Y. and J.N. analyzed the data, and L.Y. and Y.Z. (Yan Zhu) wrote the paper. All authors have read and approved the final manuscript submitted to the editor.

Acknowledgments: The authors thank all those who helped in the course of this research.

References

1. Pan, J.; Liu, Y.; Zhong, X.; Lampayan, R.M.; Singleton, G.R.; Huang, N.; Liang, K.; Peng, B.; Tian, K. Grain yield, water productivity and nitrogen use efficiency of rice under different water management and fertilizer-N inputs in South China. *Agric. Water Manag.* **2017**, *184*, 191–200. [CrossRef]
2. Cambouris, A.N.; Zebarth, B.J.; Ziadi, N.; Perron, I. Precision Agriculture in Potato Production. *Potato Res.* **2014**, *57*, 249–262. [CrossRef]
3. Guo, J.H.; Wang, X.; Meng, Z.J.; Zhao, C.J.; Zhen-Rong, Y.U.; Chen, L.P. Study on diagnosing nitrogen nutrition status of corn using Greenseeker and SPAD meter. *Plant Nutr. Fertil. Sci.* **2008**, *14*, 43–47.
4. Grohs, D.S.; Bredemeier, C.; Mundstock, C.M.; Poletto, N. Model for yield potential estimation in wheat and barley using the GreenSeeker sensor. *Eng. Agrícola* **2009**, *29*, 101–112. [CrossRef]
5. Ali, A.M.; Thind, H.S.; Sharma, S.; Varinderpal-Singh, V. Prediction of dry direct-seeded rice yields using chlorophyll meter, leaf color chart and GreenSeeker optical sensor in northwestern India. *Field Crop. Res.* **2014**, *161*, 11–15. [CrossRef]
6. Moya, I.; Camenen, L.; Evain, S.; Goulas, Y.; Cerovic, Z.G.; Latouche, G.; Flexas, J. A new instrument for passive remote sensing: 1. Measurements of sunlight-induced chlorophyll fluorescence. *Remote Sens. Environ.* **2004**, *91*, 186–197. [CrossRef]
7. Youngryel, R.; Dennisd, B.; Joseph, V.; Ma, S.; Matthias, F.; Ilse, R.M.; Ted, H. Testing the performance of a novel spectral reflectance sensor, built with light emitting diodes (LEDs), to monitor ecosystem metabolism, structure and function. *Agric. For. Meteorol.* **2011**, *150*, 1597–1606.
8. Martin, D.E.; Lan, Y.B. Laboratory evaluation of the GreenSeekerTM handheld optical sensor to variations in orientation and height above canopy. *Int. J. Agric. Biol. Eng.* **2012**, *5*, 43–47.
9. Ali, A.M.; Abouamer, I.; Ibrahim, S.M. Using GreenSeeker Active Optical Sensor for Optimizing Maize Nitrogen Fertilization in Calcareous Soils of Egypt. *Arch. Agron. Soil Sci.* **2018**, *64*, 1083–1093. [CrossRef]
10. Vohland, M.; Besold, J.; Hill, J.; Fründ, H.C. Comparing different multivariate calibration methods for the determination of soil organic carbon pools with visible to near infrared spectroscopy. *Geoderma* **2011**, *166*, 198–205. [CrossRef]
11. Kusumo, B.H.; Hedley, M.J.; Hedley, C.B.; Hueni, A.; Arnold, G.C. The use of Vis-NIR spectral reflectance for determining root density: Evaluation of ryegrass roots in a glasshouse trial. *Eur. J. Soil Sci.* **2010**, *60*, 22–32. [CrossRef]
12. Nawar, S.; Buddenbaum, H.; Hill, J.; Kozak, J.; Mouazen, A.M. Estimating the soil clay content and organic matter by means of different calibration methods of vis-NIR diffuse reflectance spectroscopy. *Soil Tillage Res.* **2016**, *155*, 510–522. [CrossRef]
13. Cao, Q.; Miao, Y.; Wang, H.; Huang, S.; Cheng, S.; Khosla, R. Non-destructive estimation of rice plant nitrogen status with Crop Circle multispectral active canopy sensor. *Field Crop. Res.* **2013**, *154*, 133–144. [CrossRef]
14. Bonfil, D.J. Wheat phenomics in the field by RapidScan: NDVI vs. NDRE. *Isr. J. Plant Sci.* **2016**, 1–14. [CrossRef]

15. Lu, J.; Miao, Y.; Wei, S.; Li, J.; Yuan, F. Evaluating different approaches to non-destructive nitrogen status diagnosis of rice using portable RapidSCAN active canopy sensor. *Sci. Rep.* **2017**, *7*, 14073. [CrossRef] [PubMed]
16. Miller, J.J.; Schepers, J.S.; Shapiro, C.A.; Arneson, N.J.; Eskridge, K.M.; Oliveira, M.C.; Giesler, L.J. Characterizing soybean vigor and productivity using multiple crop canopy sensor readings. *Field Crop. Res.* **2018**, *216*, 22–31. [CrossRef]
17. Zhou, Z.; Andersen, M.N.; Plauborg, F. Radiation interception and radiation use efficiency of potato affected by different N fertigation and irrigation regimes. *Eur. J. Agron.* **2016**, *81*, 129–137. [CrossRef]
18. Krienke, B.; Ferguson, R.B.; Schlemmer, M.; Holland, K.; Marx, D.; Eskridge, K. Using an unmanned aerial vehicle to evaluate nitrogen variability and height effect with an active crop canopy sensor. *Precis. Agric.* **2017**, *18*, 900–915. [CrossRef]
19. Shafian, S.; Rajan, N.; Schnell, R.; Bagavathiannan, M.; Valasek, J.; Shi, Y. Unmanned aerial systems-based remote sensing for monitoring sorghum growth and development. *PLoS ONE* **2018**, *13*, e0196605. [CrossRef]
20. Schirrmann, M.; Giebel, A.; Gleiniger, F.; Pflanz, M.; Lentschke, J. Monitoring Agronomic Parameters of Winter Wheat Crops with Low-Cost UAV Imagery. *Remote Sens.* **2016**, *8*, 706. [CrossRef]
21. Zheng, H.; Zhou, X.; Cheng, T.; Yao, X.; Tian, Y.; Cao, W.; Zhu, Y. Evaluation of a UAV-based hyperspectral frame camera for monitoring the leaf nitrogen concentration in rice. In Proceedings of the Geoscience and Remote Sensing Symposium, Beijing, China, 10–15 July 2016; pp. 7350–7353.
22. Stroppiana, D.; Villa, P.; Sona, G.; Ronchetti, G.; Candiani, G.; Pepe, M.; Busetto, L.; Migliazzi, M.; Boschetti, M. Early season weed mapping in rice crops using multi-spectral UAV data. *Int. J. Remote Sens.* **2018**, *39*, 5432–5452. [CrossRef]
23. Wu, M.; Huang, W.; Niu, Z.; Wang, Y.; Wang, C.; Li, W.; Hao, P.; Yu, B. Fine crop mapping by combining high spectral and high spatial resolution remote sensing data in complex heterogeneous areas. *Comput. Electron. Agric.* **2017**, *139*, 1–9. [CrossRef]
24. Sandwell, D.T. Biharmonic spline interpolation of GEOS-3 and SEASAT altimeter data. *Geophys. Res. Lett.* **2013**, *14*, 139–142. [CrossRef]
25. Ni, J.; Yao, L.; Zhang, J.; Cao, W.; Zhu, Y.; Tai, X. Development of an Unmanned Aerial Vehicle-Borne Crop-Growth Monitoring System. *Sensors* **2017**, *17*, 502. [CrossRef] [PubMed]
26. Feng, W.; Zhu, Y.; Tian, Y.; Cao, W.; Yao, X.; Li, Y. Monitoring leaf nitrogen accumulation in wheat with hyper-spectral remote sensing. *Eur. J. Agron.* **2008**, *28*, 23–32. [CrossRef]
27. Yan, Z.; Dongqin, Z.; Xia, Y.; Yongchao, T.; Weixing, C. Quantitative relationship between leaf nitrogen accumulation and canopy reflectance spectra in wheat. *Aust. J. Agric. Res.* **2007**, *58*, 1077–1085. [CrossRef]
28. Yan, Z.; Yingxue, L.; Wei, F.; Yongchao, T.; Xia, Y.; Weixing, C. Monitoring leaf nitrogen in rice using canopy reflectance spectra. *Can. J. Plant Sci.* **2006**, *86*, 1037–1046.
29. Szuvandzsiev, P.; Helyes, L.; Lugasi, A.; Szántó, C.; Baranowski, P.; Pék, Z. Estimation of antioxidant components of tomato using VIS-NIR reflectance data by handheld portable spectrometer. *Int. Agrophys.* **2014**, *28*, 521–527. [CrossRef]
30. Battay, A.E.; Mahmoudi, H. Linear spectral unmixing to monitor crop growth in typical organic and inorganic amended arid soil. *IOP Conf. Ser. Earth Environ. Sci.* **2016**, *37*, 012046. [CrossRef]
31. Fang, H.; Li, W.; Wei, S.; Jiang, C.J.A.; Meteorology, F. Seasonal variation of leaf area index (LAI) over paddy rice fields in NE China: Intercomparison of destructive sampling, LAI-2200, digital hemispherical photography (DHP), and AccuPAR methods. *Agric. For. Meteorol.* **2014**, *198–199*, 126–141. [CrossRef]
32. Sandmann, M.; Graefe, J.; Feller, C. Optical methods for the non-destructive estimation of leaf area index in kohlrabi and lettuce. *Sci. Hortic.* **2013**, *156*, 113–120. [CrossRef]
33. Kobayashi, H.; Ryu, Y.; Baldocchi, D.D.; Welles, J.M.; Norman, J.M. On the correct estimation of gap fraction: How to remove scattered radiation in gap fraction measurements? *Agric. For. Meteorol.* **2013**, *174–175*, 170–183. [CrossRef]
34. Jiradilok, V.; Gidaspow, D.; Damronglerd, S.; Koves, W.J.; Mostofi, R. Kinetic theory based CFD simulation of turbulent fluidization of FCC particles in a riser. *Chem. Eng. Sci.* **2006**, *61*, 5544–5559. [CrossRef]
35. Blocken, B.; Stathopoulos, T.; Carmeliet, J. CFD simulation of the atmospheric boundary layer: Wall function problems. *Atmos. Environ.* **2007**, *41*, 238–252. [CrossRef]

36. Nakata, T.; Liu, H.; Bomphrey, R.J. A CFD-informed quasi-steady model of flapping wing aerodynamics. *J. Fluid Mech.* **2015**, *783*, 323–343. [CrossRef] [PubMed]
37. Li, S.; Ding, X.; Kuang, Q.; Ata-UI-Karim, S.T.; Cheng, T.; Liu, X.; Tian, Y.; Yan, Z.; Cao, W.; Cao, Q. Potential of UAV-Based active sensing for monitoring rice leaf nitrogen satus. *Front. Plant Sci.* **2018**, *9*, 1834. [CrossRef]

Development and Field Evaluation of a Spray Drift Risk Assessment Tool for Vineyard Spraying Application

Georgios Bourodimos [1,2,*], Michael Koutsiaras [1], Vasilios Psiroukis [1], Athanasios Balafoutis [3] and Spyros Fountas [1]

[1] Department of Natural Resources Management & Agricultural Engineering, Agricultural University of Athens, Iera Odos 75, 11855 Athens, Greece
[2] Department of Agricultural Engineering, Institute of Soil and Water Resources, Hellenic Agricultural Organization "DEMETER", Democratias 61, 13561 Aghii Anargiri Attikis, Greece
[3] Institute for Bio-Economy & Agri-Technology, Centre of Research & Technology Hellas, Dimarchou Georgiadou 118, 38221 Volos, Greece
* Correspondence: g.bourodimos@swri.gr

Abstract: Spray drift is one of the most important causes of pollution from plant protection products and it puts the health of the environment, animals, and humans at risk. There is; thus, an urgent need to develop measures for its reduction. Among the factors that affect spray drift are the weather conditions during application of spraying. The objective of this study was to develop and evaluate a spray drift evaluation tool based on an existing model by TOPPS-Prowadis to improve the process of plant protection products' application and to mitigate spray drift for specific meteorological conditions in Greece that are determined, based on weather forecast, by reassessing the limits for wind speed and direction, temperature, and air relative humidity set in the tool. The new limits were tested by conducting experimental work in the vineyard of the Agricultural University of Athens with a trailed air-assisted sprayer for bush and tree crops, using the ISO 22866:2005 methodology. The results showed that the limits set are consistent with the values of the spray drift measured and follows the tool's estimates of low, medium, and high risk of spray drift.

Keywords: drift risk assessment tool; sedimenting spray drift; airborne spray drift; weather conditions; spray drift reduction

1. Introduction

Chemical crop protection is one of the most important factors in agricultural production, as global potential crop yield is diminished by pests up to 40%, a figure that would be twice as large if no plant protection products (PPPs) were used [1]. The benefits from crop protection are also undeniably significant in regards to improved food security and the reduction of labor [2]. However, as most PPPs are applied in the field by spraying, due to low cost and efficient performance [3], serious PPP losses are a side effect of chemical crop protection due to run-off, leaching, evaporation, and spray drift, putting the health of the environment, animals, and humans at risk [4].

Spray drift is the quantity of PPPs that is carried out by air currents from the sprayed area during the spraying application [5]. Spray drift is an important and costly (environmentally and economically) problem that is hard to control and may cause the PPPs to be deposited in off-target areas. The consequences can be serious, such as surface water contamination, air pollution, damage to sensitive nearby crops and other susceptible off-target areas, residues of chemical substances in food and feed commodities, health risks for animals and people (farm workers, bystanders, and passers-by), nearby urban or natural area contamination, and reduced PPP effectiveness due to lower dose than intended on the targeted crops [6–10]. Additionally, the financial burden resulting from spray drift due to increased inputs is very high [11].

There are many factors that contribute to spray drift and several of them are interrelated. Some factors can be controlled by the sprayer operator, while others cannot be controlled. These factors can be grouped into the following categories: (i) Equipment and application techniques (i.e., sprayer type, size and type of nozzles, spray pressure, spray volume rate, air flow rate, driving speed, sprayer's setup, etc.); (ii) weather conditions during application (i.e., wind speed and direction, temperature, relative humidity, and stability of air at the application site); (iii) spray characteristics, (i.e., volatility and viscosity of the PPP formulation; (iv) operator's care, attitude, and skill; and (v) characteristics and geometry of the crop (i.e., foliage, density, dimensions, etc.) [12–17].

Among the technical factors that affect spray drift, the most important is droplet size [18] and more particularly the percentage of fine spray droplets [19–22]. Considering the pressure atomization, droplet size depends on the nozzle design, orifice size, operating pressure, and the physical properties of the PPP formulation and spray additives [23,24]. The smaller a spray droplet, the longer it remains airborne, and the higher the possibility for it to be carried away by crosswind [25]. Droplets with diameter smaller than 100 μm contribute significantly to drift losses [26,27]. Forward speed has also a clear effect on spray drift, where the higher the driving speed the greater the spray drift, both for airborne drift and for ground deposition [28]. Regarding bush and tree sprayers, among the various application operating parameters that affect spray drift, those that concern spray generation (nozzle type and pressure) and droplet transport to the canopy (air fan volume, speed, orientation) are of great importance [29].

Environmental conditions influence spray drift and cannot be controlled by the sprayer operator. These factors need to be taken into consideration and be monitored before and during the PPP application. Among the meteorological factors affecting spray drift, wind velocity has the greatest impact, while wind direction plays also an important role. There is a strong positive correlation between wind speed and spray drift deposition [30]. Higher wind speeds result in more drift at greater distances [31]. Summer [32] pointed out that wind speed should not exceed 16 km h^{-1} (4.44 m s^{-1}) to continue spraying. According to da Cunha et al. [33], the maximum permissible wind speed for spraying is 12 km h^{-1} (3.33 m s^{-1}), while Maciel et al. [34] also proposed that spraying operation should be performed with wind speeds ranging between 2 and 12.8 km h^{-1} (0.55 to 3.55 m s^{-1}). It should be noted that different wind speeds and directions have also been reported to contribute to uneven spray distribution between the left and right side of the sprayer [35]. In addition, large fluctuations in wind direction increase the unpredictability of droplet travel direction and the amount of dilution due to atmospheric turbulence [36].

Air temperature and relative humidity during the spraying operation play also key role in spray drift. Low relative humidity and/or high air temperature can reinforce evaporation by decreasing the droplet size, especially small droplets, having, as a result, decreased sedimentation velocity and droplets more prone to drift [37]. Low relative humidity combined with high temperatures contributes to the evaporation of the spray liquid with an impact both on the environment and on the economic viability of the farm [34]. ISO 22866 standard [5] considers acceptable conditions for field measurement of spray drift at temperatures ranging from 5 to 35 °C. Da Cunha et al. [33], pointed out that spraying should be avoided when temperature is above 30 °C and relative air humidity is below 55%. As a rule, if the relative humidity is above 70%, the conditions are ideal for spraying, and if the relative humidity is below 50%, it is quite critical and requires special attention [32]. Generally, spray drift can be significantly reduced by spraying at low wind speed, low temperature, with low turbulence, at times of low sun radiation and at high relative humidity [38].

Moreover, it should be noted that during rain or shortly before it occurs, PPP application should be avoided due to leaching risk from the crop canopy before the active substance is able to act [39], polluting the soil and the underground water resources.

Since the negative effect of spray drift has been recognized, there is a need for harmonized mitigation measures to reduce human health and environmental impact. Such measures have been developed and can be divided into three classes [40–42]: (i) The use of no-spray or even no-crop buffer zones; (ii) the reduction of exposure using vegetative or artificial windbreaks, and (iii) the application of drift-reducing technology, such as drift-reducing nozzles and spray additives to coarsen the droplet size distribution, as well as shielded and band sprayers. In addition, the EU legislation has been adjusted in this direction, with Directive 2009/128/EC [43] establishing a framework to achieve sustainable use of PPPs, while Directive 2009/127/EC [44] specifies that sprayers should be designed and constructed to ensure that PPPs are deposited on target areas, to minimize losses to other areas and to prevent drift. USA has also set certain measures for spray drift prevention that are in the same direction as the EU legislation [45].

In recent years, several attempts have been made to reduce spray drift through predicting weather conditions and creating automated spray drift reduction systems. Such systems may collect meteorological data from either ground meteorological stations or geostationary satellites and predict drift or provide information to the user on how to treat and regulate the sprayer. The frequency of meteorological data gathering is crucial to optimize weather prediction and develop a reliable spray drift model. As an example, Huang and Thomson [46] highlighted the importance of knowing meteorological data on 15 min basis than on 1 h basis, as climate conditions can change rapidly, making hourly forecasts unreliable. On-line applications are also used for spray drift prediction and reduction helping farmers make decisions for best practice implementation. For example, Nansen et al. [47] created an online decision support tool for farmers and agricultural advisors. This tool allows the prediction, measurement, and archiving of spraying coverage, which is quantified using water sensitive filter papers. Spray drift investigation and prediction under a wide range of conditions have been conducted [48,49]. A comprehensive model which accurately predicts the downwind movement of spray for given circumstances, including spray liquid characteristics, spray nozzle characteristics, and meteorological conditions, was developed by the US Spray Drift Task Force [50]. Hong et al. [51] presented a software application for spray drift estimation using an orchard air-assisted sprayer, through their research study on spray drift prediction. Another case of a spray drift prediction model, which simulates spray drift taking meteorological conditions into consideration, is the one produced by Nsibande et al. [52] in South Africa.

The aim of this research was to evaluate spray drift in vineyards using a drift risk assessment tool in order to improve the process of PPP application and mitigate spray drift. The spray drift risk assessment tool was developed taking into consideration a similar drift evaluation tool developed in the framework of the TOPPS-Prowadis [53]. The factors that the tool considers are wind speed and direction, air temperature, and relative humidity. The ultimate goal was to examine the reliability of the tool by measuring the meteorological conditions in the field and assessing ground and airborne spray drift. The evaluation used field trials in a vineyard under the standardized test methodology of ISO 22866:2005 [5]. The selected vineyard variety for spray drift trials was Savatiano, the most widespread winemaking variety in the Attica region, which has been cultivated for about 4000 years [54].

2. Materials and Methods

2.1. Spay Drift Risk Assessment Tool

The tool evaluates the potential drift risk from PPP applications using air-assisted sprayers for different meteorological conditions under field conditions in vineyards. The meteorological data input to the tool is obtained from a weather forecast website for the geographic coordinates of the field. The drift evaluation tool can help the farmers or spray contractors to make decisions for PPP applications with high spray efficiency and low spray drift risk to the environment.

The tool is based on the methodology developed from the TOPPS-Prowadis drift evaluation tool [53], and presents three categories depending on the risk of spray drift to occur:

- **"Low"** indicates that it is possible to apply spraying as the drift will have small/acceptable extent;
- **"Medium"** indicates that there is medium risk of spray drift due to conditions and, the use of drift-reducing technology and/or setting drift-reducing application parameters should be considered;
- **"High"** indicates that there is high risk of spray drift and; therefore, spraying should not be applied.

The tool takes into account limitations related to air temperature, relative humidity, wind speed, and wind direction. Considering the meteorological conditions in Greece and a series of experimental measurements carried out in Greek vineyards, the tool's limits on weather conditions varied from those used by the TOPPS-Prowadis drift evaluation tool as follows:

a. Air temperature: The limits between the three categories were set at 25 and 30 °C, while the TOPPS-Prowadis tool uses, as limits, 15 and 25 °C. This is because during the spraying period of vineyards in Greece temperatures vary from 20 °C to above 30 °C.
b. Relative humidity: The TOPPS-Prowadis tool's limits between the three categories were maintained and they are 40% and 60%.
c. Wind speed: The 5 classes of the TOPPS-Prowadis tool were made 3, with the limits between them set at 3 and 4.5 m s^{-1}. The classes of the TOPPS tool which refer to low and medium wind speeds were merged, because during the spraying period the high temperatures do not allow weak winds to occur.

In order to evaluate and categorize the drift risk, the tool combines the values of the meteorological parameters by giving priority to the wind speed.

The above limits and categories apply with the following basic assumptions:

a. The wind direction is towards the sensitive area.
b. The canopy crop density is greater than 50%.
c. There is no rainfall.
d. The rows are sprayed from two sides and air is blown from two sides to each row.

2.2. Test Site and Crop Characteristics

The experimental field was an organic vineyard at the Agricultural University of Athens farm in Athens, Greece (37°59'06" N, 23°54'21" E).

The vineyard has 2.0 m row spacing with 1.6 m spacing of vines along the row to result in a density of 3125 vines ha^{-1}. The average vine height was about 1.1 m, with the leaves and grapes occupying the zone above ground between 0.3 and 1.2 m. At spraying times, the vines were in full leaf stage (BBCH 83 "Berries developing colors" and BBCH 91 "After harvest; end of wood maturation") [55], and the leaf area index (LAI) values were 1.58 and 1.46, respectively.

2.3. Experimental Design

The field experiments were carried out based on the ISO 22866:2005 standard [5], which set the criteria on the conditions for spray drift measurements. In accordance to this, the directly-sprayed area shall be at least 20 m wide upwind of the edge of the cropped area and the length of the spray track at least twice as the largest downwind sampling distance, and should be symmetrical to the axis of the sampling array. Therefore, every trial was carried out by spraying the ten outer downwind rows of the vineyard along a distance of 60 m, in order to treat a surface of 1200 m^2 (60 × 20 m) (Figure 1).

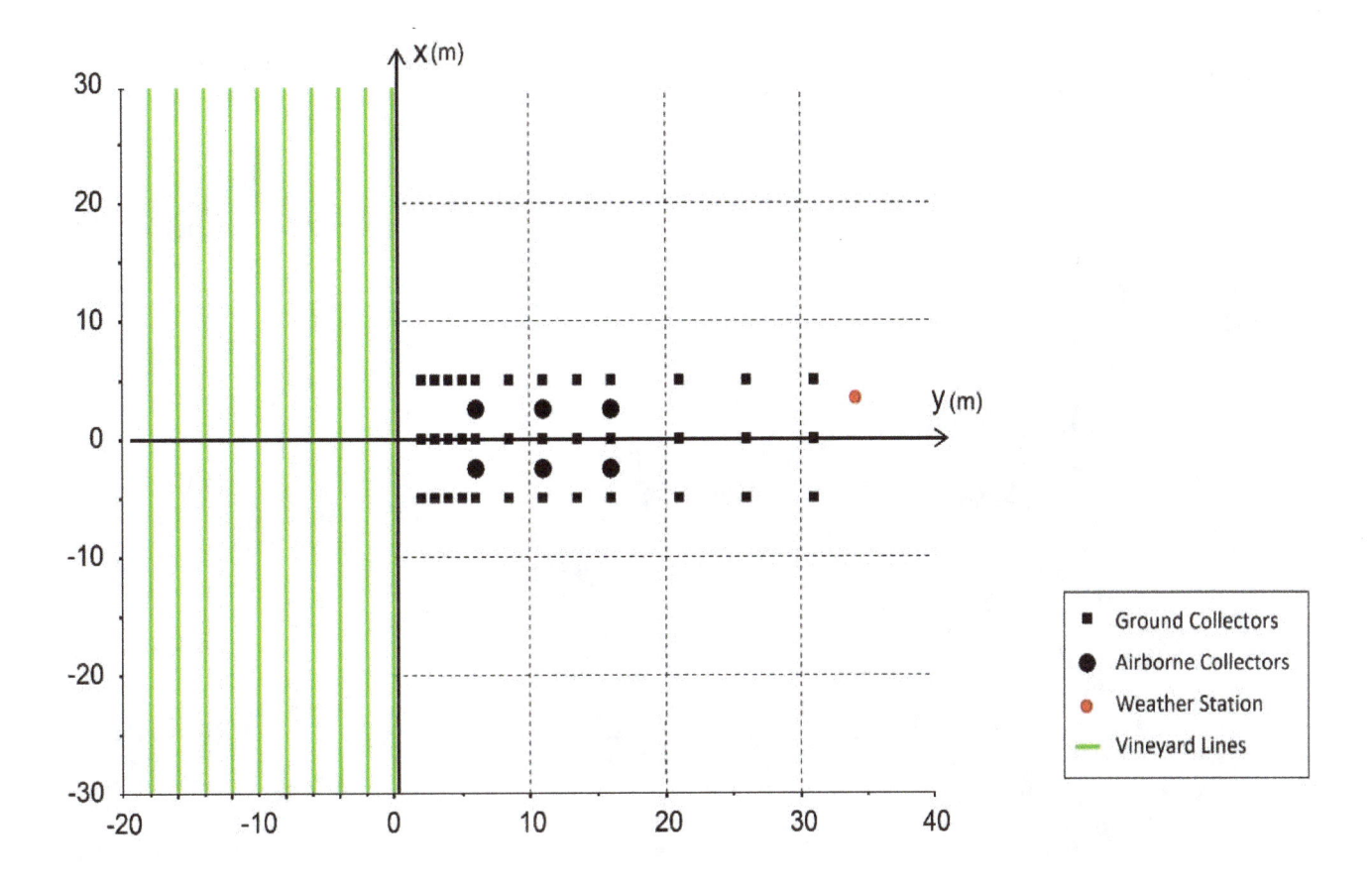

Figure 1. Test site layout according to the ISO 22866:2005 standard.

The experimental area was open and free of obstructions, other than the target crop (vineyard), as these may affect the airflow in the sampling area [5]. On the downwind side of the directly-sprayed area, there was bare soil or short vegetation (maximum height 7.5 cm), on which collectors were placed for the estimation of airborne spray drift and sedimenting spray drift [5] (Figure 2a).

In each trial, both ground sediment and airborne spray drift downwind to the directly-sprayed area were sampled. The ground collectors were placed at 12 different sampling distances in bare soil at 1, 2, 3, 4, 5, 7.5, 10, 12.5, 15, 20, 25, and 30 m from the edge of the directly-sprayed area (Figure 1). These distances started from the parallel straight line in front of the last plant row spaced 1 m (half of the row spacing). At each sampling distance, three wooden laths were placed with an upper surface covered in filter paper, Whatman Grade−1, 46 × 8 cm (Figure 2a,b), counting 36 soil samples per trial (Figure 1). So, each ground collector had a surface area of 368 cm^2, summing up to a total of 1104 cm^2 collector surface at each distance (the minimum collector area at any distance must be 1000 cm^2 [5]).

The airborne spray drift was monitored on cylindrical polyethylene lines with an external diameter of 2 mm, length 1 m, and collection area 62.8 cm^2 (Figure 2c). The measurements were taken at 3 distances, 5, 10, and 15 m, downwind from the edge of the directly-sprayed area (Figure 1). At each sampling distance, two 6 m high columns were placed, each of which had support structures per meter for the polyethylene lines, thus forming an array of 6 sampling collectors (Figure 2a,c). Therefore, at each distance there were 12 collectors summing up to a total sampling area of 753.6 cm^2, resulting in 36 collectors per trial.

After each spraying repetition, the ground and air collectors were stored into plastic sachets and sealed in refrigerators for maintenance at the appropriate temperature (4 °C) until their spectrophotometric analysis in the laboratory.

Figure 2. (a) The vineyard and sampling area; (b) detail showing the ground collector; (c) detail showing the pole holding the polythene lines.

2.4. Meteorological Conditions

Local weather conditions were measured following ISO 22866:2005 [5], but also to check the reliability of the weather forecast used by the drift risk assessment tool. Wind speed, wind direction, air temperature, and relative humidity were measured during the trials using a meteorological station placed at the edge of the downwind area in the center of the sampling area, 30 m from the sprayed area. Wind speed and direction were measured at a distance of 3 m from the ground, using an ultrasonic anemometer (Campbell Scientific WindSonic1 Gill 2D, Logan, UT, USA). Temperature and relative humidity were measured at two different heights, 2 and 3 m above ground, using two thermo-hygrometer probes (Rotronic HC2A-S3, Hauppauge, NY, USA). All measurements were taken at a frequency of 1 Hz sampling rate and all data were recorded automatically by a data logger (Campbell Scientific CR850, Logan, UT, USA).

The following parameters were calculated for each trial [5]:

a. Percentage of wind speed measurements less than 1 m s^{-1} (must be <10%);
b. Mean wind direction shall be at 90° ± 30° to the spray track (in this experiment between 0° and 60°) and no more than 30% of results shall be >90° ± 45° to the spray track;
c. Mean temperatures must be between 5 and 35 °C.

2.5. Treatments and Equipment Application Parameters

A total of six trials were carried out, two were related to the assessment of spray drift with a low risk tool indicator, two with medium, and two tests with high risk tool indicator.

Trials were performed using a trailed air-assisted sprayer for bush and tree crops Archimedes Turbo FS 1000 ("Archimedes" G. Roumeliotis, Aridaia Pellas, Greece), equipped with a 1000 L polyester tank, an axial fan of 800 mm in diameter with a two-speed gearbox and 7 nozzles for each side of the sprayer. The nozzles used were conventional hollow cone Teejet TXA 8002VK, yellow signed, with nominal nozzle flow rate of 1.40 L min^{-1} at 1.0 MPa. The real flow rate was closest to the nominal one. It was measured in the Department of Agricultural Engineering, Institute of Soil and Water Resources of the Hellenic Agricultural Organization "DEMETER", using an electronic measuring device (AAMS-Salvarani BVBA, Maldegem, Belgium) (Figure 3a). During testing, after adjusting the spray profile to target characteristics by means of water-sensitive paper spread on the vineyard canopy and poles, it was decided to activate a total of 6 nozzles (3 on each side of the sprayer). For all treatments, the working pressure was 1.0 MPa, driving speed 1.61 m s^{-1} (5.8 km h^{-1}), and volume application rate 434 L ha^{-1}. In all trials the PTO revolution speed was 56.55 rad s^{-1} (540 rev min^{-1}), fan speed 169.65 rad s^{-1} (1620 rev min^{-1}) (fan ratio 1:3), and fan airflow rate 10,000 m^3 h^{-1}. Fan air flow rate was measured in the Department of Agricultural Engineering, using a measuring tunnel (AAMS-Salvarani BVBA, Maldegem, Belgium) (Figure 3b).

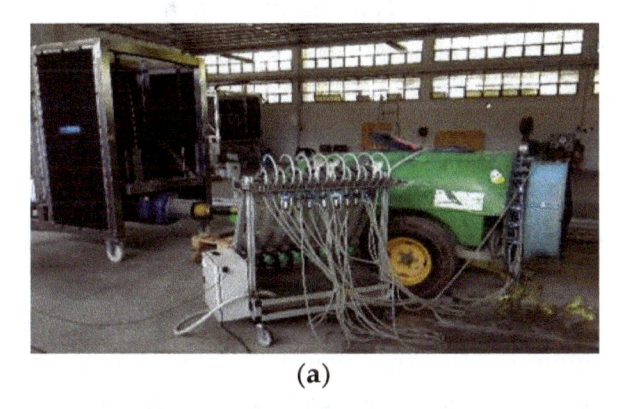

Figure 3. (**a**) Nozzles flow rate measurement; (**b**) fan air flow rate measurement.

2.6. Spray Liquid, Sample Extraction, and Spray Drift Estimation

The spray liquid was a solution of clean water and E−102 Tartrazine yellow dye tracer 85% (*w/w*) at a concentration of about 4 g L^{-1} [56,57].

Before each test, a blank sample of filter paper was placed in the sprayed area and collected just before the start of spraying. Two samples of the spray liquid were also collected from the spray tank directly from a nozzle, one at the beginning and one at the end of the test, to determine the precise tracer concentration at the nozzle outlet at each test.

After collecting the samples, they were transferred for analysis to the laboratory. Tartrazine's concentration in soil and air collectors was studied and quantified using a Shimadzu UV−1800 spectrophotometer, functioning at a wavelength of 426 nm.

Deposits of the spray tracer were extracted from samples using deionized water. For the respective samples the following volumes of deionized water were applied: A total of 40 mL for filter paper from 1 to 5 m distance, 20 mL for filter paper from 7.5 to 30 m distance, and 10 mL for polyethylene lines.

The reading of the spectrophotometer is related to the amount of tracer in solution through a calibration curve. From the reading of the spectrophotometer, the calibration factor, the collector surface area, the spray concentration and the volume of dilution liquid, and the amount of spray deposit per unit area were calculated as follows [5]:

$$drift_{dep} = \frac{(\rho_{smpl} - \rho_{blk}) \cdot F_{cal} \cdot V_{dil}}{\rho_{spray} \cdot A_{col}}, \quad (1)$$

$drift_{dep}$ = spray drift deposit (μL cm^{-2})
ρ_{smpl} = spectrophotometer reading of the sample (Abs)
ρ_{blk} = spectrophotometer reading of the blanks (collector + deionized water) (Abs)

F_{cal} = calibration factor (μg L^{-1})
V_{dil} = volume of dilution liquid (L)
ρ_{spray} = spray concentration of tracer (g L^{-1})
A_{col} = collection area of the spray drift collector (cm^2)

From this spray drift deposition figure, the percentage of spray drift on a collector can be calculated relating spray drift deposition to the amount applied in the field on the same unit of area, with the following formula:

$$drift_\% = \frac{drift_{dep} \cdot 10^4}{\beta_v}, \quad (2)$$

where β_v is the spray application volume in liters per hectare (L ha^{-1}) and given by the following equation:

$$\beta_v = \frac{Total\ nozzle\ flow\ rate \cdot Time}{Area} = \frac{Total\ nozzle\ flow\ rate \cdot 60}{Row\ Spacing \cdot Velocity/10} = \frac{Total\ nozzle\ flow\ rate \cdot 600}{Row\ Spacing \cdot Velocity},$$

Total nozzle flow rate = number of nozzles used multiplied by the nozzle nominal flow rate (L min^{-1})
Row spacing = distance between lines (m)
Velocity = velocity of the tractor (km h^{-1})
Time = 60 min
Area = 10,000 m^2

For drift ground sediment, once the tracer amount on each collector was measured, the mean of values derived from the three samples placed at each downwind distance was calculated. For airborne drift, the mean tracer amount derived from the two samples placed at each sampling height above the ground was calculated separately for each of the three sampled downwind distances (5, 10, and 15 m from the sprayed area).

While the drift is precisely defined by ISO 22866:2005 [5], for mitigation estimate a 1% line was calculated, which is the distance from the sprayer where the drift equaled only 1% of the original applied rate. If the reference parameter is the distance of the 1% line from the sprayed area, the mitigation of sedimenting drift from one treatment to another is given in the following equation [58]:

$$M(\%) = \frac{D_i - D_j}{D_i} \cdot 100, \quad (3)$$

where,

M = mitigation of sedimenting drift from (i) to (j) treatment (%)
Di = distance of the 1% line from the sprayed area at (i) treatment

The 1% line was selected because its distance is detected with good precision in field trials and has been used by other researchers [58]; however, the results can be extended to other parameters (e.g., the 0.5% line).

2.7. Data Analysis

STATGRAPHICS Centurion XVI Version 16.1.15 software for Windows was used for all statistical analyses [59]. In all tests a confidence level of 95% was considered. The effect of each treatment on the sedimenting drift was studied using two-way analysis of variance (ANOVA) considering drift risk indication from the tool and distance from the sprayed area as sources of variation. Airborne drift was evaluated with three-way ANOVA considering drift risk indication, distance, and height above the ground as factors. Spearman's correlation was used to identify the correlations between the above parameters [60]. Fisher's least significant difference (LSD) procedure was applied for pair-by-pair

3. Results and Discussion

3.1. Weather Conditions

Trials were carried out during July, August, and October 2018 in order to achieve alignment with the weather conditions that respond to the three different categories of drift risk assessment tool. The weather forecasts used in the tool are shown in Table 1.

The weather conditions during trials were monitored (Table 2) to evaluate the tool and weather forecasts and to ensure following limitations set in ISO22866:2005 standard [5]. Tables 1 and 2 indicate the validity of the prediction tool, as the measurements from the weather station have very small differences compared to the forecasts used by the tool. This outcome is considered sufficient given that the drift risk assessment tool used data from an open data source which uses a weather station that is not located in the experimental site, in contrast to the local weather station.

During trials the mean temperature fell within ISO 22866 [5] requirements. The maximum differences (Δ) in air temperature and relative humidity measured for the two heights (2 and 3 m from ground) were 0.38 °C and 0.46%, respectively. The mean wind speed was from 1.71 to 4.79 m s^{-1}, and the mean wind direction from 12.86° to 60.56° (ideal direction was 30° ± 30°). Wind direction deviation for treatment L1 was 7.30% over the limit prescribed in [5], meaning 24 records out of the 337 recorded during the trial. Considering the deviation slight, all data were used in the statistical analysis.

Table 1. Meteorological data used by the tool.

Date	Treatments	Replicates	Temperature °C	Relative Humidity %	Wind Speed m s^{-1}	Wind Direction	Rainfall
07.10.2018	"Low"	L1	21.50	69.00	2.60	North-East	No
20.10.2018		L2	23.10	62.00	1.65	North-East	No
21.07.2018	"Medium"	M1	33.00	46.10	2.00	North-East	No
29.07.2018		M2	29.20	56.00	2.50	North-East	No
02.08.2018	"High"	H1	31.00	55.00	4.00	North/North-East	No
02.08.2018		H2	31.00	55.00	4.00	North/North-East	No

Table 2. Meteorological data collected during trials.

Treatments	Replicates	Temperature Mean °C	Δ °C	Relative Humidity Mean %	Δ %	Wind Speed Min m s^{-1}	Max m s^{-1}	Mean m s^{-1}	Deviation [a] %	Wind Direction Mean ° az	Range °	Deviation [b] %
"Low"	L1	20.67	0.14	67.17	0.24	0.70	5.91	2.84	1.50	60.56	120	37.30
	L2	22.38	0.24	60.25	0.43	0.27	4.58	1.71	9.61	59.30	139	29.95
"Medium"	M1	32.67	0.38	44.76	0.46	0.26	4.40	2.12	10.19	47.80	281	29.21
	M2	28.91	0.16	54.92	0.40	0.77	5.30	2.77	0.24	31.27	110	0.72
"High"	H1	30.58	0.19	51.47	0.20	1.68	7.21	3.68	0.00	12.86	75	0.99
	H2	30.40	0.12	55.23	0.23	1.92	9.51	4.79	0.00	19.69	92	2.87

[a] Percentage of measurements <1 m s^{-1} (must be <10%); [b] percentage of measurements ∉ [30° − 45°, 30° + 45°] (must be <30%).

Temperature and relative humidity may vary much during the year, but they are stable during short periods of time (time of trial). This does not apply to wind speed and wind direction, which can drastically change during each trial, resulting in high influence on spray drift measurements [63]. For that reason, wind speed and wind direction correlation were analyzed indicating that high speed results in lower wind direction variance (Figure 4), in agreement with the result of previous research [64]. Studying the relationship between minimum wind speed, mean wind speed, and range of wind directions during trials, it appears that for minimum wind speeds less than 1 m s^{-1} and mean wind speeds less than 3 m s^{-1}, the range of wind directions was greater than 110°. Simultaneously, when the minimum wind speed was greater than 1.5 m s^{-1}, and the mean wind speed greater than 3.5 m s^{-1}, the wind direction was more uniform and range of the wind direction was less than about 90° (Table 2).

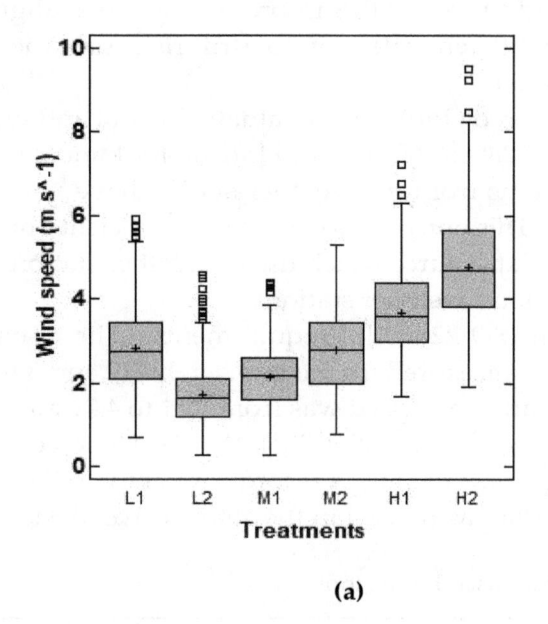

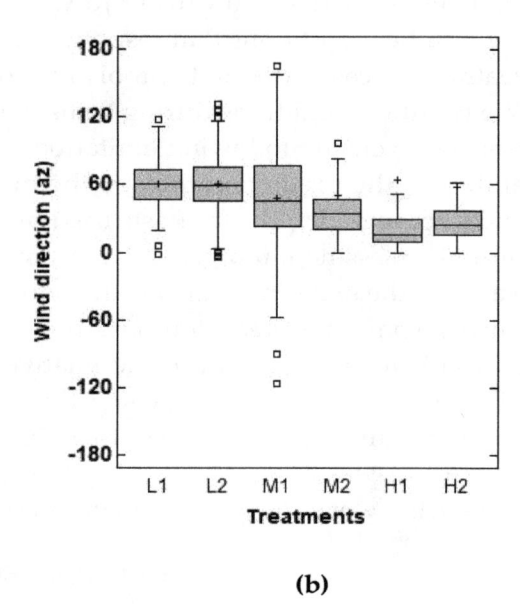

Figure 4. Boxplots of all wind speed and wind direction measurements; (a) wind speed; (b) wind direction.

3.2. Sedimenting (Fallout) Spray Drift

The spray drift deposits on ground collectors measured at different distances downwind of the sprayed area for all treatments are presented in Table 3, and the mean data curves are shown in Figure 5.

Table 3. Ground deposit of spray drift.

Distance m	Treatments					
	"Low"		"Medium"		"High"	
	L1	L2	M1	M2	H1	H2
	% Application Rate	% Application Rate	% Application Rate	% Application Rate	% Application Rate	% Application Rate
1	13.07	13.88	14.44	14.95	14.31	23.93
2	9.86	10.50	10.04	11.79	12.85	15.02
3	7.70	9.63	9.79	9.21	10.82	13.06
4	5.33	7.16	6.07	6.97	7.87	8.61
5	3.59	5.13	3.35	6.47	7.47	6.97
7.5	1.80	2.25	2.42	3.21	4.19	3.64
10	1.05	1.12	1.17	2.13	2.62	2.73
12.5	0.54	0.67	0.77	1.54	1.63	2.02
15	0.52	0.52	0.41	1.30	1.36	1.65
20	0.29	0.36	0.30	0.97	0.89	1.39
25	0.16	0.15	0.19	0.62	0.46	0.89
30	0.14	0.14	0.17	0.44	0.36	0.69

The results indicate a significant amount of ground sediment at all sampled distances. In all treatments the greatest deposition was measured in the first few meters of the downwind area. Additionally, all curves showed continued decreased deposition as distance increased from 1 to 30 m, but the rate of decrease varied among the three treatments, keeping, though, a common trend for all treatments. Similar results in both vineyard and orchard experiments have also been found by other researchers [65–68].

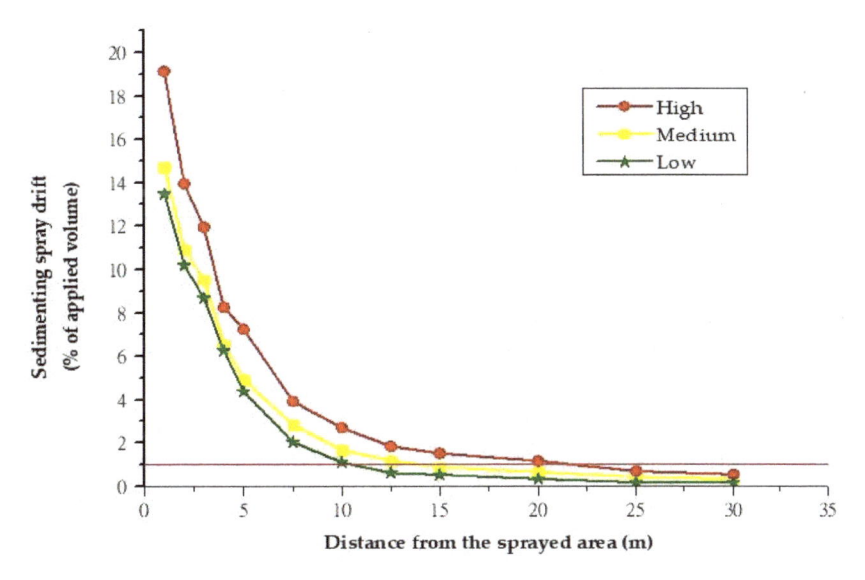

Figure 5. Spray drift deposit on the ground collectors.

It is also apparent that the three treatments (three categories of drift risk) generated different amounts of sedimenting drift over distance. As expected, the greatest amounts of ground sediment were observed when spraying was performed under worst-case conditions of high wind speed, high temperature, and low relative humidity ("High" treatment), while the lowest amounts were found under favorable weather conditions of low wind speed, low temperature, and high relative humidity ("Low" treatment). This finding comes in accordance with previous studies [69,70].

More precisely, spray drift deposit corresponding to about 1% of the spray volume was achieved at 10, 13.5, and 20 m from the sprayed area during "Low", "Medium", and "High" treatments, respectively (Figure 5).

Therefore, the mitigation of sedimenting drift, according to Equation (3), was 32.5% between "High" and "Medium" treatments, 50% between "High" and "Low" treatments, and 25.93% for "Medium" and "Low" treatments.

For the ground collectors the two-way ANOVA test showed that there is a statistically highly significant effect of the collector's placement distance, downwind of the sprayed area, and drift risk indication of the tool on the spray drift deposits ($p < 0.001$). However, there is no interaction between distance and drift risk indication ($p > 0.05$) (Table 4).

Table 4. Results of ANOVA for sedimenting spray drift as affected by distance and drift risk indication (Df: degrees of freedom).

Source	Df	Sum Sq	Mean Sq	F-Ratio	*p*-Value
Distance	11	5338.32	485.302	107.47	0.00001
Drift Risk	2	167.041	83.5206	18.50	0.00001
Distance × Drift Risk	22	92.4256	4.20116	0.93	0.5555
Residual	180	812.82	4.51566		

Spearman's rank correlation test indicated (Table 5) that there is very strong negative correlation between distance and drift ($r_s = -0.9392$), meaning that drift decreased drastically with increasing distance ($p < 0.001$), while drift risk category and drift have weak positive correlation ($r_s = 0.2047$), meaning that drift slightly increased when drift risk category changed following "Low"–"Medium"–"High" sequence ($p < 0.01$). These results are aligned with previous research trials, in which spray drift amounts are directly influenced by distance of the sprayed area and weather parameters, especially with the wind speed [48,64,71,72].

Table 5. Results of Spearman rank correlation test.

		Distance	Drift Risk
Drift (%)	Correlation	−0.9392	0.2047
	Sample size	216	216
	p-Value	0.00001	0.0027

Fisher's least significant difference (LSD) procedure was applied for pair-by-pair comparison among the means of the three treatments for all distances, showing statistically significant differences between "High"–"Medium" and "High"–"Low" treatments at downwind distances from 5 to 30 m, while no statistically significant differences were observed between "Medium"–"Low" treatments ($p < 0.05$) (Table 6). This finding enhances the fact, that the wind speed, which is less than 3 m s^{-1} in "Low" and "Medium" treatments, while in the "High" treatment is greater than 3.5 m s^{-1}, has the most significant effect on the spray drift from other meteorological parameters [30,73]. Additionally, the effect of meteorological parameters on ground drift is shown in Figure 6, where the means of sedimenting drift indicate that "High" treatment gave the highest ground drift deposition and the other treatments decreased from "Medium" to "Low".

Table 6. Multiple range LSD tests for ground collectors (LSD: least significant difference; Sig: significant).

Distance	+/−Limits	Low–Medium		Medium–High		Low–High	
		Sig.	Difference	Sig.	Difference	Sig.	Difference
1	5.44014		−1.22583		−4.4235	*	−5.64933
2	3.95954		−0.736167		−3.02083		−3.757
3	4.70071		−0.832333		−2.446		−3.27833
4	2.77943		−0.277167		−1.7155		−1.99267
5	2.12408		−0.546667	*	−2.31417	*	−2.86083
7.5	1.03509		−0.7955	*	−1.0965	*	−1.892
10	0.718107		−0.5655	*	−1.02733	*	−1.59283
12.5	0.541351	*	−0.554	*	−0.670167	*	−1.22417
15	0.506432		−0.338667	*	−0.644167	*	−0.982833
20	0.466114		−0.31	*	−0.507	*	−0.817
25	0.264212		−0.247667	*	−0.2675	*	−0.515167
30	0.172946		−0.162667	*	−0.216833	*	−0.3795

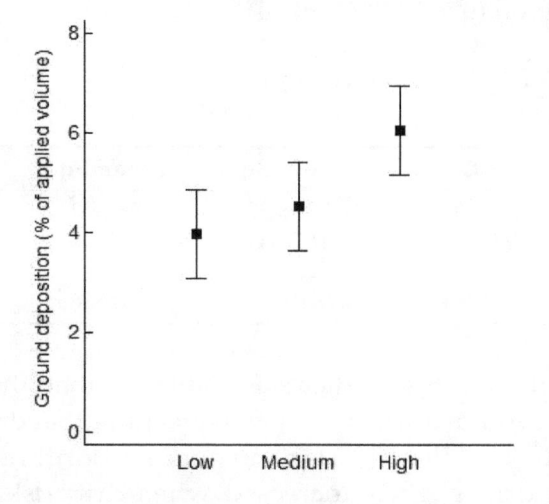

Figure 6. Sedimenting spray drift mean ± SE of the mean.

3.3. Airborne Spray Drift

The spray drift deposits on vertical samplers measured at six different heights for three different distances, downwind of the sprayed area, are presented in Table 7, and the mean data curves are shown in Figure 7.

Table 7. Airborne spray drift deposition.

Distance m	Height m	"Low"		"Medium"		"High"	
		L1	L2	M1	M2	H1	H2
		% Application Rate	% Application Rate	% Application Rate	% Application Rate	% Application Rate	% Application Rate
5	1	5.42	6.71	9.93	11.69	14.54	17.93
	2	5.53	5.39	9.63	10.41	14.98	15.55
	3	4.73	4.41	8.29	9.14	13.40	12.10
	4	3.61	2.72	7.43	6.31	9.83	9.83
	5	2.81	2.09	5.66	4.21	6.50	5.35
	6	2.40	1.48	3.83	3.51	4.33	3.75
10	1	3.70	2.34	3.39	8.81	10.31	10.16
	2	3.25	2.24	3.25	8.56	9.78	10.60
	3	2.77	1.72	3.89	7.12	10.78	8.50
	4	2.20	1.87	3.86	5.27	7.75	7.12
	5	2.06	1.24	3.07	3.59	7.04	5.39
	6	1.91	1.76	3.03	3.07	4.85	3.47
15	1	2.39	1.42	2.32	7.88	8.35	8.26
	2	1.82	2.31	1.84	7.21	7.80	7.99
	3	1.68	1.38	2.35	5.99	7.36	7.22
	4	1.62	1.63	2.02	5.01	6.60	6.35
	5	1.82	1.04	2.54	4.08	5.29	4.67
	6	1.45	0.91	1.78	3.58	3.95	3.41

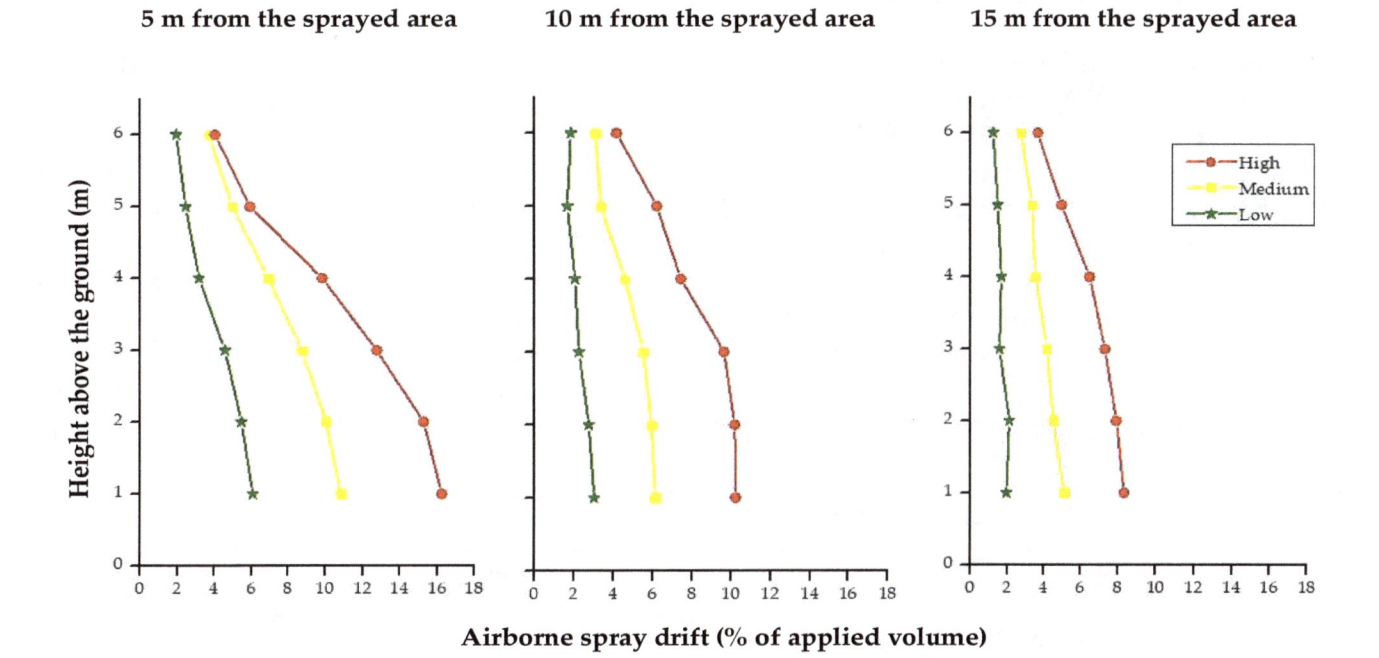

Figure 7. Airborne spray drift deposition profile at three distances from the sprayed area.

The results indicate that, in all treatments, the greatest deposition was measured at 5 m distance from the sprayed area, followed by 10 and 15 m distances. The three treatments (three categories of drift risk) generated different amounts of airborne drift over distance, having the greatest amount for "High" treatment and the lowest for "Low" treatment. The meteorological parameters had more influence on airborne spray drift than ground spray drift, since at 5, 10, and 15 m distances the amounts of airborne drift were greatest than the corresponding ground sediment.

Furthermore, in all treatments, spray drift deposition decreased with increasing height above ground, especially at 5 m distance from the sprayed area. This shape of the airborne drift profile is similar to the profile described by other researches in orchards [67,74], while there are also studies in vineyards showing that the airborne spray drift deposition increased with increasing height [64]. This discrepancy is justified by the fact that spray drift amount depends heavily on the architecture and geometry of the canopy [75,76]. The experimental vineyard had particularly poor canopy and small height, resulting in larger drift depositions in lower heights.

For the vertical samplers the three-way ANOVA test showed that there is a statistically significant effect of the collector's placement distance, downwind of the sprayed area, height from the ground, and drift risk indication of the tool on the spray drift deposits at level $p < 0.001$. Moreover, it was found that there are statistically highly significant interactions between distance and height ($p < 0.001$), height and drift risk indication ($p < 0.001$), and distance and drift risk indication ($p < 0.05$) (Table 8).

Table 8. Results of ANOVA for airborne spray drift as affected by distance, height, and drift risk indication.

Source	Df	Sum Sq	Mean Sq	F-Ratio	p-Value
Distance	2	436.733	218.366	91.09	0.00001
Height	5	614.543	122.909	51.27	0.00001
Drift Risk	2	1194.12	597.058	249.05	0.00001
Distance × Height	10	161.672	16.1672	6.74	0.00001
Distance × Drift Risk	4	24.7385	6.18463	2.58	0.0393
Height × Drift Risk	10	158.014	15.8014	6.59	0.00001
Distance × Height × Drift Risk	20	23.0276	1.15138	0.48	0.9709
Residual	162	388.369	2.39734		

Spearman's rank correlation test indicated that there is weak negative correlation between distance and drift ($r_s = -0.3599$), moderate negative correlation between height and drift ($r_s = -0.4130$), while drift risk category and drift have a strong positive correlation ($r_s = 0.6993$), meaning that drift strongly increased when wind speed and air temperature increased and relative humidity decreased, and the drift risk category followed the "Low"–"Medium"–"High" succession ($p < 0.001$) (Table 9). These results are largely consistent with other field studies. Fox et al. [77], in experiments on apple trees, found that deposits on floss decreased with height at 7.5 m distance downwind, but were more uniform across all heights at 15, 30, and 60 m downwind. Kasner et al. [78] reported that the distance from sprayed area, the height above ground, and the wind speed were significantly associated with drift level. Grella et al. [64] found a good significant relationship between airborne drift and wind speed variables, especially for the maximum and mean wind speed.

Table 9. Results of Spearman rank correlation test.

		Distance	Height	Drift Risk
	Correlation	−0.3599	−0.4130	0.6993
Drift (%)	Sample size	216	216	216
	p-Value	0.00001	0.00001	0.00001

Fisher's least significant difference (LSD) procedure was applied for pair-by-pair comparison among the means of the three treatments for the distances of 5, 10, and 15 m from the sprayed area and for all heights above the ground ($p < 0.05$). The results showed statistically significant differences between all treatment pairs, except for "Medium"–"High" pair at the heights of 5 and 6 m, at 5 m distance, and 6 m high at 15 m distance (Table 10). Furthermore, in Figure 8 the means of airborne drift for all treatments at distances 5, 10, and 15 m are presented. The highest airborne drift was registered in the distance closest to the sprayed area, and in each distance airborne drift was reduced following the "High"–"Medium"–"Low" order.

Table 10. Multiple-range LSD tests for vertical samplers.

Distance	Height	+/−Limits	Low-Medium Sig.	Low-Medium Difference	Medium-High Sig.	Medium-High Difference	Low-High Sig.	Low-High Difference
5	1	4.56578	*	−4.7505	*	−5.42075	*	−10.1713
	2	3.73621	*	−4.56775	*	−5.2415	*	−9.80925
	3	2.71095	*	−4.148	*	−4.03025	*	−8.17825
	4	2.08707	*	−3.70825	*	−2.95925	*	−6.6675
	5	1.65814	*	−2.48625		−0.98875	*	−3.475
	6	0.989128	*	−1.73475		−0.369	*	−2.10375
10	1	3.07183	*	−3.087	*	−4.13	*	−7.217
	2	3.10751	*	−3.1645	*	−4.2845	*	−7.449
	3	2.82998	*	−3.25975	*	−4.134	*	−7.39375
	4	1.44807	*	−2.5295	*	−2.86675	*	−5.39625
	5	1.27078	*	−1.67875	*	−2.8825	*	−4.56125
	6	0.897631	*	−1.2185	*	−1.10525	*	−2.32375
15	1	3.06148	*	−3.19225	*	−3.209	*	−6.40125
	2	2.949	*	−3.156	*	−3.3745	*	−5.8305
	3	2.069	*	−2.64175	*	−3.1235	*	−5.76525
	4	1.86667	*	−1.89125	*	−2.9625	*	−4.85375
	5	1.11037	*	−1.88125	*	−1.66925	*	−3.5505
	6	1.21974	*	−1.49475		−1.002	*	−2.49675

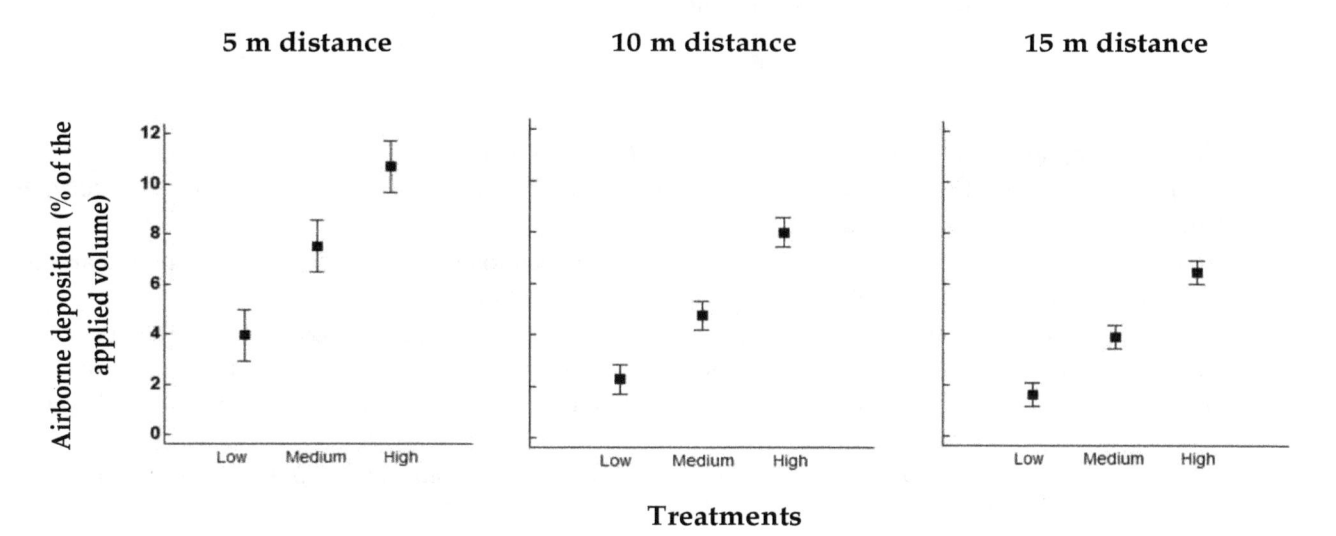

Figure 8. Airborne spray drift mean ± SE of the mean.

4. Conclusions

It is undeniable that there is great need to develop measures for spray drift reduction. Weather conditions during spraying application have considerable impact on sediment and airborne spray drift. Therefore, this work has developed a drift risk assessment tool based on the methodology derived from the TOPPS-Prowadis drift evaluation tool, which uses weather forecast data to predict spray drift extend. This tool is adapted to the Greek meteorological conditions and has three classification categories (low, medium, high) based on air temperature, relative humidity, and wind speed.

This tool and its limits between categories were evaluated successfully in real conditions, by conducting spray drift measurements in the vineyard of the Agricultural University of Athens, according to the ISO 22866:2005 methodology. Results showed that there are significant differences in airborne as well as in sediment drift among the treatments that correspond to the three drift risk classifications. The highest amount of spray drift deposits was observed within "High risk" treatments, which relate to unfavorable weather conditions; and the lowest within "Low risk" treatments, which are related to ideal weather conditions for spraying application.

The experimental results make evident that fine-tuning of the limits provides an optimized tool for Greek conditions that allows farmers to spray their vineyards with limited spray drift, resulting in higher spraying efficacy and minimized environmental impact. This tool shows that the limits classifying spray drift risk can be adjusted based on local experience for optimized results in spraying, but at the same time strengthens the importance of the TOPPS-Prowadis tool as a basis for such local optimization. Such a tool could be converted to software to assist farmers to predict and plan future spraying activities avoiding PPP application in "High" spray drift risk days.

This research produced the first set of data on spray drift amounts in vineyards when working with a conventional air-assisted sprayer and proved the efficiency of the tool. However, since the assessment of the effect of uncontrolled environmental conditions is objectively very difficult, the developed drift assessment tool will have to be tested in a wider range of environmental conditions and under different spraying techniques and crop characteristics.

Author Contributions: G.B., A.B. and S.F. conceived and designed the experiments; G.B., M.K. and V.P. performed the experiments; G.B., M.K. and V.P. analyzed the samples in the laboratory; G.B. and M.K. analyzed the data; G.B. wrote the paper; and M.K., A.B. and S.F. contributed in the analysis and presentation of data.

Acknowledgments: We would like to thank Nuri Bezolli, Giannis Kalliakmanis, Charalampos Miliotis and Dinos Grivakis for their collaboration in conducting our field experiments.

References

1. Oerke, E.C. Crop losses to pests. *J. Agric. Sci.* **2006**, *144*, 31–43. [CrossRef]
2. Cooper, J.; Dobson, H. The benefits of pesticides to mankind and the environment. *Crop Prot.* **2007**, *26*, 1337–1348. [CrossRef]
3. Giles, D.K.; Akesson, N.B.; Yates, W.E. Pesticide application technology: Research and development and the growth of the industry. *Trans. ASABE* **2008**, *51*, 397–403. [CrossRef]
4. Garcera, C.; Roman, C.; Molto, E.; Abad, R.; Insa, J.A.; Torrent, X.; Planas, S.; Chueca, P. Comparison between standard and drift-reducing nozzles for pesticide application in citrus: Part II. Effects on canopy spray distribution, control efficacy of Aonidiella aurantii (Maskell), beneficial parasitoids and pesticide residues on fruit. *Crop Prot.* **2017**, *94*, 83–96. [CrossRef]
5. ISO 22866:2005. *Equipment for Crop Protection-Methods for Field Measurement of Spray Drift*; International Organization for Standardization: Geneva, Switzerland, 2005; pp. 1–17.
6. Nuyttens, D.; De Schampheleire, M.; Baetens, K.; Sonck, B. The influence of operator- controlled variables on spray drift from field crop sprayers. *Trans. ASABE* **2007**, *50*, 1129–1140. [CrossRef]
7. Butler Ellis, M.C.; Lane, A.G.; O'Sullivan, C.M.; Miller, P.C.H.; Glass, C.R. Bystander exposure to pesticide spray drift: new data for model development and validation. *Biosyst. Eng.* **2010**, *107*, 162–168. [CrossRef]
8. Felsot, A.S.; Unsworth, J.B.; Linders, J.B.H.J.; Roberts, G. Agrochemical spray drift; assessment and mitigation-a review. *J. Environ. Sci. Health Part B* **2011**, *46*, 1–23. [CrossRef]
9. Hilz, E.; Vermeer, A.W.P. Spray drift review: the extent to which a formulation can contribute to spray drift reduction. *Crop Prot.* **2013**, *44*, 75–83. [CrossRef]

10. Benbrook, C.M.; Baker, B.P. Perspective on dietary risk assessment of pesticide residues in organic food. *Sustainability* **2014**, *6*, 3552–3570. [CrossRef]
11. Kruger, R.G.; Klein, N.R.; Ogg, L.C. *Spray Drift of Pesticides*; University of Nebraska-Lincoln Extension: Nebraska, NE, USA, 2013; p. G1773.
12. Balsari, P.; Grella, M.; Marucco, P.; Matta, F.; Miranda-Fuentes, A. Assessing the influence of air speed and liquid flow rate on the droplet size and homogeneity in pneumatic spraying. *Pest Manag. Sci.* **2019**, *75*, 366–379. [CrossRef]
13. Ozkan, H.E. *New Nozzles for Spray Drift Reduction*; AEX-523-98; Ohio State University Extension Fact Sheet Food Agricultural and Biological Engineering: Columbus, OH, USA, 1998.
14. Hofman, V.; Solseng, E. *Reducing Spray Drift*; North Dakota State University NDSU Extension Service AE–1210: Fargo, ND, USA, 2001.
15. Farooq, M.; Salyani, M. Modeling of spray penetration and deposition on citrus tree canopies. *Trans. ASABE* **2004**, *47*, 619–627. [CrossRef]
16. Da Silva, A.; Sinfort, C.; Tinet, C.; Pierrot, D.; Huberson, S. A lagrangian model for spray behaviour within vine canopies. *Aerosol Sci.* **2006**, *37*, 658–674. [CrossRef]
17. Yi, C. Momentum transfer within canopies. *J. Appl. Meteorol. Climatol.* **2008**, *47*, 262–275. [CrossRef]
18. Take, M.E.; Barry, J.W.; Richardson, B. *An FSCBG Sensitivity Study for Decision Support Systems*; ASAE Annual Meeting: Phoenix, AZ, USA, 1996; p. 961037.
19. Arvidsson, T.; Bergström, L.; Kreuger, J. Spray drift as influenced by meteorological and technical factors. *Pest. Manag. Sci.* **2011**, *67*, 586–598. [CrossRef] [PubMed]
20. Nuyttens, D.; Schampheleire, M.D.; Verboven, P.; Sonck, B. Comparison between indirect and direct spray drift assessment methods. *Biosyst. Eng.* **2010**, *105*, 2–12. [CrossRef]
21. Nuyttens, D.; de Schampheleire, M.; Baetens, K.; Brusselman, E.; Dekeyser, D.; Verboven, P. Drift from field crop sprayers using an integrated approach: results of a five-year study. *Trans. ASABE* **2011**, *54*, 403–408. [CrossRef]
22. Miranda-Fuentes, A.; Marucco, P.; Gonzalez-Sanchez, E.J.; Gil, E.; Grella, M.; Balsari, P. Developing strategies to reduce spray drift in pneumatic spraying vineyards: Assessment of the parameters affecting droplet size in pneumatic spraying. *Sci. Total Environ.* **2018**, *616–617*, 805–815. [CrossRef]
23. Miller, P.C.H.; Butler Ellis, M.C. Effects of formulation on spray nozzle performance for applications from ground-based boom sprayers. *Crop Prot.* **2000**, *19*, 609–615. [CrossRef]
24. Stainier, C.; Destain, M.F.; Schiffers, B.; Lebeau, F. Effect of the entrained air and initial droplet velocity on the release height parameter of a Gaussian spray drift model. *Commun. Agric. Appl. Biolog. Sci.* **2006**, *71*, 197–200.
25. De Ruiter, H.; Holterman, H.J.; Kempenaar, C.; Mol, H.G.J.; de Vlieger, J.J.; van de Zande, J. Influence of Adjuvants and Formulations on the Emission of Pesticides to the Atmosphere. In *A Literature Study for the Dutch Research Programme Pesticides and the Environment (DWK) Theme C-2*; Plant Research International B.V.: Wageningen, The Netherlands, 2003; Report 59.
26. Hobson, P.A.; Miller, P.C.H.; Walklate, P.J.; Tuck, C.R.; Western, N.M. Spray drift from hydraulic spray nozzles: the use of a computer simulation model to examine factors influencing drift. *J. Agric. Eng. Res.* **1993**, *54*, 293–305. [CrossRef]
27. Miller, P.C.H. The measurement of spray drift. *Pestic. Outlook* **2003**, *14*, 205–209. [CrossRef]
28. van de Zande, J.C.; Stallinga, H.; Michielsen, J.M.G.P.; van Velde, P. Effect of sprayer speed on spray drift. *Annu. Rev. Agric. Eng.* **2005**, *4*, 129–142.
29. Grella, M.; Marucco, P.; Manzone, M.; Gallart, M.; Balsari, P. Effect of sprayer settings on spray drift during pesticide application in poplar plantations (Populus spp.). *Sci. Total Environ.* **2017**, *578*, 427–439. [CrossRef] [PubMed]
30. Arvidsson, T. *Spray Drift as Influenced by Meteorological and Technical Factors. A Methodological Study*; Swedish University of Agricultural Sciences, Acta Universitatis Agriculturae Sueciae: Agraria Sweden, 1997; Volume 71, p. 144.
31. Nuyttens, D.; Sonck, B.; De Schampheleire, M.; Steurbaut, W.; Baetens, K.; Verboven, P.; Nicolai, B.; Ramon, H. Spray drift as affected by meteorological conditions. *Commun. Agric. Appl. Biol. Sci.* **2005**, *70*, 947–959. [PubMed]

32. Sumner, P.E. Reducing Spray Drift. In *Cooperative Extension Service*; The University of Georgia College of Agricultural and Environmental Sciences: Athens, GA, USA, 1997.
33. da Cunha, J.P.A.R.; Pereira, J.N.P.; Barbosa, L.A.; da Silva, C.R. Pesticide Application Windows in the Region of Uberlândia-MG, Brazil. *Biosci. J. Uberlândia* **2016**, *32*, 403–411.
34. Maciel, C.F.S.; Teixeira, M.M.; Fernandes, H.C.; Zolnier, S.; Cecon, P.R. Droplet Spectrum of a Spray Nozzle under Different Weather Conditions. *Revista Ciência Agronômica* **2018**, *49*, 430–436. [CrossRef]
35. Al-Jumaili, A.; Salyani, M. *Wind Effect on the Deposition of an Air-Assisted Sprayer*; University of Florida: Gainesville, FL, USA, 2014.
36. Thistle, H. Meteorological concepts in the drift of pesticide. In Proceedings of the International Conference on Pesticide Application for Drift Management, Washington State University, Waikoloa, HI, USA, 27–29 October 2004; pp. 156–162.
37. Holterman, H.J. *Kinetics and Evaporation of Water Drops in Air*; IMAG Report 2003-2012; Institute of Agricultural and Environmental Engendering: Wageningen, The Netherlands, 2003.
38. Carlsen, S.C.K.; Spliid, N.H.; Svensmark, B. Drift of 10 herbicides after tractor spray application. Primary drift (droplet drift). *Chemosphere* **2006**, *64*, 778–786. [CrossRef]
39. Fishel, F.M. *When a Pesticide Doesn't Work*; Agronomy Department: Florida, FL, USA, 2008.
40. FOCUS, 2004. *Focus Surface Water Scenarios in the EU evaluation process under 91/414/EEC*; Report prepared by the FOCUS working group on Surface Water Scenarios; European Commission: Brussels, Belgium, March 2004; p. 238.
41. FOCUS, 2007a. Landscape and mitigation factors. In *Aquatic Risk Assessment. Extended Summary and Recommendations, vol. 1, Report of the FOCUS Working Group on Landscape and Mitigation Factors in Ecological Risk Assessment*; EC Document Reference SANCO/10422/2005 V.2.0; European Commission: Brussels, Belgium, 2007; p. 169.
42. FOCUS, 2007b. Landscape and mitigation factors. In *Aquatic Risk Assessment. Detailed Technical Reviews, vol. 2, Report of the FOCUS Working Group on Landscape and Mitigation Factors in Ecological Risk Assessment*; EC Document Reference: SANCO/10422/2005 V.2.0; European Commission: Brussels, Belgium, September 2007; p. 436.
43. Directive 2009/128/EC of the European parliament and the council of 21 October 2009 establishing a framework for community action to achieve the sustainable use of pesticides. *Off. J. Eur. Union* **2009**, *309*, 71–86.
44. Directive 2009/127/EC of the European parliament and of the council of 21 October 2009 amending Directive 2006/42/EC with regard to machinery for pesticide application. *Off. J. Eur. Union L* **2009**, *310*, 29–33.
45. EPA-United States Environmental Protection Agency. 2015; Reducing Pesticide Drift. Available online: http://www.epa.gov/reducing-pesticide-drift (accessed on 22 December 2015).
46. Huang, Y.; Thomson, S.J. Atmospheric Stability Determination Using Fine Time-Step Intervals for Timing of Aerial Application. In Proceedings of the ASABE, 2016 Annual International Meeting 162461162, Orlando, FL, USA, 17–20 July 2016. [CrossRef]
47. Nansen, C.; Ferguson, J.C.; Moore, J.; Groves, L.; Emery, R.; Garel, N.; Hewitt, A. Optimizing Pesticide Spray Coverage Using a Novel Web and smartphone Tool, SnapCard. *Agron. Sustain. Dev.* **2015**, *35*, 1075–1085, INRA and Springer Verlag France, 2015. [CrossRef]
48. Baetens, K.; Nuyttens, D.; Verbovena, P.; De Schampheleire, M.; Nicolai, B.; Ramona, H. Predicting drift from field spraying by means of a 3D computational fluid dynamics model. *Comput. Electron. Agric.* **2007**, *56*, 161–173. [CrossRef]
49. Kruckeberg, J.; Hanna, M.; Darr, M.; Steward, B. An interactive spray drift simulator. In Proceedings of the American Society of Agricultural and Biological Engineers Annual International Meeting, ASABE, Dallas, TX, USA, 13–16 September 2010; Volume 3, pp. 2297–2311.
50. Maber, J.; Dewar, P.; Praat, J.P.; Hewitt, A.J. Real Time Spray Drift Prediction. *Acta Hortic.* **2001**, *566*, 493–498. [CrossRef]
51. Hong, S.W.; Zhaoa, L.; Zhuc, H. SAAS, a computer program for estimating pesticide spray efficiency and drift of air-assisted pesticide applications. *Comput. Electron. Agric.* **2018**, *155*, 58–68. [CrossRef]
52. Nsibande, S.A.; Dabrowski, J.M.; van der Walt, E.; Venter, A.; Forbes, P.B.C. Validation of the AGDISP model

for predicting airborne atrazine spray drift: A South African ground application case study. *Chemosphere* **2015**, *138*, 454–461. [CrossRef] [PubMed]
53. TOPPS-Prowadis Project. Best Management Practices to Reduce Spray Drift. 2014. Available online: http://www.topps-life.org/ (accessed on 24 June 2019).
54. Stavrakas, E.D. *Ampelographia*; Ziti Publications: Thessaloniki, Greece, 2010.
55. Meier, U. *Growth Stages of Mono-and Dicotyledonous Plants: BBCH Monograph*, 2nd ed.; Uwe Meier Federal Biological Research Centre for Agriculture and Forestry: Braunschweigh, Germany, 2001; pp. 1–158.
56. Pergher, G. Recovery rate of tracer dyes used for spray deposit assessment. *Trans. ASABE* **2001**, *44*, 787–794. [CrossRef]
57. Gil, E.; Balsari, P.; Gallart, M.; Llorens, J.; Marucco, P.; Andersen, P.G.; Fàbregas, X.; Llop, J. Determination of drift potential of different flat fan nozzles on a boom sprayer using a test bench. *Crop Prot.* **2014**, *56*, 58–68. [CrossRef]
58. Otto, S.; Loddo, D.; Baldoin, C.; Zanin, G. Spray drift reduction techniques for vineyards in fragmented landscapes. *J. Environ. Manag.* **2015**, *162*, 290–298. [CrossRef] [PubMed]
59. StatPoint Technologies Inc. *STATGRAPHICS Centurion XVI Version 16.1.15*; StatPoint Technologies Inc.: Warrenton, VA, USA, 1982–2011.
60. Spearman, C. The Proof and Measurement of Association between Two Things. *Am. J. Psychol.* **1904**, *15*, 72–101. [CrossRef]
61. Fisher, R.A. *The Design of Experiments*; Oliver and Boyd: Edinburg, TX, USA; London, UK, 1935.
62. Levene, H. Robust tests for equality of variances. In *Contributions to Probability and Statistics: Essays in Honor of Harold Hotelling*; Olkin, I., Ghyrye, S.G., Hoeffding, W., Madow, W.G., Mann, H.B., Eds.; Stanford University Press: Palo Alto, CA, USA, 1960; pp. 278–292.
63. Bode, L.E.; Butler, B.J.; Goering, C.E. Spray drift and recovery as affected by spray thickener, nozzle type, and nozzle pressure. *Trans. ASAE* **1976**, *19*, 213–218. [CrossRef]
64. Grella, M.; Gallart, M.; Marucco, P.; Balsari, P.; Gil, E. Ground Deposition and Airborne Spray Drift Assessment in Vineyard and Orchard: The Influence of Environmental Variables and Sprayer Settings. *Sustainability* **2017**, *9*, 728. [CrossRef]
65. Balsari, P.; Marucco, P.; Grella, M.; Savoia, S. Spray drift measurements in Italian vineyards and orchards. In Proceedings of the 13th Workshop on Spray Application in Fruit Growing (SuproFruit 2015), Lindau/Lake Costance, Germany, 15–18 July 2015; pp. 30–31.
66. Balsari, P.; Marucco, P. *Sprayer Adjustment and Vine Canopy Parameters Affecting Spray Drift: The Italian Experience*; DEIAFA-University of Turin: Torino, Italy, 2004.
67. Torrent, X.; Garcera, C.; Molto, E.; Chueca, P.; Abad, R.; Grafulla, C.; Roman, C.; Planas, S. Comparison between standard and drift-reducing nozzles for pesticide application in citrus: Part, I. Effects on wind tunnel and field spray drift. *Crop Prot.* **2017**, *96*, 130–143. [CrossRef]
68. Grella, M.; Marucco, P.; Balsari, P. Toward a new method to classify the airblast sprayers according to their potential drift reduction: comparison of direct and new indirect measurement methods. *Pest Manag. Sci.* **2019**, *75*, 2219–2235. [CrossRef] [PubMed]
69. Threadgill, E.D.; Smith, D.B. Effects of physical and meteorological parameters on the drift of controlled-size droplets. *Trans. ASAE* **1975**, *18*, 51–56.
70. De Schampheleire, M.; Baetens, K.; Nuyttens, D.; Spanoghe, P. Spray drift measurements to evaluate the Belgian drift mitigation measures in field crops. *Crop Prot.* **2008**, *27*, 577–589. [CrossRef]
71. Nuyttens, D.; Zwertvaegher, I.; Dekeyser, D. Comparison between drift test bench results and other drift assessment techniques. *Asp. Appl. Biol. Int. Adv. Pestic. Appl.* **2014**, *122*, 293–302.
72. Combellack, J.H.; Westen, N.M.; Richardson, R.G. A comparison of the drift potential of a novel twin fluid nozzle with conventional low volume flat fan nozzles when using a range of adjuvants. *Crop Prot.* **1996**, *15*, 147–152. [CrossRef]
73. Huang, Y.; Zhan, W.; Fritz, B.; Thomson, S.; Fang, A. Analysis of impact of various factors on downwind deposition using a simulation method. *J. ASTM Int.* **2010**, *7*, 1–11. [CrossRef]
74. van de Zande, J.C.; Butler Ellis, M.C.; Wenneker, M.; Walklate, P.J.; Kennedy, M. Spray drift and bystander risk from fruit crop spraying. *Asp. Appl. Biol.* **2014**, *122*, 177–185.
75. Bird, S.L.; Esterly, D.M.; Perry, S.G. Atmospheric pollutants and trace gases. Off-target deposition of pesticides from agricultural aerial spray applications. *J. Environ. Qual.* **1996**, *25*, 1095–1104. [CrossRef]

76. Duga, A.T.; Ruysen, K.; Dekeyser, D.; Nuyttens, D.; Bylemans, D.; Nicolai, B.M.; Verboven, P. Spray deposition profiles in pome fruit trees: Effects of sprayer design, training system and tree canopy characteristics. *Crop Prot.* **2015**, *67*, 200–213. [CrossRef]
77. Fox, R.; Hall, F.; Reichard, D.; Brazee, R.D.; Krueger, H.R. Pesticide tracers for measuring orchard spray drift. *Appl. Eng. Agric.* **1993**, *9*, 501–505. [CrossRef]
78. Kasner, E.J.; Fenske, R.A.; Hoheisel, G.A.; Galvin, K.; Blanco, M.N.; Seto, E.Y.W.; Yost, M.G. Spray Drift from a Conventional Axial Fan Airblast Sprayer in a Modern Orchard Work Environment. *Ann. Work Expos. Health* **2018**, *62*, 1134–1146. [CrossRef] [PubMed]

12

Non-Invasive Tools to Detect Smoke Contamination in Grapevine Canopies, Berries and Wine: A Remote Sensing and Machine Learning Modeling Approach

Sigfredo Fuentes [1,*], Eden Jane Tongson [1], Roberta De Bei [2], Claudia Gonzalez Viejo [1], Renata Ristic [2], Stephen Tyerman [2] and Kerry Wilkinson [2]

[1] School of Agriculture and Food, Faculty of Veterinary and Agricultural Sciences, The University of Melbourne, Parkville, VIC 3010, Australia
[2] School of Agriculture, Food and Wine, The University of Adelaide, PMB 1, Glen Osmond, SA 5064, Australia
* Correspondence: sfuentes@unimelb.edu.au.

Abstract: Bushfires are becoming more frequent and intensive due to changing climate. Those that occur close to vineyards can cause smoke contamination of grapevines and grapes, which can affect wines, producing smoke-taint. At present, there are no available practical in-field tools available for detection of smoke contamination or taint in berries. This research proposes a non-invasive/in-field detection system for smoke contamination in grapevine canopies based on predictable changes in stomatal conductance patterns based on infrared thermal image analysis and machine learning modeling based on pattern recognition. A second model was also proposed to quantify levels of smoke-taint related compounds as targets in berries and wines using near-infrared spectroscopy (NIR) as inputs for machine learning fitting modeling. Results showed that the pattern recognition model to detect smoke contamination from canopies had 96% accuracy. The second model to predict smoke taint compounds in berries and wine fit the NIR data with a correlation coefficient (R) of 0.97 and with no indication of overfitting. These methods can offer grape growers quick, affordable, accurate, non-destructive in-field screening tools to assist in vineyard management practices to minimize smoke taint in wines with in-field applications using smartphones and unmanned aerial systems (UAS).

Keywords: bushfires; infrared thermography; near-infrared spectroscopy; smoke taint; artificial intelligence

---

1. Introduction

A recent report from the Victorian government of Australia concluded that bushfires have increased in number and severity since the 1970s across the east and south of the country [1]. The main contributing factor to this environmental disaster is climate change, specifically the increased frequency of recurrent heat waves (i.e., prolonged periods of hotter weather) and drought conditions, which have increased the window of risk for bushfires, as well as their number, and severity. Recently, Chile (central region), USA (California), Greece, South Africa (Stellenbosch) and Australia (various states) have suffered some of the worst bushfires experienced in each country's history. These countries are major producers of wines, and their grape growers and winemakers are similarly affected by global warming with detrimental effects in drought, vine phenological changes, shifting of suitable grapevine growing regions towards the north and south, and increased bush fire events near wine growing regions [2–4].

When bushfires occur in close proximity to vineyards, smoke can contaminate leaves and fruit. One of the main physiological effects of bush fire smoke in grapevines is the reduction of stomatal conductance (g_s) [5]. Decreased g_s may be explained by the combination of the main smoke components carbon dioxide (CO_2) and carbon monoxide (CO), with water vapor (100% RH in the substomatal cavity) producing carbonic acid (H_2CO_3) that reduces pH in the stomata, thereby causing partial or complete stomatal closure [5]. In berries, smoke contamination results in adsorption of smoke-derived volatile phenols (which accumulate in glycoconjugate forms), that are extracted into the final wine during the winemaking process [6]. Several mitigating measures have been evaluated to minimize smoke taint in berries or to remove volatile phenols (and their glycoconjugates) from wine, including defoliation [7] or foliar application of kaolin [8] in the vineyard, reverse osmosis treatment [9] and the addition of fining agents in wines [10]. However, implementation of defoliation or kaolin applications are often indiscriminate and broadly applied irrespective of the degree of grapevine exposure to smoke. Furthermore, the removal of smoke taint from wine is not selective and may inadvertently remove important wine compounds, thereby affecting the desirable organoleptic characteristics of wine.

Physiological assessment of control (non-smoked) and smoke affected grapevine cultivars have shown that some cultivars are more susceptible than others in terms of photosynthesis and stomatal conductance, in particular, Merlot and Cabernet Sauvignon. In contrast, Sauvignon Blanc was not significantly affected by smoke contamination from a physiological perspective. However, berries exposed to smoke resulted in wines with significantly higher concentrations of volatile phenols and guaiacol glycoconjugates compared to wines made from uncontaminated fruit [6].

The need to assess smoke contaminated fruit and wine has led to the implementation of new laboratory-based analytical methods [11], including liquid chromatography-tandem mass spectrometry for quantification of volatile phenol glycoconjugates [12,13]. However, these techniques require expensive laboratory instrumentation, and specialized technical skills to prepare samples (i.e., to extract the analytes of interest), operate the instruments and data analysis. Spectral methods, both mid-infrared (MIR) reflectance spectroscopy and chemometric techniques have been evaluated for rapid detection of smoke-taint in grapes [8] and bottled wines [14], but also with limitations. Spectral reflectance measurements of berries were affected by fruit maturity, while MRI-based classification of wines was influenced by cultivar, oak maturation and the level of smoke taint. Thus, reliable and rapid in-field techniques available to determine whether vines and fruits have been contaminated with smoke from bushfires are not yet available. This paper presents pattern recognition and regression models based on machine learning algorithms developed for the identification of smoke contaminated grapevine canopies and fruit. The first machine learning model generated used infrared thermography data from grapevine canopies as inputs to predict smoke contamination, as a target for four grapevine cultivars. Near infrared (NIR) spectroscopy readings from berries were used as inputs for regression machine learning algorithms to assess specific smoke-related compounds in berries and final wines from seven cultivars. These models combined with affordable geo-referenced NIR spectroscopy measurements of berries could allow growers to map contaminated areas of a vineyard to facilitate decision making at harvest. Finally, potential applications of these models using proximal and mid-range remote sensing using unmanned aerial systems (UAS) are discussed.

2. Materials and Methods

2.1. Experimental Site and Application of Smoke to Grapevines

Grapevine smoke exposure experiments were conducted in the 2009/10 season using seven different cultivars grown at two locations: (i) Sauvignon Blanc, Pinot Gris, Chardonnay and Pinot Noir grown in a commercial vineyard in Adelaide Hills region, South Australia, Australia (35°00′ S, 138°49′ E) and (ii) Shiraz, Cabernet Sauvignon and Merlot vines grown in a vineyard located at the University of Adelaide's Waite campus in Adelaide, South Australia (34°58′ S, 138°38′ E). Grapevines (three per replicate) were exposed to smoke (for 1 h) using a purpose-built smoke tent and experimental conditions described previously (Figure 1) [15]. Smoke was applied to vines at a phenological stage corresponding to approximately seven days post-veraison; when total soluble solids (TSS) concentrations were approximately 15 Brix, determined using a digital handheld refractometer (PAL-1, Atago, Tokyo, Japan).

Experiment 1 consisted of the physiological assessment of smoke contamination at the canopy level using porometry, infrared thermography, and pattern recognition machine learning using four cultivars two hours after smoke exposure: Chardonnay, Merlot, Sauvignon Blanc, and Shiraz. Experiment 2 assessed smoke taint in berries and wine at harvest for all seven cultivars. In this experiment, control (unsmoked) and smoke-affected berry samples (two berries taken from the mid-section of two bunches from two replicates per treatment, per cultivar; $n = 112$ berries) were collected at harvest. Morphometry of berries was measured using a caliper to obtain diameter (equatorial length in cm); length (cm) and calculated radius (cm), area (cm^2), and perimeter (cm).

Figure 1. Grapevines were enclosed in a tent, and smoke from combustion of straw was blown into the tent using a fan. Pre-installation of the tent (**A**), and installed and operational tent (**B**). Photos obtained from the 2009/10 trial in Adelaide, Australia.

2.2. Experiment 1

2.2.1. Physiological Measurements Using Leaf Porometry

Leaf conductance to water vapor was measured as stomatal conductance (g_s) using a porometer (AP4, Delta-T Devices, Cambridge, UK). Porometer readings used were obtained from the cultivars Shiraz, Sauvignon Blanc, Chardonnay, and Merlot. Measurements were performed two hours after smoke treatments using nine mature, fully expanded sunlit leaves, for each of the two middle vines of two replicates per treatment per cultivar ($n = 72$) under natural leaf orientation with natural light intensity. Leaves were chosen to ensure measurements were performed on three leaves from the top, middle, and bottom parts of the canopies from each vine in a 3 × 3 matrix arrangement.

2.2.2. Infrared Thermal Imagery of Canopies

Thermal images were acquired from grapevine canopies using an infrared thermal camera FLIR® T-series (Model B360) (FLIR Systems, Portland, OR, USA), with a resolution of 320 × 240 pixels. The camera measures temperature in the range of −20 to +1200 °C. The thermal sensitivity of the camera is <0.08 °C @ +30 °C/80 mK with a spatial resolution of 1.36 milliradians. Each pixel is considered an effective temperature reading in degrees Celsius (°C). Infrared thermal images were acquired from the same side and in parallel to porometer measurements (shaded side of the canopy to reduce variability) in the estimation of the infrared index (I_g), which is proportional to g_s [16]. One thermal image from the canopy of each of the middle vines of two replicates per treatment per cultivar was obtained from a constant distance of 2.5 m perpendicular to the row direction (distance between rows being 3 m; Figure 2A). The calculated infrared thermal index (I_g) was compared with porometry measurements acquired immediately after obtaining each thermal image from corresponding vines. All thermal images were acquired on a clear day. The smoke treatments were applied with minimal wind; a requirement for undertaking the field trials implemented to avoid the risk of fire spreading from accidental burning of interrow dry plant material and to secure representativeness of thermal images [16,17].

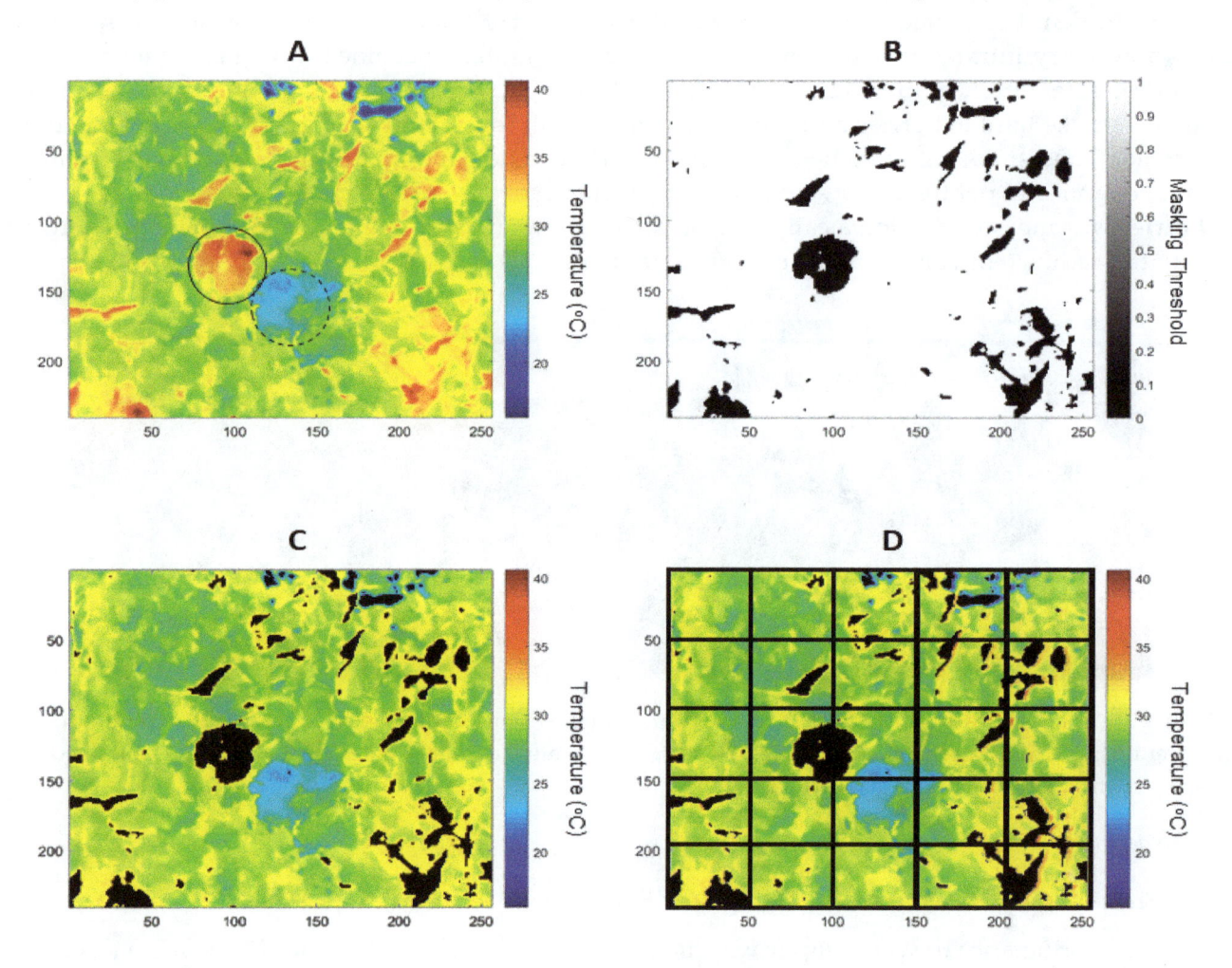

Figure 2. Examples of a radiometric thermal image (**A**) processing for data extraction of T_{dry} (**A**, solid circle) and T_{wet} (**A**, dotted circle) by painting leaves with petroleum jelly and water, respectively. Binary image obtained by thresholding T_{dry} and T_{wet} (**B**); masked radiometric image extracting non-leaf material, such as overheated elements and sky (**C**); and subdivision of thermal image to extract information from sections of the canopy in a 5 × 5 sub-division (**D**).

2.2.3. Algorithms Used to Calculate Crop Water Stress Indices (CWSI) and Infrared Index (I_g)

Crop water stress index (CWSI) was calculated using the following equation, after determining T_{dry} and T_{wet} [18]:

$$\text{CWSI} = \frac{T_{canopy} - T_{wet}}{T_{dry} - T_{wet}} \qquad (1)$$

where T_{canopy} is the actual canopy temperature extracted from the thermal image at determined positions, and T_{dry} and T_{wet} are the reference temperatures (in °C), obtained using the method of painting both sides of reference leaves with petroleum jelly and water, respectively [16].

An infrared index (I_g), proportional to leaf conductance to water vapor transfer (g_s), can be obtained using the relationship as follows [19]:

$$I_g = \frac{T_{canopy} - T_{wet}}{T_{dry} - T_{wet}} = g_s \left(r_{aw} + \left(\frac{s}{\gamma}\right) r_{HR}\right) \qquad (2)$$

where r_{aw} = boundary layer resistance to water vapor, r_{RH} = the parallel resistance to heat and radiative transfer, Υ = psychrometric constant and s = slope of the curve relating saturation vapor pressure to temperature [17,19].

2.2.4. Infrared Thermography Data Extraction

The T_{dry} and T_{wet} values were obtained on a per image basis using a customized code written in Matlab® R2019a (Mathworks Inc. Natick, MA, USA) to crop the radiometric data from the areas within the respective painted leaves with water (T_{wet}) and petroleum jelly (T_{dry}) (Figure 2A). To filter non-leaf material from the radiometric image using the determined threshold, a second customized code was written in Matlab® to binarize a masked image (Figure 2B) and to extract these values from the original image (Figure 2C). For automatic extraction of data within a canopy, a pre-defined subdivision of $3 \times 3 = 9$; $5 \times 5 = 25$; $7 \times 7 = 49$ and $10 \times 10 = 100$ was automatically implemented (Figure 2D; for the case of 5×5). From these subdivisions, data were extracted for T_{canopy} per image (Figure 2D), I_g Equation (2) and CWSI Equation (1).

The image sub-divisions (Figure 2D) represent the matrix (A) with $m \times n$ (m = rows and n = columns) extraction points represented as per the following matrix:

$$A(m,n) = \begin{pmatrix} T_{1,1} & \cdots & T_{1,n} \\ \vdots & \ddots & \vdots \\ T_{m,1} & \cdots & T_{m,n} \end{pmatrix} \qquad (3)$$

Since m, n represent the pre-determined subdivision for automatic cropping sections from the infrared thermal image (A), every sub-image is processed for automatic canopy extraction by filtering non-leaf temperatures using the T_{dry} and T_{wet} values extracted (Figure 2B) as minimum and maximum possible temperatures for the canopy. The calculated T value then corresponds to the averaged T_{canopy} for each sub-division.

2.2.5. Pattern Recognition of Infrared Thermal Imagery using Machine Learning for Smoke Contamination Prediction

Pattern recognition models were developed using a customized Matlab® code, which is able to test 17 different training algorithms, two from Backpropagation with Jacobian derivatives, 11 from Backpropagation with gradient derivatives and four from Supervised weight and bias training functions, in loop to select the best model. This model was constructed using the infrared thermal image output values as inputs to classify the samples into smoked and non-smoked (control). The infrared thermal images were analyzed with the methodology described in Figure 2 to obtain T_{canopy}, I_g, and CWSI data obtained using Equations (1) and (2) with sub-divisions of $3 \times 3 (n = 27$ per image); $5 \times 5 (n = 75$, per image; 7×7 ($n = 147$, per image) and 10×10 ($n = 300$ per image). All algorithms tested used a random data division. However, for the algorithms such as scaled conjugate gradient, which consist of three stages—training, validation and testing, the data was divided as 60% ($n = 28$ images) for training, 20% ($n = 10$ images) for validation with a cross-entropy performance algorithm, and 20% ($n = 10$ images) for testing with a default derivative function. For the algorithms such as sequential order weights and bias, which only consist of training and testing stages, the data was divided as 70% ($n = 34$) for training and 30% ($n = 14$) for testing with a cross-entropy performance algorithm. A trimming exercise was conducted using 3, 7 and 10 neurons to select the best model with no signs of overfitting (Figure 3).

Input: i) 27, ii) 75, iii) 147, iv) 300

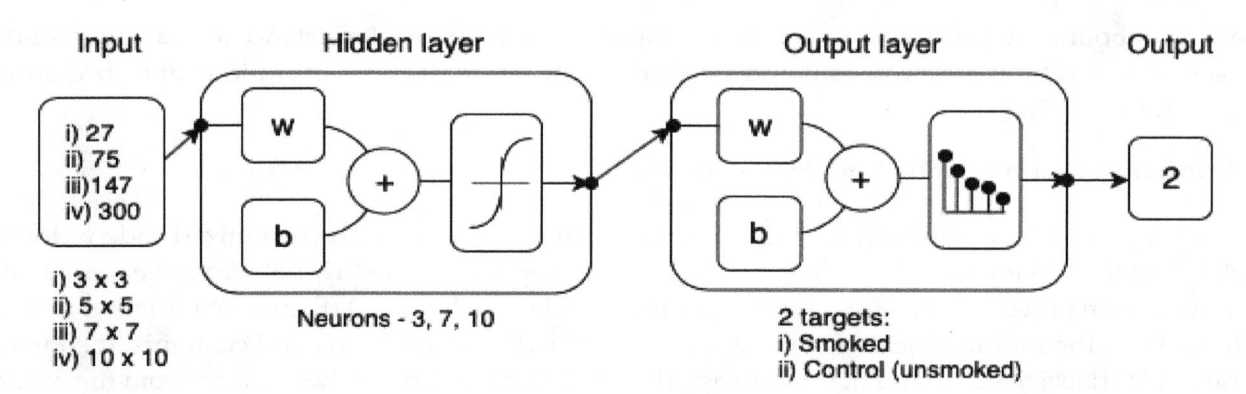

i) 3 x 3, ii) 5 x 5, iii) 7 x 7, iv) 10 x 10

Neurons - 3, 7, 10

2 targets:
i) Smoked
ii) Control (unsmoked)

Figure 3. Diagram of the two-layer feedforward network with a tan-sigmoid function in the hidden layer and a Softmax transfer function in the output layer. For hidden and output layers, w = weights and b = biases. Input volume and neuron trimming exercises are included in the diagram.

2.3. Experiment 2

2.3.1. Berry Near Infrared (NIR) Spectroscopy Measurements

Full berries were scanned using a spectrophotometer (ASD FieldSpec®3, Analytical Spectral Devices, Boulder, CO, USA) equipped with the ASD contact probe, built for contact measurements, attached by fiber optic cable to the instrument. A total of 112 berries collected at harvest from seven cultivars (16 berries per cultivar) were scanned by putting the probe's lens in contact with the berries and a total of 401 spectra were recorded for each berry. The instrument records spectra with a resolution of 1.4 nm for the region 350–1000 nm and 2 nm for the region 1000–1850 nm. The instrument was used in reflectance mode and data was then transformed into absorbance values (absorbance = log (1/reflectance)). A reference tile (Spectralon®, Analytical Spectral Devices, Boulder, CO, USA) was used as a white reference, for scatter correction. A new reference was taken every ten spectra acquisitions.

2.3.2. Winemaking and Chemical Analysis of Berries and Wine

Small scale winemaking of control and smoke-affected fruit from this trial has been described previously in detail by Ristic et al. [6]. Guaiacol glycoconjugates were measured in fruit and wine by HPLC–MS/MS using a stable isotope dilution analysis (SIDA) method developed by Dungey et al. (2011) [12]. Volatile phenols, including guaiacol, were determined in berries and wine by the Australian Wine Research Institute's (AWRI) Commercial Services Laboratory (Adelaide, Australia). Volatile phenols were measured by GC–MS according to SIDA methods reported previously [13].

2.3.3. Fitting Modeling of Near-Infrared (NIR) Spectroscopy of Berries Using Machine Learning Modeling to Predict Smoke Taint in Berries and Wine

A regression model was developed using a customized Matlab® code, which is able to test 17 different training algorithms, two from Backpropagation with Jacobian derivatives, 11 from Backpropagation with gradient derivatives and four from Supervised weight and bias training functions, in loop to select the best model. NIR absorbance values corresponding to the range of wavelengths within 700 and 1100 nm with a second derivative transformation, which were used as inputs in the machine learning algorithms, since that range corresponds to alcohol and alcohol-based compounds to predict (i) guaiacol glycoconjugates in berries (μg Kg^{-1}), (ii) guaiacol glycoconjugates in wines (μg L^{-1}) and iii) guaiacol in wine (μg L^{-1}). All algorithms tested used a random data division. However, for the algorithms, which consist of three stages—training, validation and testing, the data was divided as 60% (n = 28) for training, 20% (n = 10) for validation with a means squared error (MSE) performance algorithm and 20% (n = 10) for testing with a default derivative function (data not shown). For the algorithms such as sequential order weights and bias, which only consist of training and testing stages, data was divided as 70% (n = 34) for training and 30% (n = 14) for testing with a means squared error performance algorithm. A trimming exercise was conducted using 3, 7, and 10 neurons to select the best model with no signs of overfitting (Figure 4).

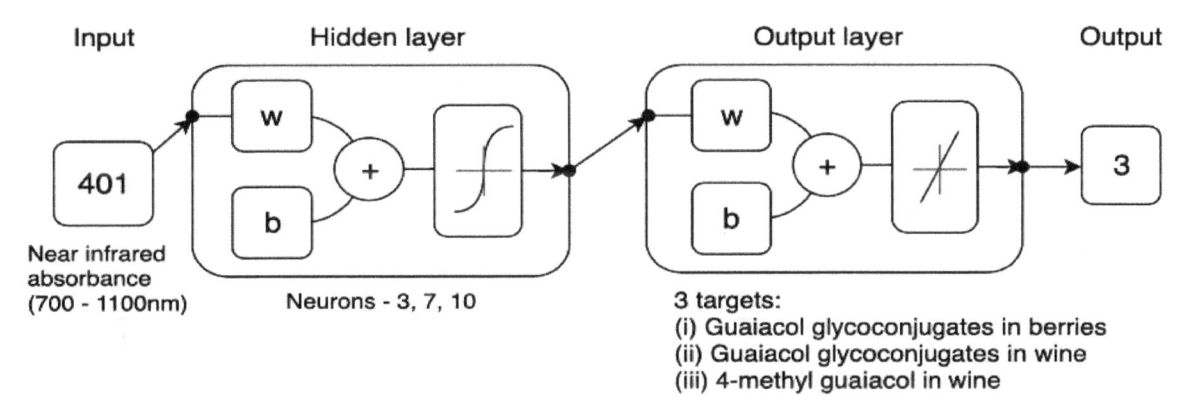

Figure 4. Diagram of the two-layer feedforward network with a tan-sigmoid function in the hidden layer and a linear transfer function in the output layer. For hidden and output layers, w = weights and b = biases. Neuron trimming exercise is included in the diagram.

2.4. Statistical Analysis

Data from chemometry and morphometry of berries, wine compounds, and I_g and g_s were analyzed through ANOVA using SAS® 9.4 software (SAS Institute Inc., Cary, NC, USA) with Tukey's studentized range test (HSD; $p < 0.05$) as post-hoc analysis for multiple comparisons to assess significant differences. Statistical data such as means and standard deviation (SD) were obtained from the replicates of each cultivar and treatment.

3. Results

3.1. Experiment 1

3.1.1. Grapevine Physiological Data Relationships between Porometry and Infrared Thermal Imagery

Table 1 shows the mean values for g_s and I_g with respective standard deviations (SD) for the four cultivars from Experiment 1. The general trend for the mean values of the control treatments follows a positive linear relationship ($R^2 = 0.99$; $I_g = 0.0027\ g_s$). On the contrary, the trend for the mean values of the smoke treatments have lower linearity and relationship, but still showed a positive linear pattern ($R^2 = 0.23$; $I_g = 0.0023\ g_s$; data not shown). In the control samples, the mean I_g values per cultivar did not show significant change, as reflected by the SD values, but Merlot showed a significantly higher I_g ($p < 0.05$). This trend was similar for the mean g_s values showing Merlot with the highest mean ($p < 0.05$). The mean I_g values for smoked samples were more variable, while g_s showed higher mean values, except Sauvignon Blanc, and more variability as reflected by the higher SD values compared to control. The I_g mean values for both treatments were not very sensitive, as seen in Table 1.

Figure 5 shows the relationships between g_s and I_g for different sections of canopies (top, middle, and bottom) of the grapevines monitored for both non-smoked (control) and smoked treatments. The graph (Figure 5A) shows a strong and significant linear relationship between g_s and I_g ($R^2 = 0.85$; $I_g = 0.0026\ g_s$). However, there was no relationship observed for smoke treatments, with the data presenting high variability, which is consistent with results shown in Table 1. Figure 5A shows that regardless of the measurement position within the canopy for I_g, there is a broader distribution of values between top, middle, and bottom of the canopy along the linear relationship found. On the contrary, Figure 5B shows that the bottom readings for g_s are more concentrated towards the lower values (<200 mmol m^2 s^{-1}). Furthermore, the Ig values become less sensitive (spread between 0 and 1). The same pattern can be seen for most of the top readings with the middle readings having a wider spread distribution.

Table 1. Means and standard deviation (SD) per variety and treatment for the infrared index (Ig, unitless) and stomatal conductance (g_s in mmol m² s⁻¹) calculated for all the images without sub-divisions.

Variety	I_g Control Mean	SD	g_s (mmol m² s⁻¹) Control Mean	SD	I_g Smoked Mean	SD	g_s (mmol m² s⁻¹) Smoked Mean	SD
Chardonnay	0.32 [b]	0.22	112.66 [c]	60.55	0.60 [a]	0.34	203.02 [ab]	145.72
Merlot	1.06 [a]	0.29	384.93 [a]	102.68	0.43 [ab]	0.28	251.00 [a]	131.44
Sauvignon Blanc	0.34 [b]	0.18	130.60 [c]	60.12	0.32 [b]	0.23	135.89 [b]	46.49
Shiraz	0.52 [b]	0.26	211.40 [b]	88.97	0.59 [a]	0.10	235.35 [ab]	148.71

Means followed by different superscript letters are statistically significant between treatments based on Tukey's studentized range test (HSD, $p < 0.05$).

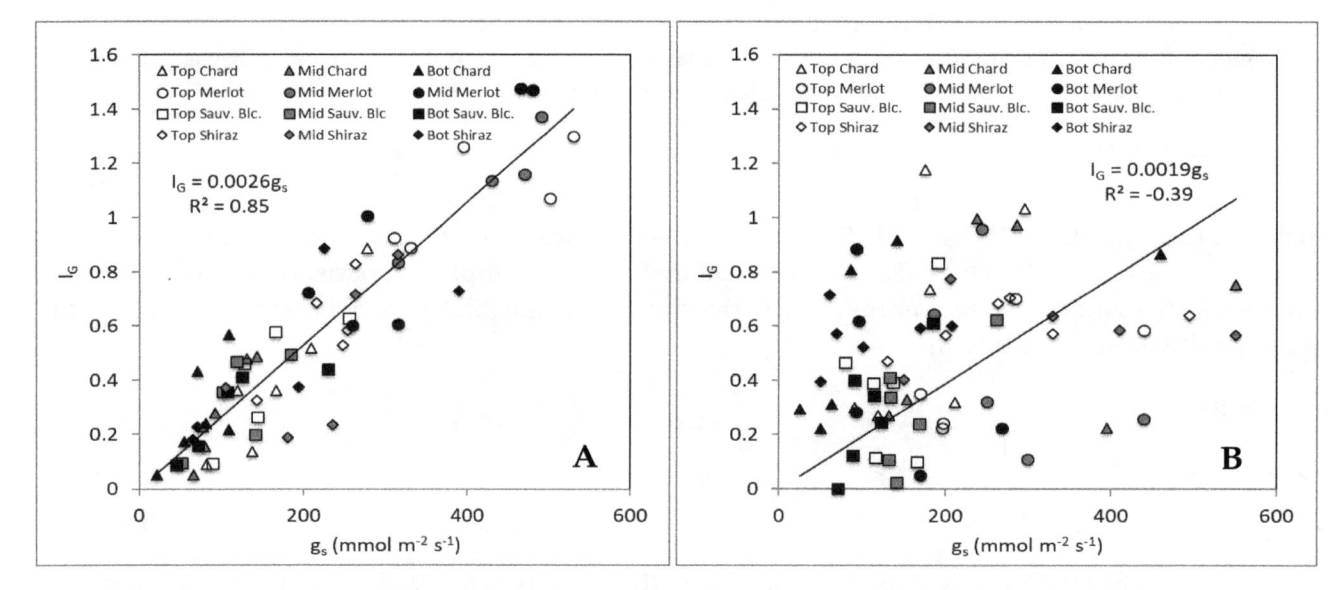

Figure 5. Relationship between g_s and I_g for Experiment 1 in the four cultivars with data separated between canopy sections: top (Top), middle (Mid) and bottom (Bot) measurements for control treatments (**A**) and smoked treatments (**B**).

3.1.2. Pattern Recognition Using Machine Learning Modeling of Physiological and Infrared Thermal Data

Table 2 shows the results of the pattern recognition modeling for the data extracted from infrared thermal images from the canopies of four different cultivars combined for Experiment 1. The best performing algorithm for the 3 × 3 sub-division and extraction of T_{canopy}, I_g, and CWSI used as inputs and classification of smoked and non-smoked as target was the scaled conjugate gradient algorithm. The training, validation, and testing procedures (using 10 neurons) resulted in an overall model with 94% accuracy. In the case of the data extracted using a 5 × 5 sub-division, the overall best model (sequential order weight and bias) resulted in an accuracy of 88% (using 10 neurons) in the classification of smoked and non-smoked canopies. For the 7 × 7 sub-division, the best algorithm (also the sequential order weight and bias) resulted in an accuracy of 94% (using 7 neurons) in the classification. Finally, the 10 × 10 was the best performing algorithm overall (sequential order weight and bias) resulted in an accuracy of 96% (using 3 neurons). Furthermore, the performance of training was lower than the one for testing, and testing accuracy was close to that from the training stage, which are evidence of no overfitting [20,21].

Table 2. Best pattern recognition model developed for each set of inputs showing the best training algorithm and number of neurons to predict whether canopies are smoked or non-smoked (control). Inputs corresponds to data extracted from infrared thermal images for T_{canopy}, I_g and crop water stress index (CWSI) in matrix arrangement of 3×3 ($n = 27$), 5×5 ($n = 75$), 7×7 ($n = 147$) and 10×10 ($n = 300$) data points per thermography. Performance reported is based on cross-entropy.

Inputs	Algorithm	Neurons	Stage	Samples	Accuracy	Performance
3×3	Scaled conjugate gradient	10	Training	28	100%	0.03
			Validation	10	90%	0.16
			Test	10	80%	0.44
			Overall	48	94%	-
5×5	Sequential order weight and bias	10	Training	34	85%	0.37
			Test	14	93%	0.43
			Overall	48	88%	-
7×7	Sequential order weight and bias	7	Training	34	94%	0.72
			Test	14	93%	0.71
			Overall	48	94%	-
10×10	Sequential order weight and bias	3	Training	34	97%	0.45
			Test	14	93%	0.47
			Overall	48	96%	-

Figure 6 shows the Receiver Operating Characteristic (ROC) for the best performing model found to predict smoke contamination in grapevine canopies (10×10 sub-division; Table 2). The figure shows that results for both smoke and control pattern recognition using infrared thermography data as inputs are projected in a similar trend to the True Positive Rate prediction axis of the graph.

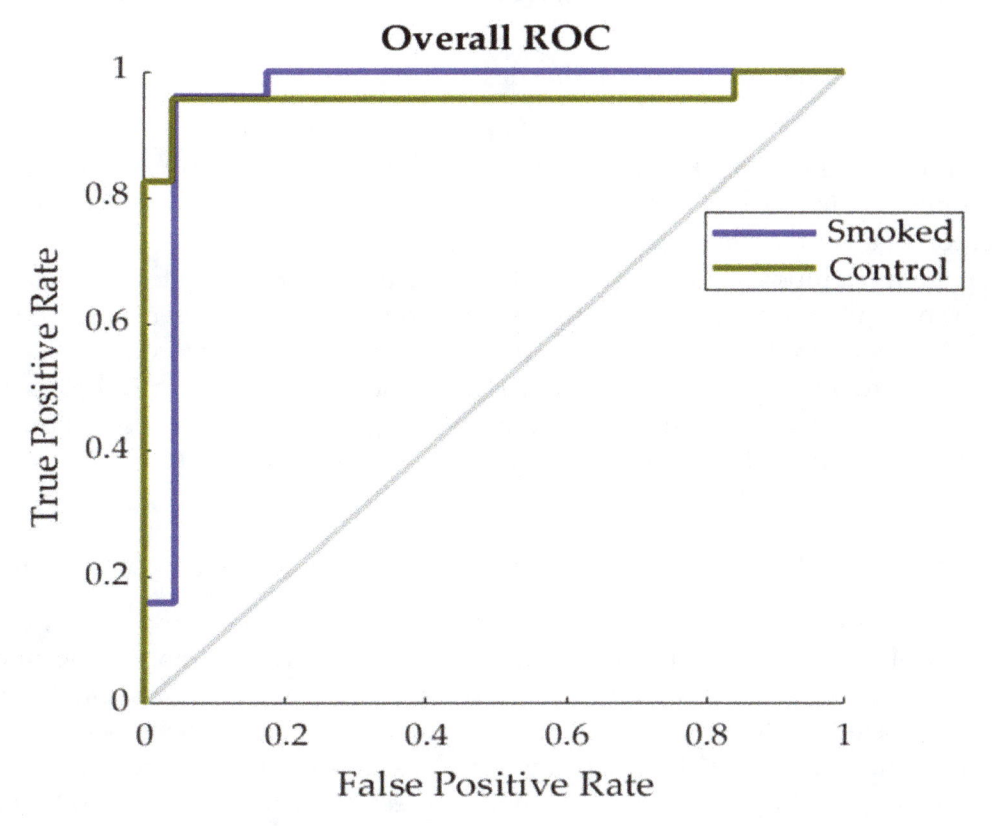

Figure 6. Receiver Operating Characteristic (ROC) showing the false positive rate (*x*-axis) and true positive (*y*-axis) for control and smoked treatments for the best performing classification model found in Table 3.

Table 3. Morphometric data obtain from berries for all seven cultivars consisting in Perimeter (P in cm), Equatorial Diameter (D in cm), calculated Area (A in cm2) and Radius (D/2 in cm). For chemometry, Total Soluble Solids (TSS) represented by Brix and Near Infrared (NIR) absorbance at 982 nm corresponding to H–O–H and O–H chemical bonds.

	P (cm) C	P (cm) S	D (cm) C	D (cm) S	Area (cm^2) C	Area (cm^2) S	R (cm) C	R (cm) S	Brix (°) C	Brix (°) S	NIR982 (nm) C	NIR982 (nm) S
Merlot	4.7 [a]	4.6 [a]	1.4 [a]	1.3 [a]	1.5 [a]	1.5 [a]	0.3 [c]	0.3 [bc]	23.9 [a]	24.2 [ab]	0.4 [abc]	0.3 [ab]
Shiraz	4.1 [b]	3.9 [bc]	1.3 [ab]	1.2 [b]	1.5 [a]	1.2 [c]	0.4 [ab]	0.3 [cd]	24.9 [a]	25.4 [a]	0.3 [cd]	0.4 [a]
PinGr	4.0 [bc]	4.2 [b]	1.3 [bc]	1.3 [ab]	1.4 [ab]	1.5 [a]	0.3 [bc]	0.4 [a]	18.7 [c]	19.8 [d]	0.3 [abc]	0.4 [a]
Char	4.0 [bcd]	3.8 [cd]	1.4 [a]	1.3 [ab]	1.5 [a]	1.3 [bc]	0.4 [a]	0.3 [abc]	19.7 [cd]	18.6 [de]	0.5 [a]	0.4 [a]
PinNoir	3.8 [cd]	3.8 [cd]	1.2 [c]	1.3 [ab]	1.2 [bc]	1.3 [abc]	0.3 [c]	0.3 [ab]	17.2 [d]	18.2 [e]	0.3 [bcd]	0.4 [a]
CabSauv	3.7 [cd]	3.6 [d]	1.1 [d]	1.1 [c]	1.1 [c]	1.0 [d]	0.3 [d]	0.3 [d]	24.1 [a]	23.1 [b]	0.4 [ab]	0.4 [a]
SauvBl	3.7 [d]	3.8 [c]	1.3 [bc]	1.3 [ab]	1.3 [b]	1.4 [ab]	0.4 [ab]	0.4 [a]	20.8 [b]	21.5 [c]	0.2 [d]	0.2 [b]

Abbreviations: C = Control, S = Smoke, PinGr = Pinot Gris, PinNoir = Pinot Noir, Char = Chardonnay, CabSauv = Cabernet Sauvignon, SauvBl = Sauvignon Blanc. Different superscript letters are statistically significant between treatments based on Tukey's studentized range test (HSD, $p < 0.05$).

3.2. Experiment 2

3.2.1. Berry Morphology and NIR Peak within the 700–1100 nm

Table 3 shows the average data of morphometric and chemometric measurements obtained from berry samples for all the seven cultivars for Experiment 2. Even though there are some significant differences between morphometric measurements of berries for the different cultivars comparing smoke and non-smoked (Control) treatments, they do not affect results and models developed.

3.2.2. Smoke-Related Compounds Found in Berries and Wines

Data for smoke-related compounds have been previously reported by Ristic et al. (2016) [6], and comprised of volatiles with statistical differences between control (non-smoked) and smoked treatments. Specifically, for purposes of modeling, guaiacol glycoconjugates found in berries ($\mu g\ Kg^{-1}$), guaiacol glycoconjugates found in wines ($\mu g\ L^{-1}$) and guaiacol found in wines ($\mu g\ L^{-1}$) were used since these are the primary compounds identified by the industry to contribute to smoke taint. In berries, the guaiacol glycoconjugates average concentration ranged for control between 37 and 602 $\mu g\ k\ g^{-1}$ and from 253 to 2452 $\mu g\ kg^{-1}$ for smoke-affected treatments. The guaiacol glycoconjugates concentrations in wines ranged from 8 to 334 $\mu g\ L^{-1}$ for control and from 111 to 1480 $\mu g\ L^{-1}$ for smoke-affected treatments. In the case of guaiacol concentration in wines, values ranged from 0 (not detected) to 9 $\mu g\ L^{-1}$ for control and from 0 (not detected) to 26 $\mu g\ L^{-1}$ [6].

3.2.3. Near-Infrared (NIR) Spectroscopy from Berries and Smoke Taint Compounds Found

Figure 7 shows the main average spectra for berries from smoke and non-smoked (control) treatments for red (Figure 7A) and white cultivars (Figure 7B). There were no significant differences in the averaged spectra between smoked and non-smoked berries for red cultivars. On the contrary, there appears to be a consistent difference for white cultivars of around 0.05 in absorbance, especially from 820 to 1100 nm for the range considered for this study. Smoke-related compounds for this trial and used for the machine learning model reported here have been previously reported by Ristic et al. [6]. In this study, statistically significant differences in the main smoke taint compounds were reported for all the seven cultivars included in Experiment 2.

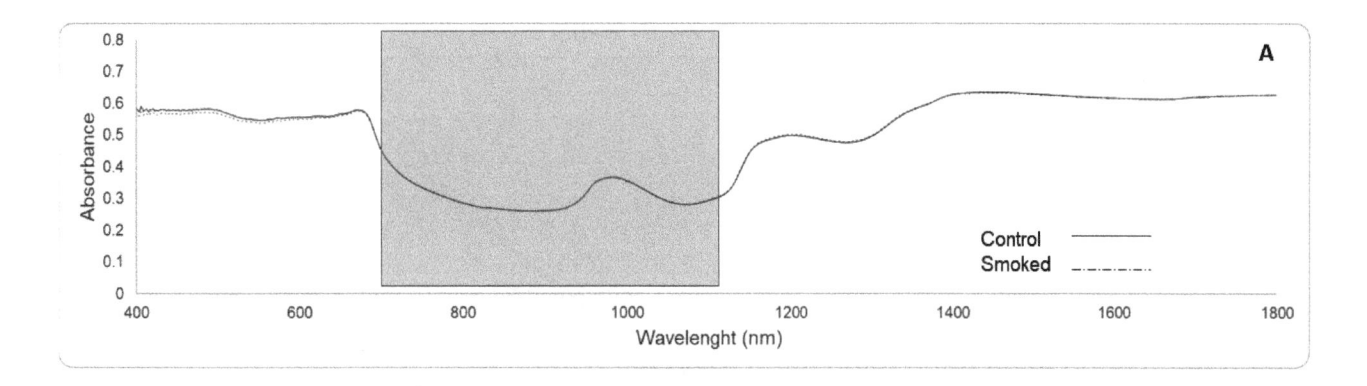

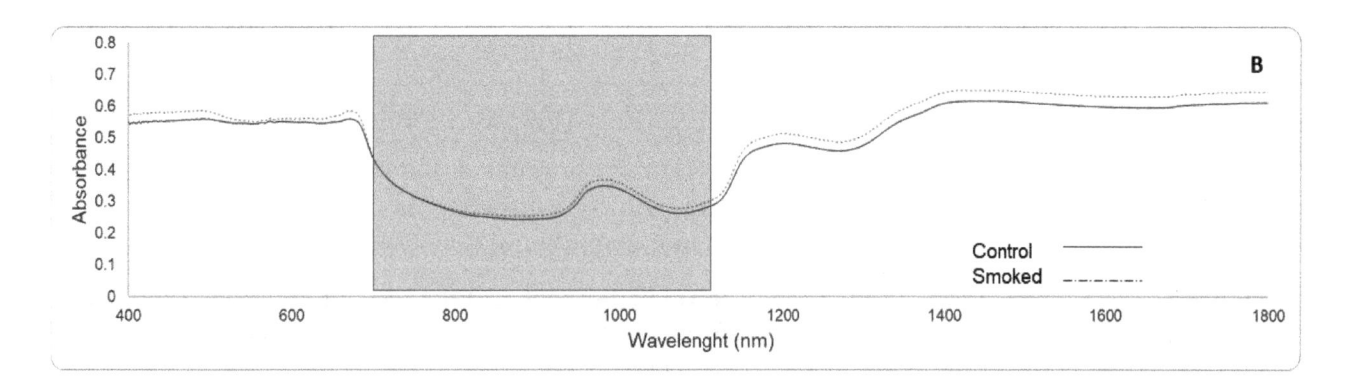

Figure 7. Average spectra for control (solid line) and smoke-affected berries (dashed line) from red cultivars (**A**): Merlot, Shiraz, Pinot Noir, Cabernet Sauvignon and white cultivars (**B**): Pinot Gris, Chardonnay and Sauvignon Blanc. The grey rectangles represent the wavelengths used for machine learning fitting modeling (700–1100 nm), with the main peak at 982 nm.

3.2.4. Machine Learning Modeling Based on NIR Spectra to Estimate Smoke Taint Compounds in Berries and Wine

Table 4 shows the best machine learning regression model obtained for the NIR data from berries (700–1100 nm using the second derivative transformation; Sequential Order Weights and Bias) as inputs and smoke taint compounds measured in berries and wine. The correlation between the estimated and observed values was R = 0.97 and slope b = 0.93 (close to unity). The same correlations and similar slopes were found for the training and the test stages. The overall model can also be seen in Figure 8, in which most of the point cloud data fits in the 1:1 line representing the accuracy of predicted versus observed data. Based on the 95% confidence bounds, the overall model had 3.6% of outliers. The performance of training was lower than the one for testing, and testing accuracy was the same as that from the training stage, which are evidence of no overfitting [20,21].

Table 4. Regression model using machine learning (Sequential Order Weights and Bias) for NIR data from berries of seven grapevine cultivars showing the correlation coefficient (R) and performance based on mean squared error (MSE) for each stage.

Stage	Samples	Observations	R	Slope	Performance (MSE)
Training	78	234	0.97	0.91	0.86
Test	33	99	0.97	0.96	0.91
Overall	111	333	0.97	0.93	-

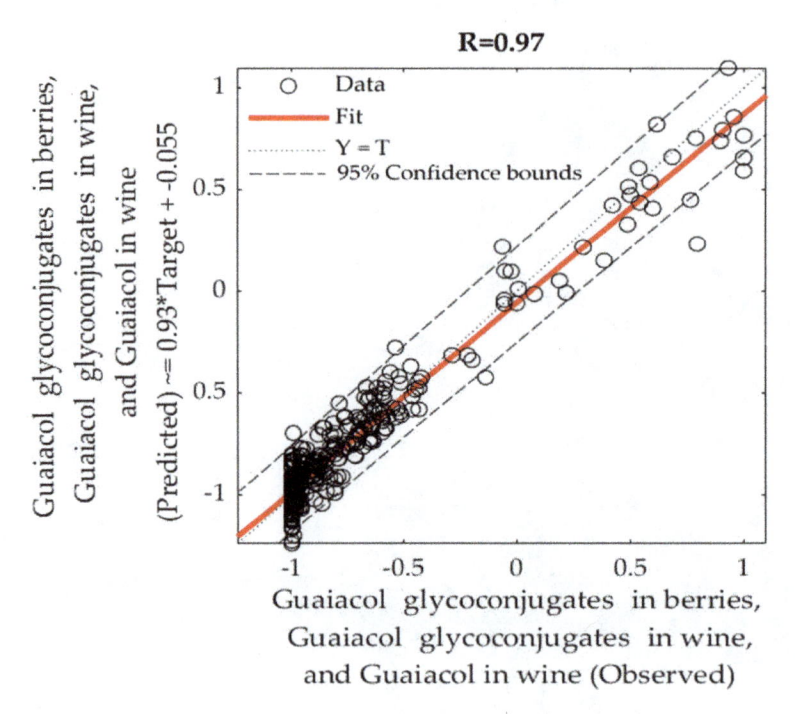

Figure 8. Overall fitting model using machine learning (Sequential Order Weights and Bias) using NIR spectra (700–1100 nm; second derivative transformation) of berries from seven grapevine cultivars as inputs and main smoke taint compounds found in berries and wine as targets.

4. Discussion

4.1. Physiological Changes within Grapevine Canopies Due to Smoke Contamination

The relationship between the I_g thermal index and g_s is linear, as shown in Table 1 and Figure 5A for non-smoked vines. These results are consistent with other studies showing the same relationships for grapevines [16,17], coffee plants [22] and olive trees [23], which are tree-like or bushy canopies. However, this relationship was not observed for smoked canopies of the four cultivars from Experiment 1 (Figure 5B). Smoke contamination is an external signal to the plant which is composed mainly of CO, CO_2 and other gases, which cause acidification of the sub-stomatal cavity due to the production of carbonic acid (H_2CO_3) when combined with water, with the resulting pH reduction causing partial or complete stomata closure [5]. This effect could explain the increased variability within g_s data amongst individual leaves that was detected in porometry data (Table 1 and Figure 5B). The reported I_g data from the whole infrared thermal images (Table 1) did not have significant differences in the variability of the data, which can be explained by the unrepresentativeness of means when using this type of high-resolution information.

It is important to note that the comparison between g_s and I_g for Figure 5 was made in this case using the methodology proposed in Figure 2 and with a sub-division of 3 × 3 for comparison purposes. Since every image was taken from 2.5 m distance, the field of view from infrared thermal images was around 140 × 110 cm of the canopy, which divided by nine gives a sub-area of 47 × 37 cm (area = 1739 cm^2). Considering that the area of an average leaf (data not shown) is of around 50–80 cm^2 [24], the I_g values represent the average of an area of approximately 25-fold of single leaves, in which porometry was conducted. This difference may explain the lower sensitivity of I_g to g_s, especially for smoked canopies with higher g_s variability expected even at the leaf level (patchy stomata behavior).

The extraction of I_g values from infrared thermal images require a T_{dry} and T_{wet} reference temperatures. In this study, the painted leaves method was implemented for more accuracy in the determination of reference temperature thresholds to separate leaf from non-leaf material in the analysis. However, this method is manual and hinders the possibility of automation. Alternatively, the leaf energy balance method could be implemented using micrometeorological weather data such as temperature, relative humidity, and solar radiation to calculate T_{dry} and T_{wet} on-the-go, while obtaining the infrared thermal images. It is common nowadays to access cheap sensor technology to measure these micrometeorological variables and dataloggers or access to the Internet of Things (IoT) for data transmission and processing. Previous research has shown that these reference temperatures can be calculated with high accuracy ($R^2 = 0.95$; RMSE = 0.85; $p < 0.001$) [16]. Furthermore, there is the requirement for infrared thermal images to be explored and assessed more in-depth at higher subdivisions and using machine learning modeling to assess the pattern variability and use it as a predictor of smoke contamination levels.

4.2. Pattern Recognition of Smoke Contamination Using Machine Learning Modeling

Considering the sub-division of infrared thermography data, the field of view of canopies and size of single leaves for this study, it is not surprising that the best pattern recognition model (96% accuracy) using machine learning (Sequential order weight and bias) was obtained with the 10 × 10 subdivision. This sub-division will render comparison areas within the canopy of 154 cm^2, which is only 2.2-fold compared to a single leaf area (70 cm^2). Furthermore, from the neuron trimming analysis, a highly accurate model was obtained for the classification of smoked and non-smoked canopies with three neurons, which makes the model more efficient and less susceptible to overfitting. The latter is also supported by the performance value obtained by this model. Results shown in this paper from pattern recognition modeling using machine learning to asses smoke contamination of canopies have excellent potential for the use in short and mid-range remote sensing based on Unmanned Aerial Vehicles (UAVs) platforms. From Figure 5B, it can be seen that the main variability within g_s values is in the bottom and top parts of the canopies, which validates obtaining infrared thermal imagery using UAVs at 0° Nadir angle. Furthermore, models developed in this study should be tested using UAV with infrared cameras that could render a 15 × 15-pixel resolution, which corresponds to an area of 225 cm^2, which is close to the 154 cm^2 area used for machine learning modeling here.

This kind of remote sensing tool can render spatial distribution maps of contaminated areas within vineyards that could aid growers to apply differential management strategies discussed before to mitigate smoke contamination of the fruit. Spatial maps of smoke contamination can also help to achieve differential harvests to avoid mixing fruit with smoke-tainted fruit. Hence, a system is proposed using these methods, which is depicted in Figure 9 for proximal and mid-distance remote sensing using infrared cameras and UAV platforms. For proximal remote sensing, the algorithms developed in this study can be implemented in smartphone devices as computer applications (Apps) connected to portable and affordable infrared thermal cameras (i.e., FLIR One®, FLIR Systems, Portland, OR, USA) and near-infrared spectroscopy devices (i.e., Lighting Passport®, AsenseTek, Taipei, Taiwan).

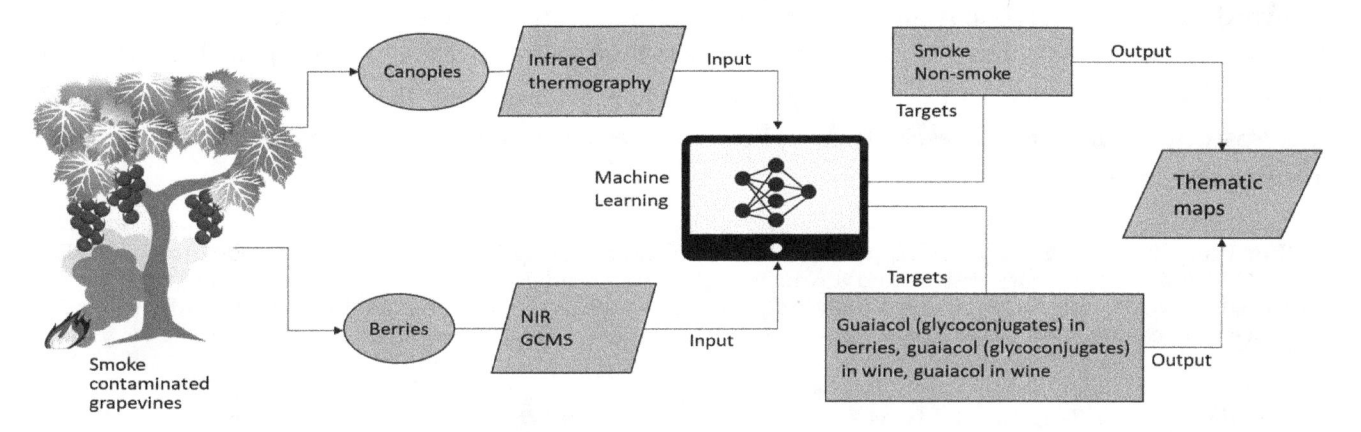

Figure 9. Diagram showing the implementation of machine learning modeling strategies proposed in this paper for proximal (using smartphones and portable infrared thermal cameras and NIR devices) and mid-distance remote sensing using unmanned aerial system (UAS) platforms.

4.3. Near-Infrared (NIR) Spectroscopy of Berries

Since NIR spectroscopy was obtained from full berries, the tool proposed in this paper is non-destructive. Furthermore, it has been shown that a higher concentration of smoke-related compounds after contamination can be found in the skin of berries, which is higher than in the pulp and higher than the seeds [12]. Furthermore, the range of 700–1100 nm was chosen since most of the available NIR instrumentation in this range can be affordable for growers compared to the instrument used in this study which can cost around 45 times more. The 982 nm overtone is associated with the OH overtone band and 1100 for the CH bands, which corresponds to alcohol and phenolic compounds [25].

The model reported using machine learning fitting algorithms can be of great assistance to growers and winemakers to obtain chemometry data in real time using the proposed methodology shown in Figure 9. Currently, growers do not have sophisticated tools to assess potential smoke contamination of berries bunches and wines. The only option available is collection of samples within a vineyard for compositional analysis by an accredited laboratory using GC-MS or HPLC-MS/MS. This process is destructive, expensive, and takes a long time, which makes it less ideal for the implementation of mitigation strategies and/or decision making before harvest. Furthermore, it may minimize smoke taint by the information provided through a spatial assessment of the contamination either through canopies or berries for informed decision making regarding palliative measures (as presented in this paper) or differential harvest.

The models developed in this study were able to predict smoke contamination in canopies, berries and wines, regardless of the cultivar. Hence, the models could be applied as a universal methodology. Further studies and data acquired could be added to models to include more cultivars. However, the seven cultivars included in this study were some of the most commercially important in Australia. Finally, it is important to note that the levels of smoke-taint compounds present in wine are in part related to the winemaking process (i.e., duration of skin contact time during fermentation), hence this model will need to be adjusted for different winemaking techniques, which can influence the extraction of smoke-related compounds from the berry.

5. Conclusions

This paper showed two main advancements for tools to detect smoke contamination in grapevine canopies and smoke-related compounds in berries and wine using remote sensing techniques. This study is the first to apply machine learning modeling techniques to assist growers confronted with vineyard exposure to smoke from bushfires, an issue which has been exacerbated in prominent wine regions around the world due to climate change. Furthermore, this paper has proposed an affordable method to implement these novel techniques using smartphones, portable thermal imagery and NIR spectroscopy devices. More research is required to assess the usage of these affordable devices in the future using the models proposed.

Author Contributions: S.F. conceived the machine learning modeling idea and practical applications; S.F., E.J.T. and C.G.V. analyzed the data and created the machine learning models; K.W., S.T. and S.F. were awarded funding for the study; K.W. and R.R. performed field trials, laboratory analysis and winemaking; S.F. and R.D.B. acquired the physiological and NIR data. All authors contributed to the writing of the paper.

Acknowledgments: This research was supported under the Australian Research Council's Linkage Projects funding scheme (LP0989138); the financial contributions of industry partners are also gratefully acknowledged. The machine learning modeling research was supported by the Digital Viticulture program funded by the University of Melbourne's Networked Society Institute, Australia.

References

1. Hughes, L.; Alexander, D. Climate Change and the Victoria Bushfire Threat: Update 2017. Climate Council Report. 2017. Available online: http://www.climatecouncil.org.au/uploads/98c26db6af45080a32377f3ef4800102.pdf (accessed on 10 July 2019).
2. Webb, L.; Whetton, P.; Bhend, J.; Darbyshire, R.; Briggs, P.; Barlow, E. Earlier wine-grape ripening driven by climatic warming and drying and management practices. *Nat. Clim. Chang.* **2012**, *2*, 259–264. [CrossRef]
3. Webb, L.B. Climate change and winegrape quality in Australia. *Clim. Res.* **2008**, *36*, 99–111. [CrossRef]
4. Webb, L.B.; Whetton, P.H.; Barlow, E.W.R. Modelled impact of future climate change on the phenology of winegrapes in Australia. *Aust. J. Grape Wine Res.* **2007**, *13*, 165–175. [CrossRef]
5. Su, B.; Xue, J.; Xie, C.; Fang, Y.; Song, Y.; Fuentes, S. Digital surface model applied to unmanned aerial vehicle based photogrammetry to assess potential biotic or abiotic effects on grapevine canopies. *Int. J. Agric. Biol. Eng.* **2016**, *9*, 119–130.
6. Ristic, R.; Fudge, A.L.; Pinchbeck, K.A.; De Bei, R.; Fuentes, S.; Hayasaka, Y.; Tyerman, S.D.; Wilkinson, K.L. Impact of grapevine exposure to smoke on vine physiology and the composition and sensory properties of wine. *Theor. Exp. Plant Physiol.* **2016**, *28*, 67–83. [CrossRef]
7. Ristic, R.; Pinchbeck, K.; Fudge, A.; Hayasaka, Y.; Wilkinson, K. Effect of leaf removal and grapevine smoke exposure on colour, chemical composition and sensory properties of Chardonnay wines. *Aust. J. Grape Wine Res.* **2013**, *19*, 230–237. [CrossRef]
8. van der Hulst, L.; Munguia, P.; Culbert, J.A.; Ford, C.M.; Burton, R.A.; Wilkinson, K.L. Accumulation of volatile phenol glycoconjugates in grapes following grapevine exposure to smoke and potential mitigation of smoke taint by foliar application of kaolin. *Planta* **2019**, *249*, 941–952. [CrossRef]
9. Fudge, A.; Ristic, R.; Wollan, D.; Wilkinson, K.L. Amelioration of smoke taint in wine by reverse osmosis and solid phase adsorption. *Aust. J. Grape Wine Res.* **2011**, *17*, S41–S48. [CrossRef]
10. Fudge, A.; Schiettecatte, M.; Ristic, R.; Hayasaka, Y.; Wilkinson, K.L. Amelioration of smoke taint in wine by treatment with commercial fining agents. *Aust. J. Grape Wine Res.* **2012**, *18*, 302–307. [CrossRef]
11. Wilkinson, K.; Ristic, R.; Pinchbeck, K.; Fudge, A.; Singh, D.; Pitt, K.; Downey, M.; Baldock, G.; Hayasaka, Y.; Parker, M. Comparison of methods for the analysis of smoke related phenols and their conjugates in grapes and wine. *Aust. J. Grape Wine Res.* **2011**, *17*, S22–S28. [CrossRef]
12. Dungey, K.A.; Hayasaka, Y.; Wilkinson, K.L. Quantitative analysis of glycoconjugate precursors of guaiacol in smoke-affected grapes using liquid chromatography–tandem mass spectrometry based stable isotope dilution analysis. *Food Chem.* **2011**, *126*, 801–806. [CrossRef]
13. Hayasaka, Y.; Parker, M.; Baldock, G.A.; Pardon, K.H.; Black, C.A.; Jeffery, D.W.; Herderich, M.J. Assessing the impact of smoke exposure in grapes: Development and validation of a HPLC-MS/MS method for the quantitative analysis of smoke-derived phenolic glycosides in grapes and wine. *J. Agric. Food Chem.* **2012**, *61*, 25–33. [CrossRef] [PubMed]
14. Fudge, A.L.; Wilkinson, K.L.; Ristic, R.; Cozzolino, D. Classification of smoke tainted wines using mid-infrared spectroscopy and chemometrics. *J. Agric. Food Chem.* **2011**, *60*, 52–59. [CrossRef] [PubMed]
15. Kennison, K.R.; Gibberd, M.R.; Pollnitz, A.P.; Wilkinson, K.L. Smoke-derived taint in wine: The release of smoke-derived volatile phenols during fermentation of Merlot juice following grapevine exposure to smoke. *J. Agric. Food Chem.* **2008**, *56*, 7379–7383. [CrossRef] [PubMed]
16. Fuentes, S.; De Bei, R.; Pech, J.; Tyerman, S. Computational water stress indices obtained from thermal image analysis of grapevine canopies. *Irrig. Sci.* **2012**, *30*, 523–536. [CrossRef]
17. Jones, H.G.; Stoll, M.; Santos, T.; Sousa, C.D.; Chaves, M.M.; Grant, O.M. Use of infrared thermography for monitoring stomatal closure in the field: Application to grapevine. *J. Exp. Bot.* **2002**, *53*, 2249–2260. [CrossRef] [PubMed]
18. Moran, M.S.; Inoue, Y.; Barnes, E. Opportunities and limitations for image-based remote sensing in precision crop management. *Remote Sens. Environ.* **1997**, *61*, 319–346. [CrossRef]
19. Jones, H.G. Use of infrared thermometry for estimation of stomatal conductance as a possible aid to irrigation scheduling. *Agric. For. Meteorol.* **1999**, *95*, 139–149. [CrossRef]
20. Beale, M.H.; Hagan, M.T.; Demuth, H.B. *Deep Learning Toolbox User's Guide*; The Mathworks Inc.: Herborn, MA, USA, 2018.

21. Gonzalez Viejo, C.; Torrico, D.D.; Dunshea, F.R.; Fuentes, S. Development of Artificial Neural Network Models to Assess Beer Acceptability Based on Sensory Properties Using a Robotic Pourer: A Comparative Model Approach to Achieve an Artificial Intelligence System. *Beverages* **2019**, *5*, 33. [CrossRef]
22. Craparo, A.; Steppe, K.; Van Asten, P.J.; Läderach, P.; Jassogne, L.T.; Grab, S. Application of thermography for monitoring stomatal conductance of Coffea arabica under different shading systems. *Sci. Total Environ.* **2017**, *609*, 755–763. [CrossRef]
23. Egea, G.; Padilla-Díaz, C.M.; Martinez-Guanter, J.; Fernández, J.E.; Pérez-Ruiz, M. Assessing a crop water stress index derived from aerial thermal imaging and infrared thermometry in super-high density olive orchards. *Agric. Water Manag.* **2017**, *187*, 210–221. [CrossRef]
24. Fuentes, S.; Hernández-Montes, E.; Escalona, J.; Bota, J.; Viejo, C.G.; Poblete-Echeverría, C.; Tongson, E.; Medrano, H. Automated grapevine cultivar classification based on machine learning using leaf morpho-colorimetry, fractal dimension and near-infrared spectroscopy parameters. *Comput. Electron. Agric.* **2018**, *151*, 311–318. [CrossRef]
25. Wang, X.; Bao, Y.; Liu, G.; Li, G.; Lin, L. Study on the best analysis spectral section of NIR to detect alcohol concentration based on SiPLS. *Procedia Eng.* **2012**, *29*, 2285–2290. [CrossRef]

Assessing Topsoil Movement in Rotary Harrowing Process by RFID (Radio-Frequency Identification) Technique

Ahmed Kayad [1,2,*], **Riccardo Rainato** [1], **Lorenzo Picco** [1,3,4], **Luigi Sartori** [1] and **Francesco Marinello** [1]

1. Department TESAF, University of Padova, viale dell'Università, 16, I-35020 Legnaro (PD), Italy
2. Agricultural Engineering Research Institute (AEnRI), Agricultural Research Centre, Giza 12619, Egypt
3. Faculty of Engineering, Universidad Austral de Chile, Campus Miraflores, Valdivia 5090000, Chile
4. Universidad Austral de Chile, RINA–Natural and Anthropogenic Risks Research Center, Campus Miraflores, Valdivia 5090000, Chile
* Correspondence: ahmed.kayad@phd.unipd.it

Abstract: Harrowing is a process that reduces the size of soil clods and prepares the field for seeding. Rotary harrows are a common piece of equipment in North Italy that consists of teeth rotating around a vertical axis with a processing depth of 5–15 cm. In this study, the topsoil movement in terms of distance and direction were estimated at different rotary harrow working conditions. A total of eight tests was performed using two forward speeds of 1 and 3 km/h, two working depths of 6 and 10 cm and two levelling bar positions of 0 and 10 cm from the ground. In order to simulate and follow topsoil movement, Radio-Frequency Identification (RFID) tags were inserted into cork stoppers and distributed in a regular pattern over the soil. Tags were distributed in six lines along the working width and repeated in three rows for each test: a total number of 144 tags was tracked. Results showed that there were no significant differences between the performed tests, on the other hand the reported tests highlight the effectiveness of the RFID monitoring approach.

Keywords: rotary harrow; secondary tillage; soil erosion; RFID

1. Introduction

Soil tillage is an agronomic practice that effects on both soil and crop properties. The main objective for tillage operations is to improve the soil environment for seed germination and, subsequently, improve crop yield [1–3]. There are several kinds of tillage methods and tools such as conventional tillage, which is commonly divided into primary and secondary tillage. In primary tillage, moldboard or chisel plows make the major part of the tillage operation. Meanwhile, soil surface after primary tillage still needs further operation in order to smooth and reduce clods size. Therefore, secondary tillage operation by rotary harrow is used for preparing suitable seedbed [4].

Tillage operations have a great impact on topsoil in terms of aggregate size and crop residue cover, which plays an important role in sustainable agriculture. A minimum amount of residue is needed to protect soil erosion, reduce greenhouse gas emissions and maintain soil carbon level [5–8]. The impact of tillage operations on topsoil depend on tillage operational depth and speed, soil characteristics, initial amount of residues, and type of used equipment [9,10].

The contribution of harrowing in soil erosion could be realized from weed dispersal studies [4]. Soil movement by tillage operations was investigated through different instruments [11] and different approaches [12,13]. Researchers used different techniques to monitor soil movement such as plastic beads [14], granite rocks [13], or aluminum cubes [4]. It is in the authors' opinion that Radio-Frequency Identification (RFID) systems can be successfully applied for this scope.

The RFID is a system that transmits the identity of an object using radio waves and consists of RFID tags and RFID readers. RFID tags are small electronic devices attached to objects and RFID reader is an antenna that identifies the tags [15]. The passive tags have no power source and discovered by antenna within 0.5–1.0 m radius while other tags may include batteries to be detected at higher distances [16,17]. In recent years, due to relatively low costs, many applications have been developed based on RFID such as location identification for many shipping and postal services, security purposes in shops and companies, retailers and supply chain for confirming that products on shelf, payment systems for payments at drive-through windows, livestock and agricultural production monitoring [18–21]. With reference to the soil, some recent advancements have been reported, mainly dealing with the assessment of erosion, landslide displacement monitoring, or sediment mobility [22–25], while the authors are not aware of RFID applications in relation to soil tillage practices.

The main objective of this study is to assess the movements of topsoil layer after the application of a tillage operation such as rotary harrowing. In addition, the study highlights how RFID technology can be effectively applied in order to simulate crop residues or even soil clods for tillage experiments as more robust technique especially in open field trials (Supplementary Materials S1).

2. Materials and Methods

This study was conducted at a 5 ha field in Agripolis experimental field (University of Padova, Italy). The soil can be defined, according to the United State Department of Agriculture (USDA), as loamy and containing 46% sand, 30% silt, and 24% clay. Primary tillage was applied at the end of the winter season and followed by a harrowing process as secondary tillage on May 2018 in order to prepare the soil for soybean sowing. It is worth noting how most of crop residues are buried after primary tillage, especially in the case of mouldboard ploughing. Such conditions were helpful in order to maximize the understanding of implemented RFID tags dynamics.

A rotary harrow with two horizontal rollers from Alpergo Co., Lonigo, VI, Italy (Model: Rotodent DP) was used to perform the harrowing process. The implement consisted of a series of 20 pairs of tines (indicated by *A* in Figure 1) which rotate about a vertical axis in order to produce soil disturbance over a working width of 5 m. The working volume is limited on the back by a levelling bar (*B* in Figure 1): its height relatively to the ground can be adjusted in order to limit the flow of clods through the machine and allow a better control on aggregates dimensions. The levelling bar thus allows the soil to be hold for a certain time (from a few tenths up to a few seconds depending on its relative position) within the shell where the tines operate, allowing aggregates mixing and reduction. Two horizontal cage rollers (*C* in Figure 1) were positioned on the back end of the rotary harrow to allow soil compaction and levelling. The implement was driven by a 160 hp tractor through the three-point hitch and Power Take Off (PTO) shaft at 1000 rpm.

A RFID package from Oregon RFID Co., Portland, OR, USA was implemented in this work. This package consists of RFID reader and passive tags. Tags are in a cylindrical shape with 3.65 mm diameter and 23 mm length. Being small in dimensions and weight (less than 1 gram) allows their integration into bigger envelopes. In fact, in order to simulate topsoil components such as crop residues and soil clods, a cork stopper was drilled laterally and one RFID tag inserted inside each stopper then closed by glue. Cork stoppers were chosen, due to their wide availability, durability, and high similarity (in terms of shape, size, and density) specially with crop residues such as dry corn stems or corncob. Furthermore, they allow to damp the effects of rotary harrow tools, minimizing the eventuality of damages to the delicate RFID tags. Each tag (and thus each cork stopper) had an identification number which allowed individual monitoring. Additionally, cork trackers were painted with shining fluorescent color and tag identification number in order to facilitate recognition after the harrowing process (Figure 2).

Assessing Topsoil Movement in Rotary Harrowing Process by RFID (Radio-Frequency Identification) Technique

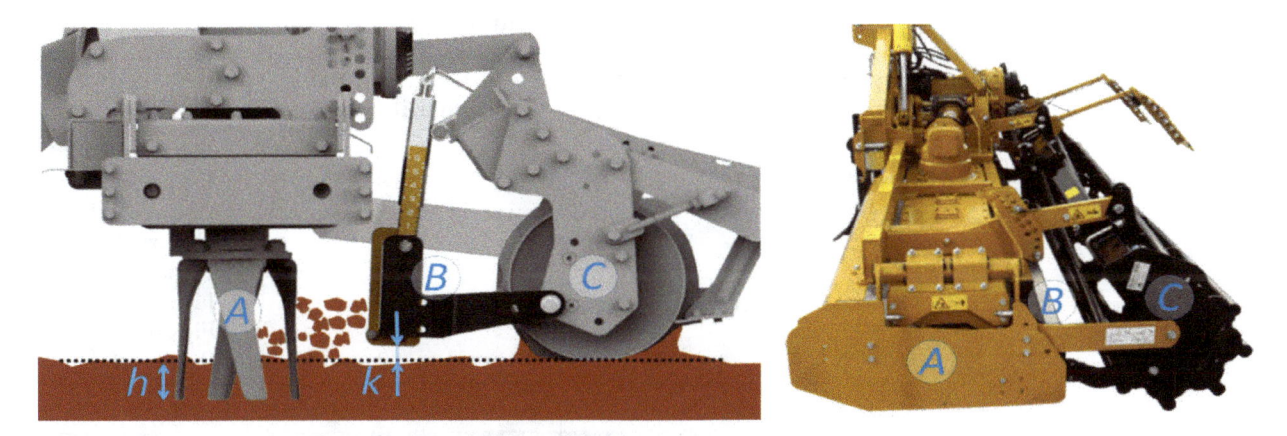

Figure 1. On the left schematic view and on the right a picture of the implemented harrowing machine, with indication of the working depth h of the rotary tines (**A**) and of the height k of the levelling bar (**B**) closing the back part of the working volume; Soil is eventually levelled by a couple of cage rollers (**C**).

Figure 2. On the left a picture if the RFID tag, and on the right a cork stopper with RFID inside, painted and numbered.

The experimental field was divided into two lateral strips (5 m × 80 m) and each strip consisted of four different working conditions, as summarized in Table 1 and depicted in Figure 3. The different applied working conditions were as follows: two forward speeds of 1 and 3 km/h, two working depths (namely h in Figure 1) of 6 and 10 cm, and two levelling bar positions (namely k in Figure 1) of 0 (i.e., at ground level) and 10 cm (i.e., lifted up from the soil). Eight working conditions were investigated by using the RFID trackers. In each treatment, 18 trackers were distributed in six lines and three rows. The distance between rows was 1 m while the distance between lines were 40, 80, 220, 280, 420, and 460 cm starting from one of the harrow sides in order to maintain a symmetric measurement between the two horizontal rollers (Figure 3). The three rows of trackers act as replicates for each treatment.

Trackers' positions were recorded in conjunction with their identification number and tested with the RFID antenna before applying the experimental treatments. After performing each treatment, the position of each tracker was recorded in order to investigate its movement in terms of distance, depth, and direction. At the start of each treatment, two stick markers were fixed to act as a datum for trackers position recognition (Figure 3C).

Table 1. Experimental conditions.

Acronym	Forward Speed (km/h)	Working Depth (cm)	Levelling Bar Position (cm)	Reference
V1D10	1	10	0	Figure 3B a
	1	10	10	Figure 3B e
V3D10	3	10	0	Figure 3B b
	3	10	10	Figure 3B f
V1D6	1	6	0	Figure 3B c
	1	6	10	Figure 3B g
V3D6	3	6	0	Figure 3B d
	3	6	10	Figure 3B h

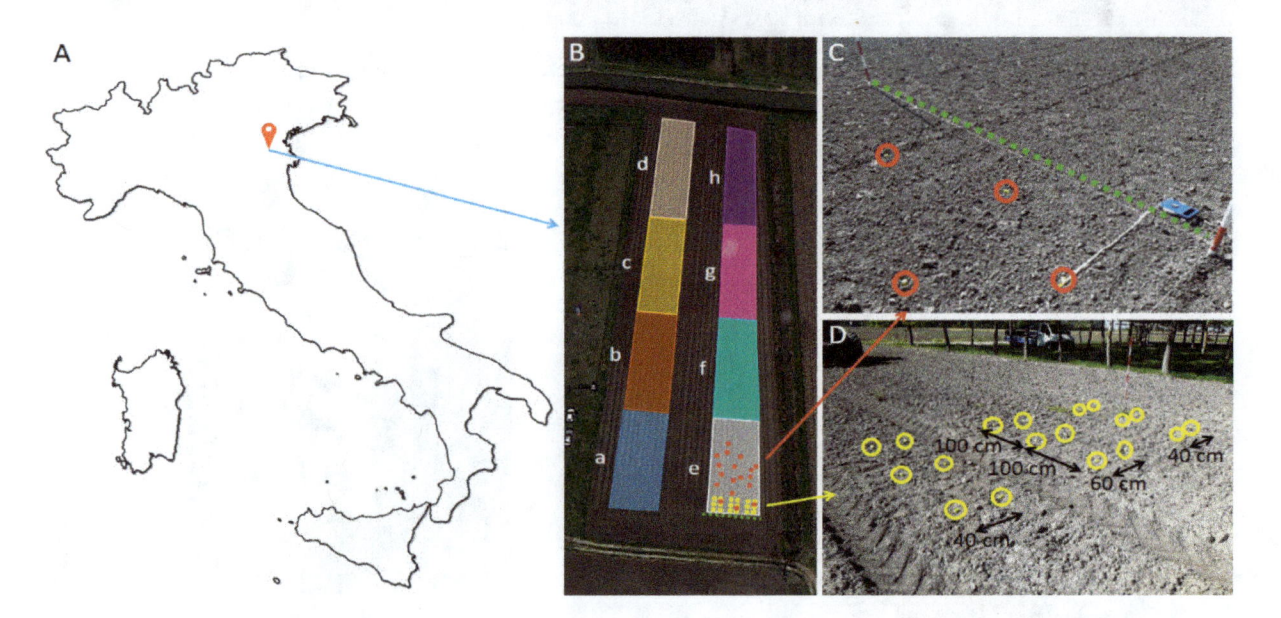

Figure 3. (**A**)—Geographical localization of the experimental site in Italy. (**B**)—Experimental area with representation of the experimental design. (**C**)—Some of the tags after tillage operation. (**D**)—Trackers distribution on undisturbed soil, before harrowing process.

3. Results and Discussion

A total number of 144 trackers were tracked after applying eight different treatments of rotary harrowing. Trackers were localized by naked eye in the case of surface positions, conversely buried trackers (more than eighty) were localized taking advantage of the RFID reader. During the experiments, only two of the trackers were lost: in such cases they were not revealed by the RFID reader most probably due to some breakage occurred during harrowing operation. Hence, the overall recovery rate was 98.6%. Results showed that there was a clear movement in the machine direction with an average value of 2.3 m, but exceeding 5 m in different situations. On the other hand, the lateral movement was limited to an average of 0.2 m. The average movements for applied treatments are shown in Figure 4.

The major direction for trackers movement was in the machine direction. The major factor that effected on movement distance was noticed from the levelling bar. The average movement were 2 and 0.8 m at bar level of 0 and 10 cm, respectively. Also, many trackers moved more than 5 m in the machine direction with the bar at ground level, compared to less than 2.5 m in the case of lifted up bar. This is ascribable to the fact that when the levelling bar is down, the amount of soil kept within the chamber between tines and levelling bar itself increases. Concurrently, there is an increase in the amount of time soil remains within such chamber: aggregates exit only after their size has been sufficiently reduced to pass under the levelling bar. Such phenomenon results in a longer distance done by clods or residues (and thus also by the RFID tags) and in a higher chance of burial effects. Moreover, as tines are rotating, there is a probability of hitting the cork trackers which may throw trackers to a forward or backward direction. In case of backward thrown, the levelling bar prevents trackers from moving beyond that barrier. In contrary, there is no limit for the forward thrown where there is no barrier and the rugged material of cork trackers may explain the long forward displacement in some cases. Figure 5 illustrates the forward movement at different rotary harrow working conditions.

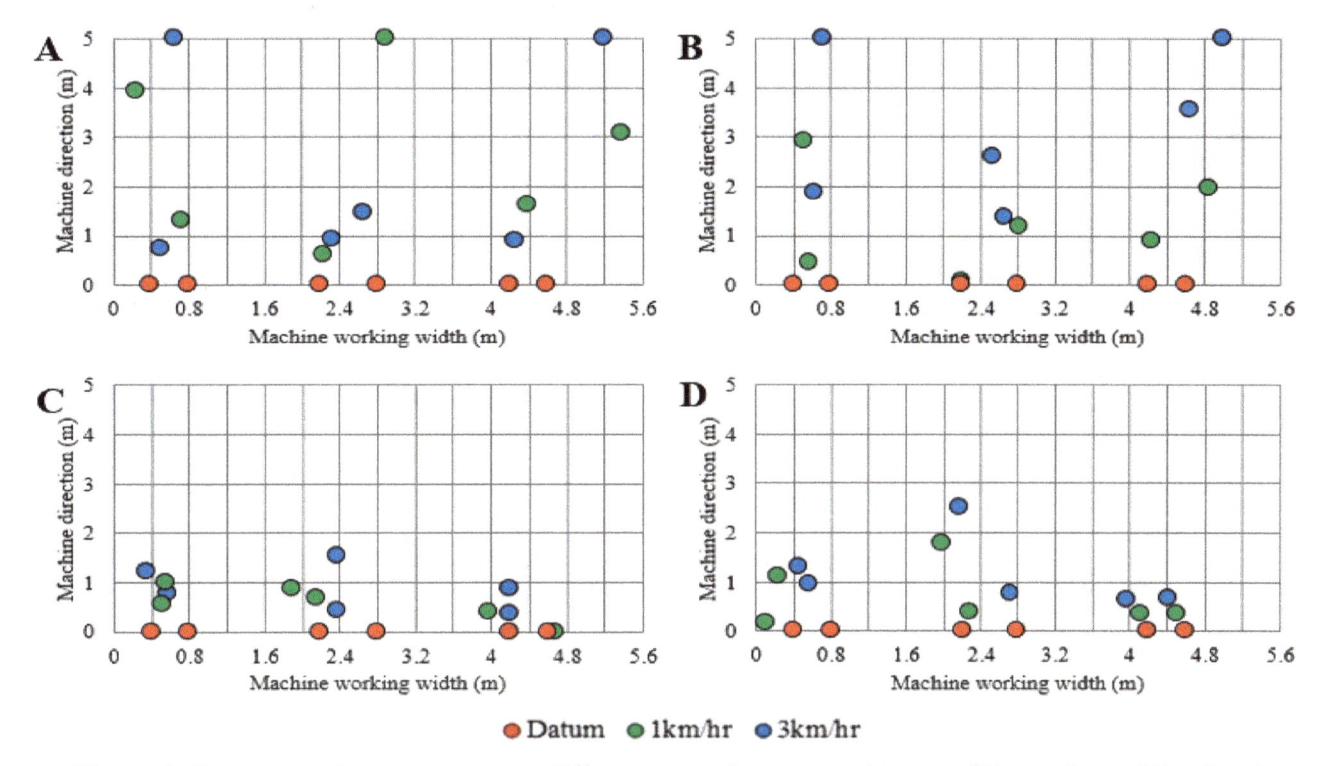

Figure 4. Average trackers movement at different rotary harrow working conditions where: (**A**) at level bar of zero and depth of 6 cm, (**B**) at level bar of zero and depth of 10 cm, (**C**) at level bar of 10 cm and depth of 6 cm, and (**D**) at level bar of 10 cm and depth of 10 cm.

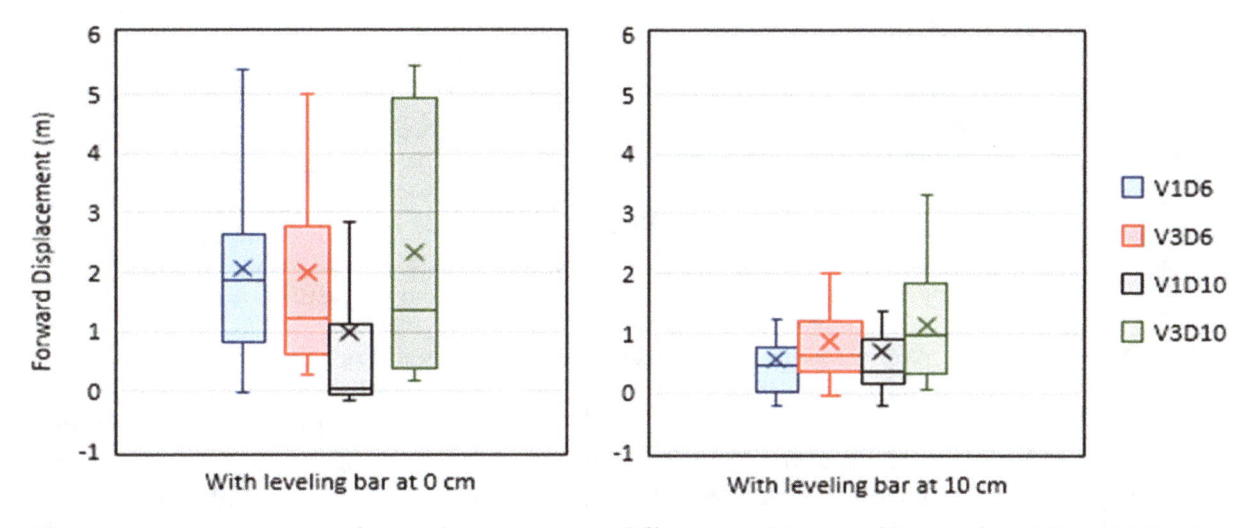

Figure 5. Average trackers forward movement at different working conditions where V1 and V3 refer to forward speed of 1 and 3 km/h. Also, D6 and D10 refer to working depth of 0.06 and 0.1 m.

Shifts in lateral direction were limited to 0.03 m left on average, with a standard deviation as high as 0.40 m. This is in agreement with the rotary harrow symmetric construction, where clockwise and anticlockwise rotating tine pairs are alternated: such working mode allows avoidance of soil piling phenomena, as confirmed by RFID distribution. In this case, the levelling bar position plays only a marginal role statistically irrelevant: 0.12 m right lateral displacement in the case of bar at ground level and 0.20 m left shift in the case of lifted up position. A slightly different situation is highlighted in the case of analyses related to tags depth. In fact, lower position of the levelling bar caused an average increase on burial phenomenon from −0.8 cm to −1.5 cm on average. Such apparently low values (compared to the working depth) are most probably ascribable to some kind of floating phenomenon the cork stoppers underwent due to their density, sensibly lower than soil clods density, but comparable to crop residues one. Figures 6 and 7, illustrates the lateral movement and burial depth at different rotary harrow working conditions.

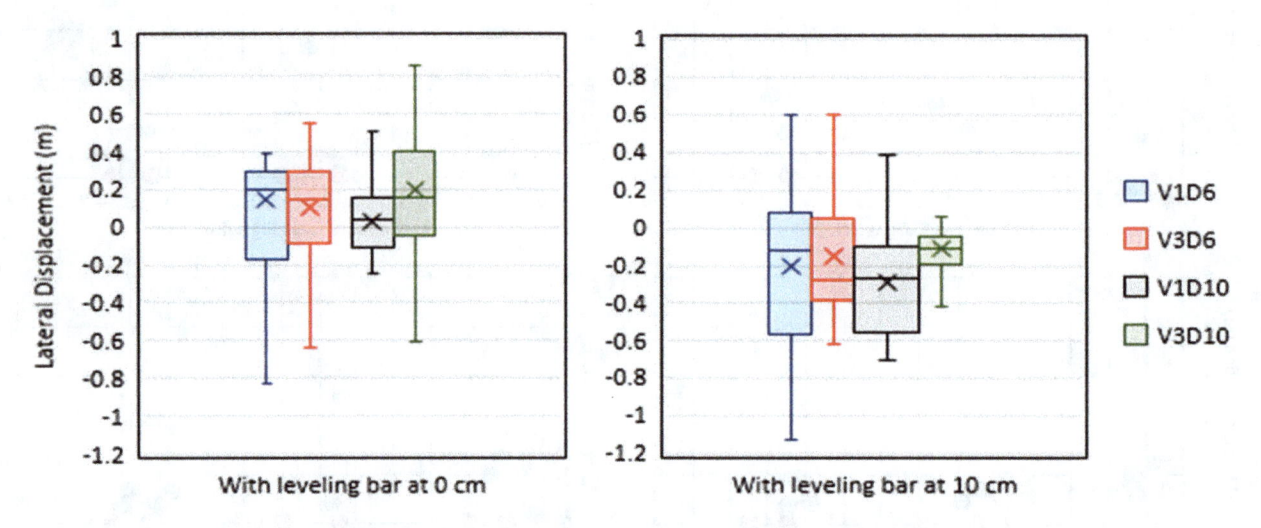

Figure 6. Average trackers lateral movement at different working conditions.

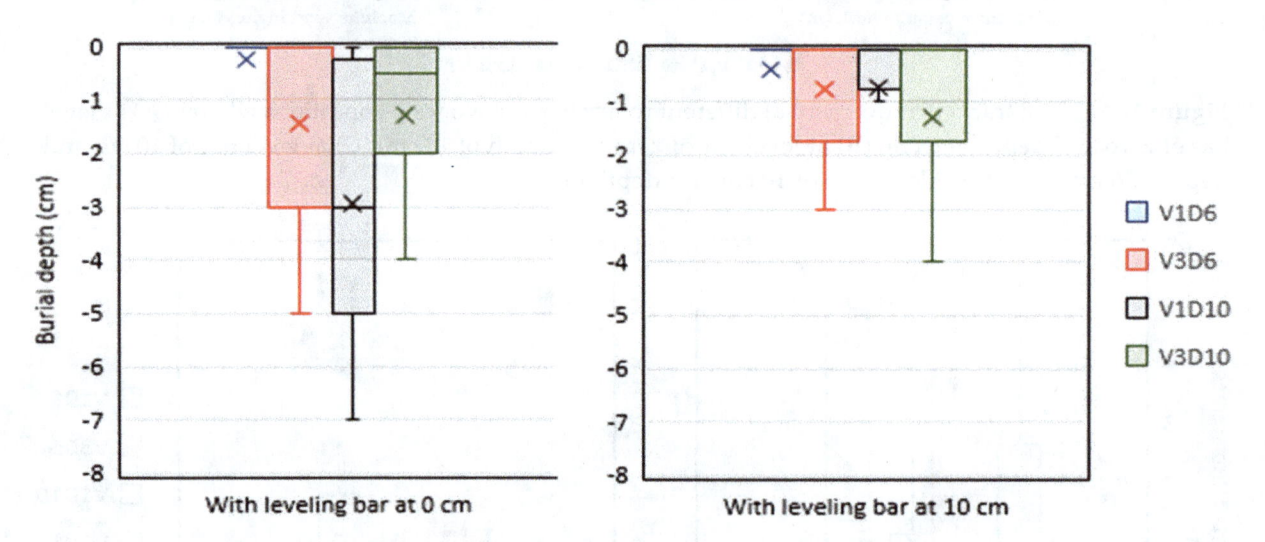

Figure 7. Average trackers burial depth at different working conditions.

Moreover, 38% of trackers were covered by soil after the harrowing process. The average cover depth was ranging between 0.01 m and 0.07 m. In addition, at the working depth of 0.1 m and levelling bar at ground level, more than 60% of trackers were covered by soil, which highlights a higher mixing rate at these conditions. Such higher mixing rate is due to the higher interaction between the soil and machine in terms of higher tines depth and longer contact compared with other studied working conditions. Also, it can be noticed how higher tillage speed resulted in higher trackers displacements, in agreement with Liu et al. [26] in sweep trials at soil bin.

The experimental results exhibited a large variability especially when the bar was set at a null height, which was the main reason for no significant differences between variables. Such large variations were due to a number of reasons such as; tracker (rigid material of cork) might be hit by tines and thrown at long distances, or tracker dragging through tillage operation, in agreement with similar results reported by different relevant research works in soil bin trials [26,27].

Besides characterization of the topsoil movement after the application of rotary harrow tillage operation, the proposed study demonstrates the effectiveness of the RFID technique soil movements monitoring operations. Differently, from metal or polymer trackers, RFID allows minimization of tags dimensions, allowing univocal identification and maximizing localization process both in terms of high detection percentage (close to 100%, and lower in case of harsh operation possibly affecting RFID tags integrity) and low detection time (a few seconds per tag). Additionally, it is also worth noting how the RFID approach can be in principle applied in order to simulate different materials in similar trials. The very tiny shape of RFID tags offers a wide range of simulations such as soil clods, stones, and crop residues of different shapes and densities. The best simulation for any studied object is the object itself where the material properties, such as density, surface roughness, and shape, will simulate and act naturally. Figure 8 illustrates the possibility of integrating the RFID tags in wheat spikes, straw, corn residues, potato, soil clods, or even small rocks for suggested future applications. In addition, it is worth mentioning that the easy recognition of RFID by the antenna and the unique ID for each tag will reduce the source of error and limitations in open field experiments, which will also lead to more reliable and understandable results.

Figure 8. Suggested applications for the use of Radio-Frequency Identification (RFID) to simulate or monitor crop residues, straw, soil clods and potato in agricultural experiments.

4. Conclusions

Field trials were performed to investigate the effect of different rotary harrow working conditions of forward speed, levelling bar, and tillage depth on topsoil aggregates displacement using RFID technique. Results showed no significant difference between treatments, while the higher displacement was noticed from the levelling bar factor. The average movement was about 2.3 m in the machine direction and exceeded 5 m in different cases. Lateral movements were limited to 0.03 m on average because of the different rotating directions between each tine pairs. Also, at the levelling bar of 0.1 m, 60% of trackers were buried because of the high mixing rate. Furthermore, using RFID tags and the antenna was a promising application in this field, since it is providing a robust way to simulate different materials.

Author Contributions: Conceptualization and data analysis F.M. and A.K.; Technical, practical support and method optimization for RFID trackers R.R. and L.P.; Writing and reviewing A.K. and F.M.; Supervision L.S. and F.M.

Acknowledgments: The authors acknowledge researchers involved in the project PRIN 2015 2015KTY5NW for their support in connection with the development of the technique and with the experiment.

References

1. Torabian, S.; Farhangi-Abriz, S.; Denton, M.D. Do tillage systems influence nitrogen fixation in legumes? A review. *Soil Tillage Res.* **2019**, *185*, 113–121. [CrossRef]
2. Zeyada, A.M.; Al-Gaadi, K.A.; Tola, E.; Madugundu, R.; Kayad, A.G. Impact of soil firmness and tillage depth on irrigated maize silage performance. *Appl. Eng. Agric.* **2017**, *33*, 491–498. [CrossRef]
3. Pezzuolo, A.; Dumont, B.; Sartori, L.; Marinello, F.; de Antoni Migliorati, M.; Basso, B. Evaluating the impact of soil conservation measures on soil organic carbon at the farm scale. *Comput. Electron. Agric.* **2017**, *135*, 175–182. [CrossRef]
4. Van Muysen, W.; Govers, G. Soil displacement and tillage erosion during secondary tillage operations: The case of rotary harrow and seeding equipment. *Soil Tillage Res.* **2002**, *65*, 185–191. [CrossRef]
5. Cherubini, F.; Ulgiati, S. Crop residues as raw materials for biorefinery systems—A LCA case study. *Appl. Energy* **2010**, *87*, 47–57. [CrossRef]
6. Graham, R.L.; Nelson, R.; Sheehan, J.; Perlack, R.D.; Wright, L.L. Current and potential U.S. corn stover supplies. *Agron. J.* **2007**, *99*, 1–11. [CrossRef]
7. Cillis, D.; Maestrini, B.; Pezzuolo, A.; Marinello, F.; Sartori, L. Modeling soil organic carbon and carbon dioxide emissions in different tillage systems supported by precision agriculture technologies under current climatic conditions. *Soil Tillage Res.* **2018**, *183*, 51–59. [CrossRef]
8. Pezzuolo, A.; Basso, B.; Marinello, F.; Sartori, L. Using SALUS model for medium and long term simulations of energy efficiency in different tillage systems. *Appl. Math. Sci.* **2014**, *8*, 129–132. [CrossRef]
9. Chen, Y.; Monero, F.V.; Lobb, D.; Tessier, S.; Cavers, C. Effects of six tillage methods on residue incorporation and crop performance in a heavy clay soil. *Trans. ASAE* **2004**, *47*, 1003–1010. [CrossRef]
10. Wagner, L.E.; Nelson, R.G. Mass reduction of standing and flat crop residues by selected tillage implements. *Trans. ASAE* **1995**, *38*, 419–427. [CrossRef]
11. Dubbini, M.; Pezzuolo, A.; de Giglio, M.; Gattelli, M.; Curzio, L.; Covi, D.; Yezekyan, T.; Marinello, F. Last generation instrument for agriculture multispectral data collection. *CIGR J.* **2017**, *19*, 158–163.
12. Lindstrom, M.J.; Nelson, W.W.; Schumacher, T.E.; Lemme, G.D. Soil movement by tillage as affected by slope. *Soil Tillage Res.* **1990**, *17*, 255–264. [CrossRef]
13. Thapa, B.B.; Cassel, D.K.; Garrity, D.P. Assessment of tillage erosion rates on steepland Oxisols in the humid tropics using granite rocks. *Soil Tillage Res.* **1999**, *51*, 233–243. [CrossRef]
14. Marshall, E.J.P.; Brain, P. The horizontal movement of seeds in arable soil by different soil cultivation methods. *J. Appl. Ecol.* **1999**, *36*, 443–454. [CrossRef]
15. Kaur, M.; Sandhu, M.; Mohan, N.; Sandhu, P.S. Technology principles, advantages, limitations and its applications. *Int. J. Comput. Electr. Eng.* **2011**, *3*, 1793–8163. [CrossRef]
16. Want, R. An introduction to RFID technology. *IEEE Pervasive Comput.* **2006**, *5*, 25–33. [CrossRef]
17. Nikitin, P.V.; Rao, K.V.S. Theory and measurement of backscattering from RFID tags. *IEEE Antennas Propag. Mag.* **2006**, *48*, 212–218. [CrossRef]
18. Tian, F. An agri-food supply chain traceability system for China based on RFID & blockchain technology. In Proceedings of the 13th International Conference on Service Systems and Service Management (ICSSSM), Kunming, China, 24–26 June 2016.
19. Vanderroost, M.; Ragaert, P.; Devlieghere, F.; de Meulenaer, B. Intelligent food packaging: The next generation. *Trends Food Sci. Technol.* **2014**, *39*, 47–62. [CrossRef]
20. Prinsloo, J.; Malekian, R. Accurate vehicle location system using RFID, an internet of things approach. *Sensors* **2016**, *16*, 825. [CrossRef]
21. Abdullah, M.F.F.; Ali, M.T.B.; Yusof, F.Z.M. Rfid application development for a livestock monitoring system. In *Bioresources Technology in Sustainable Agriculture*; Apple Academic Press: Palm Bay, FL, USA, 2018; pp. 103–116.

22. Parsons, A.; Cooper, J.; Onda, Y.; Sakai, N. Application of RFID to soil-erosion research. *Appl. Sci.* **2018**, *8*, 2511. [CrossRef]
23. Rainato, R.; Mao, L.; Picco, L. Near-bankfull floods in an Alpine stream: Effects on the sediment mobility and bedload magnitude. *Int. J. Sediment. Res.* **2018**, *33*, 27–34. [CrossRef]
24. Picco, L.; Tonon, A.; Rainato, R.; Lenzi, M.A. Bank erosion and large wood recruitment along a gravel bed river. *J. Agric. Eng.* **2016**, *47*, 72. [CrossRef]
25. Le Breton, M.; Baillet, L.; Larose, E.; Rey, E.; Benech, P.; Jongmans, D.; Guyoton, F.; Jaboyedoff, M. Passive radio-frequency identification ranging, a dense and weather-robust technique for landslide displacement monitoring. *Eng. Geol.* **2019**, *250*, 1–10. [CrossRef]
26. Liu, J.; Chen, Y.; Kushwaha, R.L. Effect of tillage speed and straw length on soil and straw movement by a sweep. *Soil Tillage Res.* **2010**, *109*, 9–17. [CrossRef]
27. Liu, J.; Chen, Y.; Lobb, D.A.; Kushwaha, R.L. Soil-straw-tillage tool interaction: Field and soil bin study. *Can. Biosyst. Eng. Genie Biosyst. Can.* **2007**, *49*, 2.

Predicting Forage Quality of Warm-Season Legumes by Near Infrared Spectroscopy Coupled with Machine Learning Techniques

Gurjinder S. Baath [1,*], Harpinder K. Baath [2], Prasanna H. Gowda [3], Johnson P. Thomas [2], Brian K. Northup [4], Srinivas C. Rao [4] and Hardeep Singh [1]

[1] Department of Plant and Soil Sciences, Oklahoma State University, 371 Agricultural Hall, Stillwater, OK 74078, USA; hardeep.singh@okstate.edu
[2] Department of Computer Science, Oklahoma State University, 219 MSCS, Stillwater, OK 74078, USA; hbaath@okstate.edu (H.K.B.); jpthomas@okstate.edu (J.P.T.)
[3] USDA-ARS, Southeast Area Branch, 114 Experiment Station Road, Stoneville, MS 38776, USA; prasanna.gowda@usda.gov
[4] USDA-ARS, Grazinglands Research Laboratory, 7207 W. Cheyenne St., El Reno, OK 73036, USA; brian.northup@usda.gov (B.K.N.); srinivas.rao@usda.gov (S.C.R.)
[*] Correspondence: gbaath@okstate.edu.

Abstract: Warm-season legumes have been receiving increased attention as forage resources in the southern United States and other countries. However, the near infrared spectroscopy (NIRS) technique has not been widely explored for predicting the forage quality of many of these legumes. The objective of this research was to assess the performance of NIRS in predicting the forage quality parameters of five warm-season legumes—guar (*Cyamopsis tetragonoloba*), tepary bean (*Phaseolus acutifolius*), pigeon pea (*Cajanus cajan*), soybean (*Glycine max*), and mothbean (*Vigna aconitifolia*)—using three machine learning techniques: partial least square (PLS), support vector machine (SVM), and Gaussian processes (GP). Additionally, the efficacy of global models in predicting forage quality was investigated. A set of 70 forage samples was used to develop species-based models for concentrations of crude protein (CP), acid detergent fiber (ADF), neutral detergent fiber (NDF), and in vitro true digestibility (IVTD) of guar and tepary bean forages, and CP and IVTD in pigeon pea and soybean. All species-based models were tested through 10-fold cross-validations, followed by external validations using 20 samples of each species. The global models for CP and IVTD of warm-season legumes were developed using a set of 150 random samples, including 30 samples for each of the five species. The global models were tested through 10-fold cross-validation, and external validation using five individual sets of 20 samples each for different legume species. Among techniques, PLS consistently performed best at calibrating (R^2_c = 0.94–0.98) all forage quality parameters in both species-based and global models. The SVM provided the most accurate predictions for guar and soybean crops, and global models, and both SVM and PLS performed better for tepary bean and pigeon pea forages. The global modeling approach that developed a single model for all five crops yielded sufficient accuracy (R^2_{cv}/R^2_v = 0.92–0.99) in predicting CP of the different legumes. However, the accuracy of predictions of in vitro true digestibility (IVTD) for the different legumes was variable (R^2_{cv}/R^2_v = 0.42–0.98). Machine learning algorithms like SVM could help develop robust NIRS-based models for predicting forage quality with a relatively small number of samples, and thus needs further attention in different NIRS based applications.

Keywords: partial least square; support vector machine; Gaussian processes; soybean; pigeon pea; guar; tepary bean

1. Introduction

Perennial warm-season grasses, such as bermudagrass (*Cynodon dactylon*), old world bluestems (*Bothriochloa* spp.), and bahiagrass (*Paspalum notatum*), serve as major summer forage resources for grazing stocker cattle in the southern United States (US). While capable of producing large amounts of biomass, these perennial grasses often show a decline in forage quality with their maturation towards the mid-late summer growing season and do not meet the nutritional needs of grazing stocker cattle for the entire season [1,2]. Legumes, being high-quality forages, can be adopted to offset the summer slump in forage quality, and enhance the efficiency of forage-based beef production systems. Further, the continued increase in the cost of nitrogen fertilizers has added to the interest of producers in utilizing legume crops as forage in many regions across the US. In response, extensive research in the southern US over the last decade has focused on evaluating warm-season annual legumes as summer forage resources that can be grown in rotation with winter-wheat (*Triticum aestivum* L.) [3–6]. In more recent years, several legumes have received increased attention due to their capabilities of generating high amounts of biomass under the limited moisture conditions that prevail in the southern US [7].

Quantifying the quality of forage in pastures is crucial for both agriculture research and forage management, including cattle grazing and harvesting. However, the determination of the different parameters of forage quality, such as crude protein (CP), neutral detergent fiber (NDF), acid detergent fiber (ADF), and in vitro true digestibility (IVTD), by classical analytical techniques is time-consuming and expensive, especially when numerous samples are required. The vast evolution of computers and multivariate statistical techniques has enabled the use of near infrared spectroscopy (NIRS) in assessing the quality parameters of many forages. The NIRS method is quick, inexpensive, and facilitates timely decision-making related to grazing periods. The technique is based on interactions between light reflectance in the wavelength ranging between 750–2500 nm and organic compounds in the plant biomass [8]. The method of applying NIRS to predict forage quality involves analyzing a particular forage with both traditional lab analysis and NIRS, and then developing a predictive equation by pairing the information in a calibration dataset (Figure 1). The NIRS has been widely used in forage quality predictions of crops including alfalfa (*Medicago sativa*) [9], maize (*Zea mays*) [10], ryegrass (*Lolium multiflorum*) [11], tall fescue (*Festuca arundinacea*) [12], and other species. However, the technique has been underutilized to provide predictions of forage quality for many warm-season legumes.

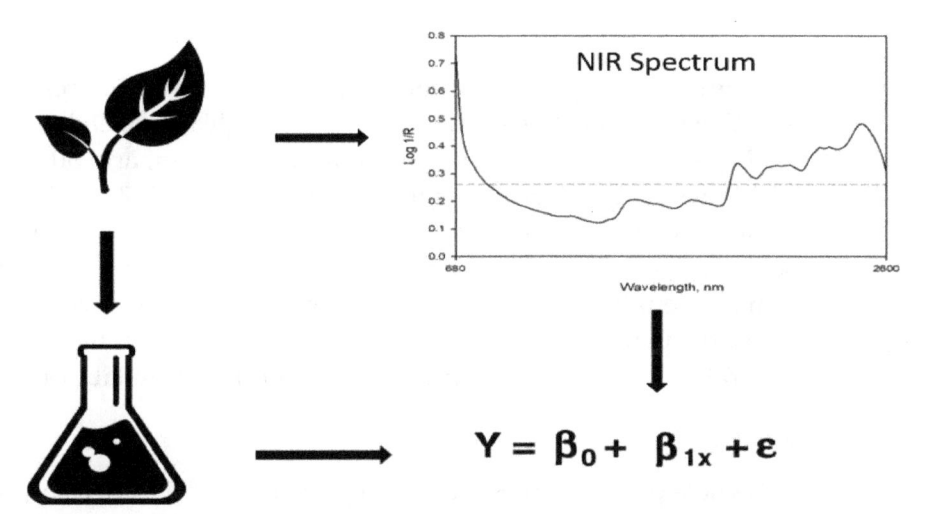

Figure 1. Illustration of the procedure used for applying near infrared spectroscopy (NIRS) technique in forage quality predictions.

As developed for other forage crops, well-calibrated NIRS species-based models for warm-season legumes could be useful tools to quickly asses the forage quality of different legume species grown under a range of environmental or management settings, or harvested at different stages of growth, or cutting or grazing height. Therefore, it is necessary to examine the effectiveness of NIRS in predicting forage characteristics of some important warm-season legumes. This work includes species such as guar (*Cyamopsis tetragonoloba*), tepary bean (*Phaseolus acutifolius*), soybean (*Glycine max*), and pigeon pea (*Cajanus cajan*), given past research, and the potential for expansion of use of these species across the southern US and other similar environments. There are also other warm-season legumes that may be capable of providing high-quality forage for summer grazing [3,13,14]. However, developing NIR calibration equations for every species can become challenging for public or private laboratories that test forage quality. Generally, accurate chemical analyses of a large number of samples is not readily available or feasible to develop calibrations, especially when novel legume species are involved. In response to challenges related to developing species-based calibrations, global models developed from samples of ranges of different warm-season legumes can prove useful, if such calibrations provide sufficiently accurate predictions.

Several calibration techniques are known to perform well in the application of NIRS in estimating forage quality and are generally available in most chemometric packages [15]. Partial least squares (PLS) is among the most commonly used methods, where least square algorithms are used to compute regressions [16]. In contrast, a comparatively novel and robust machine learning algorithm, support vector machine (SVM), has been gaining attention for NIRS calibrations [15]. Further, the Gaussian processes (GP) have provided better calibration results than PLS and SVM, in some cases [17,18]. However, tests of these calibration techniques on wide ranges of common and more novel legumes are required to define their function.

The combination of NIRS and machine learning calibration techniques could serve as an effective tool to streamline the monitoring efforts in warm-season legumes by eliminating the need for classical forage analytical methods. Therefore, the objective of this research was to evaluate the performance of NIRS in predicting the forage quality of four warm-season legumes (guar, tepary bean, pigeon pea, and soybean), using three different calibration techniques—PLS, SVM, and GP—on individual species bases. Additionally, the efficacy of global calibrations of these techniques, developed by combining datasets of all four species and mothbean (*Vigna aconitifolia*), was tested using different independent datasets of five species.

2. Materials and Methods

2.1. Materials

Forage samples used in the study ($n = 410$) were collected as parts of two different field experiments conducted at the USDA-ARS Grazinglands Research Laboratory near El Reno, Oklahoma, US (35.57° N, 98.03° W, elevation 414 m). Ninety samples each for guar and tepary bean, and 50 mothbean samples were collected from field experiments conducted during the summers of 2017 and 2018. An additional 90 samples of both soybean and pigeon pea were obtained from two long-term experiments (2001–2008) conducted in the same location [3,19]. In all three experiments, aboveground biomass was collected from randomly clipped 0.5 m row lengths from experimental plots at 15-day intervals, starting at 45 days after planting. Apart from whole plant samples, a major proportion of collected biomass samples in these experiments were separated into leaf, stem, and pods fractions before laboratory analysis.

2.2. Laboratory and NIRS Analysis

All leaf, stem, pod, and whole plant samples were oven-dried at 60 °C until a constant weight. Dry samples were ground to pass a 2-mm filter using a Wiley grinding mill. Total nitrogen concentration in each sample was determined by the flash combustion method (Model Vario Macro, Elementar Americas, Inc., Mt. Laurel, NJ, USA) and then converted into CP by multiplying with a factor of 6.25 (Table 1). The IVTD was obtained for each sample by following the Daisy Digester procedures (ANKOM Technology, Macedon, NY, USA). The NDF and ADF concentrations were only determined in samples of guar and tepary bean, in accordance with the batch fiber analyzer techniques (ANKOM Technology, Macedon, NY, USA).

Table 1. Summary statistics of lab datasets used for calibration, cross-validation and external validation of crude protein (CP), neutral detergent fiber (NDF), acid detergent fiber (ADF), and in vitro true digestibility (IVTD) of four warm season legumes.

Species	Parameter	Calibration and Cross-Validation (n = 70)				External Validation (n = 20)			
		Min	Max	Mean	SD	Min	Max	Mean	SD
		———————————— (%) ————————————								
Guar	CP	3.94	34.87	17.66	8.66	3.69	33.56	15.07	9.50
	NDF	16.83	70.80	37.57	16.95	22.95	75.78	45.94	17.82
	ADF	8.90	58.39	27.19	15.29	12.79	62.93	34.70	16.57
	IVTD	40.35	95.22	79.27	14.11	42.96	94.37	73.38	16.08
Tepary bean	CP	4.50	31.12	15.76	7.78	5.94	30.25	19.35	8.13
	NDF	22.90	71.57	48.34	12.31	25.52	60.95	43.90	10.21
	ADF	15.32	59.16	34.92	11.99	17.08	48.16	30.36	10.39
	IVTD	55.88	93.16	75.50	10.83	60.23	92.56	81.34	8.56
Soybean	CP	4.15	39.75	21.16	11.03	6.31	36.12	19.73	8.94
	IVTD	42.45	99.30	78.25	16.28	57.66	98.31	80.21	12.38
Pigeon pea	CP	4.52	32.48	16.30	8.77	6.24	28.64	15.62	7.41
	IVTD	30.71	91.08	61.55	19.28	33.31	82.89	59.76	16.40

n, number of samples; Min, minimum value; Max, maximum value; SD, standard deviation.

Aliquots of ground samples were filled into ring cups to eliminate voids. Spectral reflectance (R) of monochromatic light, averaged over 10 spectra per sample, were collected by scanning spectrophotometer (Model SpectraStar 2600 XT-R, Unity Scientific, Columbia, MD, USA). Spectral data were obtained as the logarithm of the inverse of reflectance [log(1/R)] at 1-nm interval over the range of 680–2600 nm.

2.3. Calibration Techniques

Partial least squares (PLS) is an extensively used class of statistical methods, which includes regression, classification, and dimension reduction techniques. It uses latent variables, also called score vectors, to model the relationship between input and response variables. In the case of regression problems, PLS first generates the latent variables from the given data and uses them as new predictor variables. There are different types of PLS, based on techniques employed to extract the latent variables. Two approaches are used to extend PLS for modeling non-linear relations among data. The first approach is to reformulate the linear relationship between score vectors, u and v, by a non-linear model:

$$v = g(u) + h = g(X, w) + h \qquad (1)$$

where g is the continuous function that models the existing non-linear relation. Generally, g is modeled using artificial neural networks, smoothing splines, polynomial, or radial basis functions. Remaining variables h and w denote a residual vector and a weight vector, respectively.

The second PLS approach is to apply kernel-based learning. The kernel PLS method transforms the input space data to higher dimensional feature space and linearly estimates PLS in that space. To avoid the mapping function Φ from projecting data to feature space, PLS applies the kernel trick which uses the fact that a value of the inner product of two vectors x and y in feature space can be calculated using a kernel function $k(x, y)$ [20]:

$$k(x, y) = \Phi(x)^T \Phi(y) \qquad (2)$$

By using the kernel function, score vectors (u and v) can be identified and used to define the non-linear relationship. The kernel PLS approach is used to model complex non-linear relations easily in terms of implementation and computation.

Gaussian processes (GP) are kernel-based, probabilistic, non-parametric regression models. A Gaussian process involves a set of random variables such that every finite number of those variables possess joint Gaussian distributions. A Gaussian process, $f(x)$, can be described using a mean function $m(x)$ and a covariance function $k(x, x')$. The covariance function defines the smoothness of responses, and the basis function Φ projects the input space vector x to a higher dimension feature space vector $\Phi(x)$. A Gaussian process regression (GPR) model describes the response by using latent variables from a Gaussian process. A GPR model is represented as:

$$\Phi(x_i)^T w + f(x) \tag{3}$$

where $f(x) \sim GP(0, k(x, x'))$, and $f(x)$ are from a zero mean GP having a covariance function, $k(x, x')$ [21]. The covariance is specified by kernel parameters, which are also known as hyperparameters. GPR is a probabilistic model, and an instance of response y is:

$$P(y_i | f(x_i), x_i) \sim N(y_i | \Phi(x_i)^T w + f(x_i), \sigma^2) \tag{4}$$

GPR is non-parametric as there is a latent variable $f(x_i)$ for each observation x_i. Noise variance σ^2, basis function coefficients w, and hyperparameters of the kernel can be estimated from the data while training the GPR model.

Support vector machine (SVM) is a popular machine-learning algorithm used for identifying linear as well as non-linear dependency between input vectors and outputs. SVMs are non-parametric models, which means parameters are selected, estimated, and tuned in such a way that the model capacity matches the data complexity [21]. Generally, SVM starts by observing the multivariate inputs X and outputs Y, estimates its parameters w, and then learns the performed mapping function $y = f(x, w)$, which approximates the underlying dependency between inputs and responses. The obtained function, also known as a hyperplane, must have a maximal margin (for classification) or the error of approximation (for regression) to predict the new data. In the case of SVM regression, Vapnik's error (loss) function is used with ε-insensitivity. It finds a regression function $f(x)$ that deviates from the actual responses (y) by values no more than ε and is considerably flat at the same time.

For non-linear regression problems, SVM maps the input space to feature space (a higher dimension space) using a mapping $\Phi(x)$ to find a linear regression hyperplane in that space. However, there is no need to know the mapping Φ, as the kernel function $k(x_i, x_j)$, which is the inner product of the vectors $\Phi(x_i)$ and $\Phi(x_j)$, can be used to find the optimal regression hyperplane in extended space. There are many kernel functions available to describe non-linear regressions, such as the polynomial kernel, RBF kernel, Gaussian Kernel, normalized polynomial kernel, etc. The learning problem in classification as well as in regression, leads to solving the quadratic programming (QP). The sequential minimal optimization (SMO) is considered as the most popular optimizer for solving SVM problems [22]. It divides the large QP problem into a set of small QP problems and analytically solves them.

2.4. Performance Evaluation

Apart from calibration, 10-fold cross-validations and external validations were conducted to assess the performance of the calibration techniques. The 10-fold cross-validation is a unique statistical way of performance evaluations of machine learning models in which ten repeated hold-out executions are obtained and averaged. In each execution, the model is trained with 90% of the data points and tested with the remaining 10%, and thus every data point is taken nine times for training and once for testing the model. For each species-based model, the original dataset of 90 samples for each species was split into two subsets (Figure 2). A subset of 70 samples was used for running calibration and 10-fold cross-validation. The other subset of 20 remaining samples was used only for external validation and neither used in calibration nor cross-validation of any model. For the global model, the original dataset consisted of 250 samples, involving 50 samples each of guar, tepary bean, soybean, pigeon pea, and mothbean. These samples were divided into six subsets (Figure 2). One random subset of 150 samples (30 samples per species) was employed for calibration as well as 10-fold cross-validation. Each of the remaining five subsets, comprising 20 samples of individual species, was used for external validation.

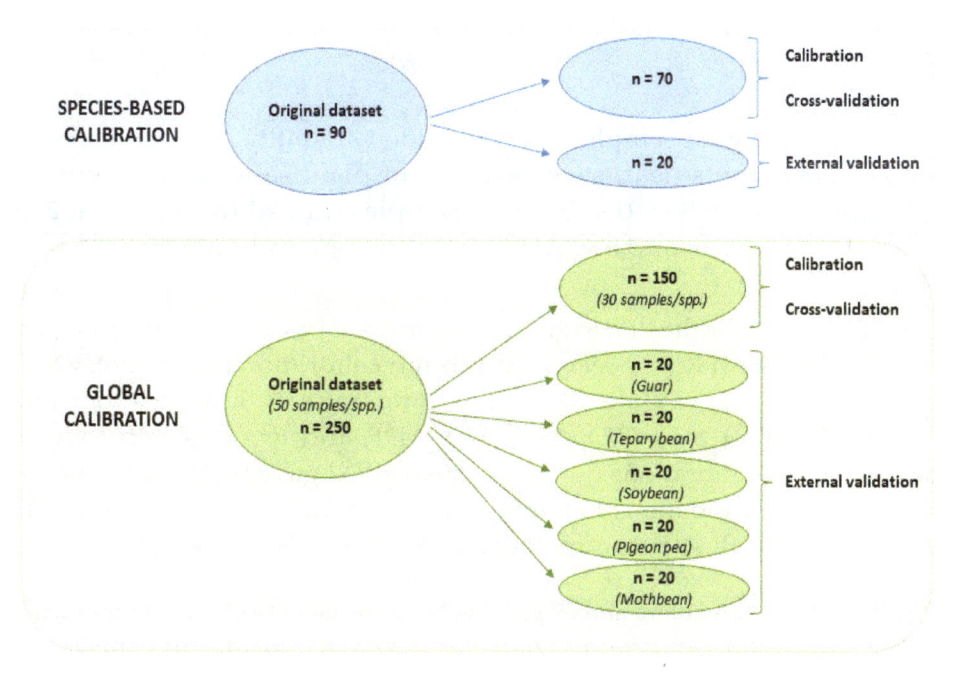

Figure 2. Diagram of the datasets, calibration, and different validation processes used in two calibration strategies.

Coefficients of determination, being upper-bounded by 1.0, are often adopted for meaningful comparisons across different models and therefore was used here as an estimate of prediction accuracy. To be precise, coefficient of determination in calibration (R^2_c), coefficient of determination in cross-validation (R^2_{cv}), and coefficient of determination in validation (R^2_v) were used for direct computation of the variance in the data captured at calibration, cross-validation, and external validation, respectively by each model. Additionally, root mean squared error estimation was also presented for comparing models, which were termed as $RMSE_c$, $RMSE_{cv}$, and $RMSE_v$ for calibration, cross-validation, and external validation, respectively.

2.5. Software

Regression models were calibrated, cross-validated, and externally validated using the *Weka* software, version 3.8 [23]. Weka is a suite of machine learning algorithms and is widely used for data mining. For implementing PLS, we used the PLS classifier package in Weka, which uses the prediction capabilities of PLSFilter. The PLSFilter runs the PLS regression on the given set of data and computes the beta matrix for prediction. By default, missing values are replaced, and the data are centered. For GP implementation, the Gaussian classifier for regression without hyperparameter-tuning was used. The kernel for the Gaussian classifier was configured as a polynomial. By default, missing values were replaced by the global mean. The SMOReg classifier was used to implement SVM in Weka. The classifier used the polynomial kernel and RegSMOImproved optimizer to learn SVM for regression. All remaining parameters, such as batch size, debugging, and filter type, which do not check capabilities, noise, etc., were kept as default.

3. Results and Discussion

The prediction accuracy of calibrated models is discussed by comparing their cross-validation (R^2_{cv}) and external validation (R^2_v) results to a scale proposed for NIRS calibrations [24]. According to the scale, the performance of a model is considered excellent if the R^2 of validations is greater than

0.95, and the resultant model can be used in any application. A model is assumed satisfactory with R^2 ranging from 0.9–0.95 and would be usable for most applications involving quality assurance. Models with R^2 ranging between 0.8–0.9 are considered moderately successful and can be used with caution for most applications, including research.

3.1. Guar

The chemical analysis of guar samples showed wide variability in parameters that define forage quality for different components (leaf, stem, or pod) of plant sampled at different growth stages (Table 1). The CP content for all 90 (70 + 20) guar samples ranged from 3.7% to 34.9%, while NDF concentrations ranged from 16.8% to 75.8%, ADF concentrations ranged from 8.9% to 62.9%, and IVTD from 40.3% to 95.2%.

Among the three techniques, the PLS technique performed best at calibrating each of the four forage quality parameters in guar with R^2_c of 0.98–0.99, though calibration results of SVM (R^2_c = 0.94–0.98) were also comparable (Table 2). While GP had a comparatively lower calibration accuracy with R^2_c ranging between 0.88–0.91 for IVTD, NDF and ADF, and R^2_c of 0.95 for CP of guar samples. Although PLS provided best calibrations out of the three, SVM gave better prediction accuracy in both cross-validation and external validation of all four indices of forage quality for guar. Thee GP approach generated the lowest R^2_{cv} for all four parameters and R^2_v for NDF and ADF.

Table 2. Calibration, cross-validation, and external validation statistics obtained for crude protein (CP), neutral detergent fiber (NDF), acid detergent fiber (ADF), and in vitro true digestibility (IVTD) in guar using three calibration techniques.

Parameter	Method	Calibration (n = 70) R^2_c	$RMSE_c$	Cross-Validation (n = 70) R^2_{cv}	$RMSE_{cv}$	External Validation (n = 20) R^2_v	$RMSE_v$
CP	GP	0.95	1.84	0.93	2.20	0.96	2.12
	PLS	0.99	0.78	0.95	1.97	0.93	2.52
	SVM	0.98	1.23	0.97	1.56	0.98	1.27
NDF	GP	0.90	5.53	0.84	6.73	0.90	6.98
	PLS	0.98	2.17	0.85	6.66	0.93	5.52
	SVM	0.94	3.98	0.91	5.08	0.94	4.67
ADF	GP	0.91	4.79	0.86	5.77	0.92	6.02
	PLS	0.99	1.18	0.95	3.36	0.94	4.23
	SVM	0.97	2.46	0.95	3.51	0.96	3.78
IVTD	GP	0.88	4.92	0.81	6.10	0.93	5.63
	PLS	0.98	2.15	0.81	6.69	0.87	5.66
	SVM	0.94	3.51	0.83	5.88	0.94	4.19

GP, Gaussian processes; PLS; partial least square; SVM, support vector machine; R^2_c, determination coefficient in calibration; $RMSE_c$, root mean square error in calibration; R^2_{cv}, determination coefficient in cross-validation; $RMSE_{cv}$, root mean square error in cross-validation; R^2_v, determination coefficient in external validation; $RMSE_v$, root mean square error in external validation.

Among forage quality parameters, the greatest prediction accuracy was recorded for CP by all three techniques with R^2_{cv} of 0.93–0.97 and R^2_v of 0.93–0.98 (Table 2). In comparison, only the SVM technique resulted in a satisfactory prediction accuracy (R^2_{cv} = 0.92; R^2_v = 0.94) for NDF, based on the proposed scale [24]. Both the SVM and PLS techniques showed excellent accuracy at predicting ADF with R^2_{cv} and R^2_v between 0.94–0.96, while GP produced R^2_{cv} of 0.86. All three techniques resulted in relatively low prediction accuracy for IVTD, with R^2_{cv} ranging from 0.81–0.83. Overall, performances of SVM was most satisfactory among the three calibration methods, and it can be employed in NIRS-based prediction of CP, ADF, and NDF of guar. In contrast, use of IVTD predictions of guar would require caution, based on the type of application.

While currently a minor crop in the southern US, guar has a proven potential to serve as a multi-purpose legume and has potential for expansion in use. Guar is a common crop in regions of the Indian subcontinent, Africa, North and South America, and Australia [25]. Guar has been gaining attention as a forage resource in the southern US due to its capability of producing high N biomass under limited water conditions [3,5]. Therefore, this first report investigating the application of NIRS in guar would encourage the utilization of the technique in its research and forage management.

3.2. Tepary Bean

Results from the laboratory analysis of tepary bean samples showed high variability in all four of the quality indices, though the observed ranges were narrower than guar (Table 1). The concentration of CP varied from 4.5–31.1%, while NDF ranged from 22.9% to 71.6%. In contrast, ADF and IVTD ranged between 15.3–59.2% and 55.9–93.2%, respectively. Best calibration results for tepary bean were recorded using the PLS technique, with R^2_c of 0.98–0.99 (Table 3). Whereas, neither SVM nor PLS clearly resulted in better predictions for all quality indices when cross-validated and externally validated.

Table 3. Calibration, cross-validation, and external validation statistics obtained for crude protein (CP), neutral detergent fiber (NDF), acid detergent fiber (ADF), and in vitro true digestibility (IVTD) in tepary bean using three calibration techniques.

Parameter	Method	Calibration ($n = 70$)		Cross-Validation ($n = 70$)		External Validation ($n = 20$)	
		R^2_c	$RMSE_c$	R^2_{cv}	$RMSE_{cv}$	R^2_v	$RMSE_v$
CP	GP	0.94	1.89	0.90	2.42	0.94	2.20
	PLS	0.99	0.68	0.93	2.03	0.98	1.35
	SVM	0.97	1.35	0.95	1.74	0.94	1.94
NDF	GP	0.85	4.96	0.75	6.22	0.75	5.10
	PLS	0.98	1.64	0.84	5.09	0.75	5.53
	SVM	0.94	2.97	0.72	7.01	0.84	4.03
ADF	GP	0.87	4.60	0.78	5.62	0.86	3.90
	PLS	0.98	1.47	0.89	3.97	0.92	3.34
	SVM	0.96	2.45	0.86	4.52	0.95	2.23
IVTD	GP	0.87	4.02	0.75	5.39	0.75	4.25
	PLS	0.98	1.55	0.79	5.00	0.88	2.89
	SVM	0.93	2.86	0.75	5.70	0.82	3.82

GP, Gaussian processes; PLS; partial least square; SVM, support vector machine; R^2_c, determination coefficient in calibration; $RMSE_c$, root mean square error in calibration; R^2_{cv}, determination coefficient in cross-validation; $RMSE_{cv}$, root mean square error in cross-validation; R^2_v, determination coefficient in external validation; $RMSE_v$, root mean square error in external validation.

All calibration techniques showed best results at predicting CP in tepary bean samples with a R^2_{cv} or R^2_v above 0.90 among the forage quality characteristics (Table 3). The SVM technique resulted in the lowest $RMSE_{cv}$ value (1.74) for cross-validation of CP, whereas PLS had the lowest $RMSE_v$ of 1.35 for external validation among the three techniques. In contrast, PLS showed the lowest $RMSE_{cv}$ values of 5.09 and 3.97 and SVM had the lowest $RMSE_v$ of 4.03 and 2.23 for NDF and ADF, respectively. Both PLS and SVM produced satisfactory results at predicting ADF concentration in tepary bean with R^2_{cv} of 0.86–0.89 and R^2_v of 0.92–0.95 compared to GP, while all three techniques had comparatively low performance at predicting NDF in tepary bean with R^2_{cv} and R^2_v of 0.72–0.84 and 0.75–0.84, respectively.

In comparison to ADF, the NDF concentration of tepary samples were less accurately predicted by all three techniques (Table 3). Similar differences between prediction accuracy of ADF and NDF were also noticed for guar in this study, and also reported earlier in NIRS studies involving *Brassica napus* [26], *Lolium multiflorum* [11], and *Oryza sativa* [27]. Though PLS performed better at predicting IVTD in tepary bean compared to other two, all three techniques resulted in relatively low prediction accuracy with R^2_{cv} and R^2_v ranging between 0.75–0.79 and 0.75–0.88, respectively. Overall, both PLS and SVM could be considered as good among three tested techniques and hence can be employed for satisfactory predictions of CP and ADF in tepary bean. While prediction results of NDF and IVTD would need some caution if calibrations are developed with similar sample sizes ($n = 70$) as used in this study.

Tepary bean is a vining, warm-season legume species originated from the areas of the southwestern United States and northwestern Mexico, that may have value for multiple uses in dryland agricultural systems. Due to its spreading growth habit, and the ability to generate high N biomass with limited soil moisture, tepary bean could be an ideal summer forage for the Southern Great Plains [14]. This first study investigating the application of NIRS to attributes of forage quality in tepary bean showed that the technique could aid in quantifying its role in meeting animal nutrition needs.

3.3. Soybean

All three techniques (PLS, SVM, and GP) gave excellent accuracies at calibrating CP and IVTD of soybean samples with PLS again performing the best out of three with a R^2_c greater than 0.98 (Table 4). Among three techniques, SVM performed best at predicting CP with $RMSE_{cv}$ and $RMSE_v$ of 1.85 and 1.78, respectively, followed closely by PLS. All three calibration techniques produced better predictions of IVTD in soybean ($R^2_{cv} > 0.84$ and $R^2_v > 0.89$), compared to prediction accuracies obtained for guar and tepary bean. As observed for CP, SVM performed better than the other techniques in cross-validation ($R^2_{cv} = 0.89$) of IVTD, while the other two techniques performed better in external validation (R^2_v of 0.92–0.93). All three techniques can be employed for rapid NIR-based predictions of CP and IVTD in soybean forage samples, with SVM would be the best choice.

Table 4. Calibration, cross-validation, and external validation statistics obtained for crude protein (CP) and in vitro true digestibility (IVTD) in soybean using three calibration techniques.

Parameter	Method	Calibration ($n = 70$)		Cross-Validation ($n = 70$)		External Validation ($n = 20$)	
		R^2_c	$RMSE_c$	R^2_{cv}	$RMSE_{cv}$	R^2_v	$RMSE_v$
CP	GP	0.92	4.63	0.87	5.78	0.92	3.78
	PLS	0.98	2.16	0.84	6.92	0.93	3.46
	SVM	0.94	3.92	0.89	5.28	0.89	4.09
IVTD	GP	0.96	2.14	0.94	2.71	0.92	2.53
	PLS	0.99	0.80	0.96	2.05	0.94	2.24
	SVM	0.99	1.26	0.97	1.85	0.96	1.78

GP, Gaussian processes; PLS; partial least square; SVM, support vector machine; R^2_c, determination coefficient in calibration; $RMSE_c$, root mean square error in calibration; R^2_{cv}, determination coefficient in cross-validation; $RMSE_{cv}$, root mean square error in cross-validation; R^2_v, determination coefficient in external validation; $RMSE_v$, root mean square error in external validation.

Soybean was initially introduced as a forage into the US in the 19th Century, but is now one of the most widely grown grain legumes in the Southern Great Plains [14]. In the last two decades, there has been increased interest from researchers in utilizing soybean as a summer forage in the US [28–30]. Hence the need for rapid and low-cost techniques for estimating forage quality. The NIRS technique has not been exploited for forage quality predictions in soybean. A single report investigated modified PLS and multiple scatter correction methods for NIR predictions of CP, NDF, and ADF concentrations, using 353 soybean samples collected at one (R6) growth stage [31]. In comparison, calibrations developed in the present study, used data on IVTD and CP with just 70 soybean samples collected across a range of different growth stages. Thus, our observed ranges for CP (4.1–39.7) and IVTD (42.4–99.3%) were more diverse (Table 1). The accuracies (R^2_{cv} or $R^2_v > 0.92$) obtained in predicting CP in soybean forage by all three techniques were higher than the values reported [31], despite large differences in sample sizes (N = 70 vs. 353) used for developing calibrations. Therefore, this study showed machine learning algorithms could develop robust NIRS calibrations for precise analysis of forage quality of soybean with small sample sizes.

3.4. Pigeon Pea

Laboratory analyses for the current study showed wide variability in both CP (4.5–32.5%) and IVTD (30.7–91.1%) for forage samples of pigeon pea (Table 1). The CP concentration in pigeon pea was accurately calibrated ($R^2_c > 0.95$) by each of the three techniques (Table 5). All three techniques resulted in CP predictions with R^2_{cv} and R^2_v greater than 0.96. Both PLS and SVM also showed greater accuracies in predicting IVTD of pigeon pea with R^2_{cv} and R^2_v ranges of 0.91–0.92 and 0.96–0.97, respectively. Although lower than PLS and SVM, the performance ($R^2_{cv} = 0.86$) of GP-based calibrations were moderately satisfactory in IVTD predictions, following the proposed scale [24]. Overall, both PLS and SVM would provide excellent options for NIR predictions of CP and IVTD in pigeon pea.

Table 5. Calibration, cross-validation, and external validation statistics obtained for crude protein (CP) and in vitro true digestibility (IVTD) in pigeon pea using three calibration techniques.

Parameter	Method	Calibration ($n = 70$)		Cross-Validation ($n = 70$)		External Validation ($n = 20$)	
		R^2_c	$RMSE_c$	R^2_{cv}	$RMSE_{cv}$	R^2_v	$RMSE_v$
CP	GP	0.98	1.37	0.96	1.73	0.96	1.69
	PLS	1.00	0.43	0.97	1.46	0.98	1.02
	SVM	0.99	0.84	0.98	1.17	0.98	1.12
IVTD	GP	0.95	4.51	0.86	7.18	0.97	2.95
	PLS	0.99	1.93	0.92	5.49	0.96	3.09
	SVM	0.97	3.31	0.91	5.86	0.97	2.85

GP, Gaussian processes; PLS; partial least square; SVM, support vector machine; R^2_c, determination coefficient in calibration; $RMSE_c$, root mean square error in calibration; R^2_{cv}, determination coefficient in cross-validation; $RMSE_{cv}$, root mean square error in cross-validation; R^2_v, determination coefficient in external validation; $RMSE_v$, root mean square error in external validation.

Pigeon pea is another legume species that has seen the development of a range of cultivars for different uses in its home range, and areas of greater cultivation. This includes research on the value of cultivars of pigeon pea in the US for forage, grain, and pasture productivity [4,32]. Pigeon pea has a high degree of heat and drought tolerance, and the capacity for high levels of forage production in the US and other tropical and sub-tropical regions.

While pigeon pea is a broadly grown crop in much of the world, there was only one preliminary report that discussed the possible use of NIRS techniques to predict forage quality of pigeon pea [33]. That report used limited numbers of samples ($n = 48$), involving leaves and branches, that were mostly collected at one growth stage for calibrations of CP, NDF, and ADF concentrations; however, no validations were performed [33]. In contrast, the present study undertook both calibrations and validations using 90 (70 + 20) pigeon pea samples, involving leaves, stems, or seed pods, collected at different growth stages during a long-term experiment. Further, we investigated the NIR-based predictions of IVTD, which is assumed as an important quality parameter in pigeon pea forage [4]. Therefore, this study confirms that NIRS techniques could be effective tools for predicting forage quality of pigeon pea.

3.5. Global Calibrations

Global calibrations for CP and IVTD of warm-season legumes were developed with 150 samples, which included 30 samples each of guar, tepary bean, soybean, pigeon pea, and mothbean (Figure 2). As observed with the species-based calibrations for the four different legumes, the PLS technique performed best out of the three techniques for global calibrations both CP and IVTD (R^2_c of 0.97 and 0.94, respectively), while the GP technique was the least accurate (Table 6). In comparison, cross-validation of global models showed the SVM approach provided the greatest prediction accuracy for both CP (R^2_{cv} = 0.94) and IVTD (R^2_{cv} = 0.86), followed closely by PLS. Therefore, based on cross-validation results, the performance of global calibrations developed using SVM and PLS were satisfactory at predicting CP, and moderately satisfactory for IVTD.

When global calibrated models were validated using different external datasets for each of the five legume species, the predictions for CP by all three techniques resulted in sufficient accuracies with R^2_v ranging between 0.91–0.97 (Table 6). The SVM technique showed higher accuracy compared to the others in predicting CP, with the exception of guar, where the PLS approach provided slight improvements. Among species, the best CP predictions were noted for pigeon pea (R^2_v values of 0.98–0.99) for all three techniques. In contrast, IVTD predictions were not consistently accurate across all five species. The greatest accuracy was observed for IVTD predictions in pigeon pea with R^2_v of 0.97–0.98 under SVM and PLS. The lowest accuracy in predicting IVTD was noted for mothbean (R^2_v between 0.65–0.69 by SVM and PLS; 0.42 by GP). The best prediction accuracies for IVTD of soybean (R^2_v = 0.82–0.86 for all three techniques) and guar (PLS; R^2_v = 0.81) were moderately satisfactory. However, the performance of all three techniques was satisfactory at predicting IVTD in tepary bean (R^2_v of 0.91–92), which was better than the specific models developed for tepary bean (Table 3).

Overall, global-calibrated models for CP have the potential to offer sufficient prediction accuracies that are comparable to species-based calibration models. Diverse calibration sets that contain different legume species may allow the creation of robust, generalized models that provide predictions similar to species-based models. In some cases, global models may be capable of providing more accurate predictions, as was observed for IVTD predictions of tepary bean in this study.

The application of accurate globally calibrated models would be extremely useful for a broad range of end-users. They would reduce or eliminate the large amounts of time and other resources required to perform chemical analyses or the development and use of separate calibration sets for every species. However, adopting the global calibration approach for IVTD may not provide satisfactory predictions for all species. Some of the issues related to the low level of performance of calibrations for IVTD may be variability associated with using techniques that rely on rumen fluids in laboratory analyses [34]. Therefore, further investigations are required to compare the performance of global calibrations developed for IVTD of warm-season legumes derived using both rumen fluid and cellulose degradation methods.

Table 6. Calibration, cross-validation, and external (species) validation statistics of global models obtained for crude protein (CP) and in vitro true digestibility (IVTD) in warm-season legumes using three calibration techniques.

	Method	Calibration (n = 150)			Cross-Validation (n = 150)			External Validation (n = 20)								
								Guar		Tepary Bean		Soybean		Pigeon Pea		Mothbean	
		R^2_c	$RMSE_c$		R^2_{cv}	$RMSE_{cv}$		R^2_v	$RMSE_v$	R^2_v	$RMSE_v$	R^2_v	$RMSE_v$	R^2_v	$RMSE_v$	R^2_v	$RMSE_v$
CP	GP	0.92	2.15		0.89	2.46		0.93	2.42	0.95	2.72	0.91	3.95	0.98	2.21	0.94	3.36
	PLS	0.97	1.15		0.92	2.02		0.94	2.36	0.94	2.49	0.94	2.47	0.98	2.03	0.94	3.10
	SVM	0.96	1.48		0.94	1.87		0.92	2.77	0.95	2.36	0.94	3.16	0.99	1.29	0.97	2.54
IVTD	GP	0.86	5.09		0.81	5.84		0.65	6.16	0.91	4.41	0.82	7.93	0.91	4.75	0.42	5.40
	PLS	0.94	3.28		0.85	5.28		0.81	5.00	0.90	5.53	0.88	5.19	0.98	2.21	0.69	4.50
	SVM	0.91	3.98		0.86	4.98		0.77	5.12	0.92	4.74	0.86	5.60	0.97	2.77	0.65	4.29

GP, Gaussian processes; PLS; partial least square; SVM, support vector machine; R^2_c, determination coefficient in calibration; $RMSE_c$, root mean square error in calibration; R^2_{cv}, determination coefficient in cross-validation; $RMSE_{cv}$, root mean square error in cross-validation; R^2_v, determination coefficient in external validation; $RMSE_v$, root mean square error in external validation.

4. Conclusions

The statistics obtained for calibration, cross-validation, and external validation in this study demonstrated that NIRS techniques could be effective for supplying rapid and accurate predictions of most attributes of forage quality (cell wall fractions, crude protein) for different warm-season legumes. Further, the applications of NIRS technique to guar, tepary bean, and mothbean represent the first reports of such tools to provide estimates of forage quality for these species. Though similar to PLS, the SVM technique performed consistently well in predicting quality parameters of five warm-season legumes under both species-based and global calibration strategies. The global calibration approach can be a useful approach for predicting CP in warm-season legumes, and reduce the time and resources required for traditional chemical analysis in the use of separate calibration equations for each species. However, the global model for IVTD was not accurate for all species. Further model development based on other analytical procedures may improve the consistency and reliability of the global approach. Machine learning algorithms like SVM could also allow the development of robust models with a relatively small number of samples. Additional research is required to refine the SVM approach for different NIRS applications.

Author Contributions: Conceptualization, G.S.B., H.K.B. and P.H.G.; methodology, G.S.B., H.K.B., P.H.G. and J.P.T.; formal analysis, G.S.B. and H.K.B.; data curation, G.S.B., B.K.N. and S.C.R.; writing—original draft preparation, G.S.B. and H.K.B.; writing—review and editing, B.K.N., P.H.G., H.S. and J.P.T.; visualization, G.S.B. and H.K.B.; supervision, P.H.G. and J.P.T.; project administration, P.H.G. All authors have read and agreed to the published version of the manuscript.

Acknowledgments: The authors would like to acknowledge ARS technicians Cindy Coy, Delmar Shantz, Kory Bollinger, and Jeff Weik for their assistance in collecting, processing, and analyzing forage samples.

Disclaimer: Mention of trademarks, proprietary products, or vendors does not constitute guarantee or warranty of products by USDA and does not imply its approval to the exclusion of other products that may be suitable. All programs and services of the USDA are offered on a nondiscriminatory basis, without regard to race, color, national origin, religion, sex, age, marital status, or handicap.

Abbreviations

PLS, partial least square; SVM, support vector machine; GP, Gaussian processes; CP, crude protein; ADF, acid detergent fiber; NDF, neutral detergent fiber; IVTD, in vitro true digestibility; R^2_c, determination coefficient in calibration; R^2_{cv}, determination coefficient in cross-validation; R^2_v, determination coefficient in external validation; $RMSE_c$, root mean square error in calibration; $RMSE_{cv}$, root mean square error in cross-validation; $RMSE_v$, root mean square error in external validation.

References

1. Phillips, W.; Coleman, S. Productivity and economic return of three warm season grass stocker systems for the Southern Great Plains. *J. Prod. Agric.* **1995**, *8*, 334–339. [CrossRef]
2. Williams, M.; HaLmmond, A. Rotational vs. continuous intensive stocking management of bahiagrass pasture for cows and calves. *Agron. J.* **1999**, *91*, 11–16. [CrossRef]
3. Rao, S.C.; Northup, B.K. Capabilities of four novel warm-season legumes in the southern Great Plains: Biomass and forage quality. *Crop Sci.* **2009**, *49*, 1096–1102. [CrossRef]
4. Rao, S.C.; Northup, B.K. Pigeon pea potential for summer grazing in the southern Great Plains. *Agron. J.* **2012**, *104*, 199–203. [CrossRef]
5. Rao, S.C.; Northup, B.K. Biomass production and quality of indian-origin forage guar in Southern Great Plains. *Agron. J.* **2013**, *105*, 945–950. [CrossRef]
6. Foster, J.; Adesogan, A.; Carter, J.; Sollenberger, L.; Blount, A.; Myer, R.; Phatak, S.; Maddox, M. Annual legumes for forage systems in the United States Gulf Coast region. *Agron. J.* **2009**, *101*, 415–421. [CrossRef]
7. Baath, G.S.; Northup, B.K.; Gowda, P.H.; Turner, K.E.; Rocateli, A.C. Mothbean: A potential summer crop for the Southern Great Plains. *Am. J. Plant Sci.* **2018**, *9*, 1391. [CrossRef]

8. Rushing, J.B.; Saha, U.K.; Lemus, R.; Sonon, L.; Baldwin, B.S. Analysis of some important forage quality attributes of Southeastern Wildrye (Elymus glabriflorus) using near-infrared reflectance spectroscopy. *Am. J. Anal. Chem.* **2016**, *7*, 642. [CrossRef]
9. Brogna, N.; Pacchioli, M.T.; Immovilli, A.; Ruozzi, F.; Ward, R.; Formigoni, A. The use of near-infrared reflectance spectroscopy (NIRS) in the prediction of chemical composition and in vitro neutral detergent fiber (NDF) digestibility of Italian alfalfa hay. *Ital. J. Anim. Sci.* **2009**, *8*, 271–273. [CrossRef]
10. Volkers, K.; Wachendorf, M.; Loges, R.; Jovanovic, N.; Taube, F. Prediction of the quality of forage maize by near-infrared reflectance spectroscopy. *Anim. Feed Sci. Technol.* **2003**, *109*, 183–194. [CrossRef]
11. Yang, Z.; Nie, G.; Pan, L.; Zhang, Y.; Huang, L.; Ma, X.; Zhang, X. Development and validation of near-infrared spectroscopy for the prediction of forage quality parameters in Lolium multiflorum. *PeerJ* **2017**, *5*, e3867. [CrossRef] [PubMed]
12. Hill, N.; Cabrera, M.; Agee, C. Morphological and climatological predictors of forage quality in tall fescue. *Crop Sci.* **1995**, *35*, 541–549. [CrossRef]
13. Muir, J.P.; Pitman, W.D.; Dubeux Jr, J.C.; Foster, J.L. The future of warm-season, tropical and subtropical forage legumes in sustainable pastures and rangelands. *Afr. J. Range Forage Sci.* **2014**, *31*, 187–198. [CrossRef]
14. Baath, G.S.; Northup, B.K.; Rocateli, A.C.; Gowda, P.H.; Neel, J.P. Forage potential of summer annual grain legumes in the southern great plains. *Agron. J.* **2018**, *110*, 2198–2210. [CrossRef]
15. Agelet, L.E.; Hurburgh, C.R., Jr. A tutorial on near infrared spectroscopy and its calibration. *Crit. Rev. Anal. Chem.* **2010**, *40*, 246–260. [CrossRef]
16. Roggo, Y.; Chalus, P.; Maurer, L.; Lema-Martinez, C.; Edmond, A.; Jent, N. A review of near infrared spectroscopy and chemometrics in pharmaceutical technologies. *J. Pharm. Biomed. Anal.* **2007**, *44*, 683–700. [CrossRef]
17. Wang, K.; Chi, G.; Lau, R.; Chen, T. Multivariate calibration of near infrared spectroscopy in the presence of light scattering effect: A comparative study. *Anal. Lett.* **2011**, *44*, 824–836. [CrossRef]
18. Cui, C.; Fearn, T. Comparison of partial least squares regression, least squares support vector machines, and Gaussian process regression for a near infrared calibration. *J. Near Infrared Spectrosc.* **2017**, *25*, 5–14. [CrossRef]
19. Rao, S.; Mayeux, H.; Northup, B. Performance of forage soybean in the southern Great Plains. *Crop Sci.* **2005**, *45*, 1973–1977. [CrossRef]
20. Rosipal, R.; Kramer, N. Subspace, latent structure and feature selection techniques. *Lect. Notes Comput. Sci. Chap. Overv. Recent Adv. Part. Least Sq.* **2006**, *2940*, 34–51.
21. Williams, C.K.; Rasmussen, C.E. *Gaussian Processes for Machine Learning*; MIT Press: Cambridge, MA, USA, 2006; Volume 2.
22. Huang, C.-L.; Wang, C.-J. A GA-based feature selection and parameters optimizationfor support vector machines. *Expert Syst. Appl.* **2006**, *31*, 231–240. [CrossRef]
23. Platt, J. Probabilistic outputs for support vector machines and comparisons to regularized likelihood methods. *Adv. Large Margin Classif.* **1999**, *10*, 61–74.
24. Frank, E.; Hall, M.A.; Witten, I.H. *The WEKA Workbench; Online Appendix for "Data Mining: Practical Machine Learning Tools and Techniques"*, 4th ed.; Morgan Kaufmann: Cambridge, MA, USA, 2016.
25. Malley, D.; Martin, P.; Ben-Dor, E. Application in analysis of soils. In *Near-Infrared Spectroscopy in Agriculture*, 1st ed.; Roberts, C.A., Workman, J., Jr., Reeves, J.B., III, Eds.; American Society of Agronomy; Crop Science Society of America; Soil Science Society of America: Madison, WI, USA, 2004; pp. 729–783.
26. Baath, G.S.; Kakani, V.G.; Gowda, P.H.; Rocateli, A.C.; Northup, B.K.; Singh, H.; Katta, J.R. Guar responses to temperature: Estimation of cardinal temperatures and photosynthetic parameters. *Ind. Crop. Prod.* **2019**. [CrossRef]
27. Wittkop, B.; Snowdon, R.J.; Friedt, W. New NIRS calibrations for fiber fractions reveal broad genetic variation in Brassica napus seed quality. *J. Agric. Food Chem.* **2012**, *60*, 2248–2256. [CrossRef]
28. Kong, X.; Xie, J.; Wu, X.; Huang, Y.; Bao, J. Rapid prediction of acid detergent fiber, neutral detergent fiber, and acid detergent lignin of rice materials by near-infrared spectroscopy. *J. Agric. Food Chem.* **2005**, *53*, 2843–2848. [CrossRef]
29. Nielsen, D.C. Forage soybean yield and quality response to water use. *Field Crop. Res.* **2011**, *124*, 400–407. [CrossRef]

30. Beck, P.; Hubbell, D., III; Hess, T.; Wilson, K.; Williamson, J.A. Effect of a forage-type soybean cover crop on wheat forage production and animal performance in a continuous wheat pasture system. *Prof. Anim. Sci.* **2017**, *33*, 659–667. [CrossRef]
31. Asekova, S.; Han, S.-I.; Choi, H.-J.; Park, S.-J.; Shin, D.-H.; Kwon, C.-H.; Shannon, J.G.; LEE, J.D. Determination of forage quality by near-infrared reflectance spectroscopy in soybean. *Turk. J. Agric. For.* **2016**, *40*, 45–52. [CrossRef]
32. Rao, S.; Coleman, S.; Mayeux, H. Forage production and nutritive value of selected pigeonpea ecotypes in the southern Great Plains. *Crop Sci.* **2002**, *42*, 1259–1263. [CrossRef]
33. Berardo, N.; Dzowela, B.; Hove, L.; Odoardi, M. Near infrared calibration of chemical constituents of Cajanus cajan (pigeon pea) used as forage. *Anim. Feed Sci. Technol.* **1997**, *69*, 201–206. [CrossRef]
34. Roberts, C.A.; Stuth, J.; Flinn, P. Analysis of forages and feedstuffs. In *Near-Infrared Spectroscopy in Agriculture*, 1st ed.; Roberts, C.A., Workman, J., Jr., Reeves, J.B., III, Eds.; American Society of Agronomy; Crop Science Society of America; Soil Science Society of America: Madison, WI, USA, 2004; pp. 231–267.

Calibration and Validation of a Low-Cost Capacitive Moisture Sensor to Integrate the Automated Soil Moisture Monitoring System

Ekanayaka Achchillage Ayesha Dilrukshi Nagahage *, Isura Sumeda Priyadarshana Nagahage and Takeshi Fujino

Graduate School of Science and Engineering, Saitama University, 255 Shimo-Okubo, Sakura-ku, Saitama 338-8570, Japan
* Correspondence: ayesha@mail.saitama-u.ac.jp

Abstract: Readily available moisture in the root zone is very important for optimum plant growth. The available techniques to determine soil moisture content have practical limitations owing to their high cost, dependence on labor, and time consumption. We have developed a prototype for automated soil moisture monitoring using a low-cost capacitive soil moisture sensor (SKU:SEN0193) for data acquisition, connected to the internet. A soil-specific calibration was performed to integrate the sensor with the automated soil moisture monitoring system. The accuracy of the soil moisture measurements was compared with those of a gravimetric method and a well-established soil moisture sensor (SM-200, Delta-T Devices Ltd, Cambridge, UK). The root-mean-square error (RMSE) of the soil water contents obtained with the SKU:SEN0193 sensor function, the SM-200 manufacturer's function, and the SM-200 soil-specific calibration function were 0.09, 0.07, and 0.06 cm^3 cm^{-3}, for samples in the dry to saturated range, and 0.05, 0.08, and 0.03 cm^3 cm^{-3}, for samples in the field capacity range. The repeatability of the measurements recorded with the developed calibration function support the potential use of the SKU:SEN0193 sensor to minimize the risk of soil moisture stress or excess water application.

Keywords: calibration function; capacitive soil moisture sensor; internet-based data acquisition; soil moisture content

1. Introduction

Readily available soil moisture is a key requirement for the growth and development of plants and depends on the physical properties of the soil and the meteorological conditions of the surrounding environment. The upper and lower limits of the readily available soil moisture are known as the field capacity and the permanent wilting percentage, respectively. At the field capacity, sufficient water and air are retained in the soil, resulting in optimum plant growth. The effect of the meteorological conditions on soil moisture is minimal in indoor systems. However, the readily available soil moisture content varies throughout the soil, owing to the differences in transpiration and moisture loss from the soil, even in controlled environments [1]. Thus, continuous monitoring of the soil moisture content at different locations is required in indoor systems. These practices are costly, time-consuming, and labor-dependent.

There are several techniques to determine the soil moisture content, including the destructive gravimetric method as well as nuclear, electromagnetic, tensiometric, hygrometric, and remote sensing processes [2]. Among them, sensors employing an electromagnetic technique are widely utilized to measure soil moisture levels. Time-domain reflectometry (TDR), time-domain transmission (TDT), and capacitance sensors are the most commonly used sensors based on an electromagnetic technique [3,4]. TDR and TDT sensors are accurate but have limited large-scale applicability owing to the cost of investment. In contrast, capacitance sensors are less expensive but require precise calibration. The capacitance of a sensor is determined by the dielectric constant [5,6] and the volume fraction of each phase (bulk water, water vapor, air, solid minerals, etc.) [7,8]. It provides real-time soil moisture data according to the changes in the moisture content of the soil.

Many researchers have focused on the calibration of low-cost sensors which are used in different sensing techniques, such as capacitance-based sensors [9,10], resistivity-based granular matrix sensors [9,11], and sensors based on a tensiometer technique [9], to measure the soil water content in fields [2,12–14] and to develop low-cost automated irrigation systems [15]. Some studies have considered the effects of the physical and chemical properties of soil on the performances of soil moisture sensors [4,13,14,16–18]. These studies provide valuable insight into the necessity of soil-specific calibration.

A variety of capacitance sensors have become increasingly popular because they are less expensive than TDR and TDT high-frequency (GHz range) sensors and sufficiently reliable. Considering the cost of investment, a low-cost sensor even with a relatively weak accuracy is preferred in agriculture [19].

The SKU:SEN0193 sensor is a commercially available, low-cost capacitive soil moisture sensor which operates in low-power consumption. However, this soil moisture sensor has not been properly investigated for its accuracy and repeatability under laboratory conditions. Hence, the present study was performed (i) to investigate the accuracy and reliability of the SKU:SEN0193 low-cost capacitive soil moisture sensor under laboratory conditions, (ii) to develop a calibration function, and (iii) to integrate this sensor with a data acquisition system.

2. Materials and Methods

The SKU:SEN0193 capacitive soil moisture sensor, (DFRobot, Shanghai, China) with dimensions of 9.8 × 2.3 cm (L × W) was used for this study. The SKU:SEN0193 sensor can be powered from a voltage source in the range of 3.3 to 5.5 V. Thus, it can be interfaced with low-power microcontrollers. In addition, the sensor is made of corrosion-resistant material which increases durability. In parallel, SM-200 (Delta-T Devices Ltd., Cambridge, UK) a well-established commercially available soil moisture sensor was used for comparison to evaluate the accuracy of the SKU:SEN0193 sensor. The SM-200 sensor operates at 100 MHz, which measures a material (soil, water, air) response to polarization in an electromagnetic field (i.e., permittivity). The permittivity of the soil can be detected as a voltage output corresponding to the soil moisture content. The measuring accuracy is ±0.3% for volumetric water content, θ from 0 to 0.50 $m^3\ m^{-3}$. The SM-200 sensor has been used in studies as a reliable soil moisture sensor to determine the soil moisture content in the plant root zone [20] and as a reference sensor to compare the accuracy of other capacitive-type soil moisture sensors [21]. In addition, the SHT30 temperature and humidity sensor was used to obtain the temperature and humidity of the air.

Commercially available organic-rich gardening soil was obtained from a local producer (Gardening. Pro, Maruki, Japan), hereafter referred to as organic-rich soil. This soil was originally obtained from the surface of the kanto loam layer (Black soil, Kanuma city, Tochigi Prefrecture, Japan). The measured organic matter content and mineral content were 24.8 and 75.2%, respectively [22]. The air-dried soil was sieved through a 2 mm mesh to remove aggregated soil clumps, and the <2 mm fraction of the soil was used in this study. In addition, a laboratory soil mixture was prepared by mixing the organic-rich soil with vermiculite in a 1:1 ratio [23,24]. This soil mixture was used as a growing medium for laboratory plants.

2.1. Data Acquisition and Analysis via an Internet-Based Platform

The data acquisition system used in this study consisted of two main components called the microcontroller unit and the Wi-Fi module. The main microcontroller STC89C52RC was operated at a speed of 11.0592 MHz. A software-implemented I^2C bus [25] was used to interface a 16 × 2 LCD (1602 character-type liquid crystal display) module, an ADS1115 16-bit analog-to-digital converter (ADC), and temperature and humidity sensor with the microcontroller. An ESP8266-12E low-cost serial-to-WiFi module was interfaced through STC89C52RC inbuilt UART. The analog data output pin of the SKU:SEN0193 sensor was connected to the ADS1115 with a full-scale range of ±4.096 V. The ADS1115 converted the voltage of SKU:SEN0193 sensor to raw counts (raw). ThingSpeak API, an open IoT (Internet of Things) platform, was used to collect and analyze data with MATLAB@ analytics. The assembly program for the microcontroller (Supplementary Material: Microcontroller program code) was written by using Keil μVision 5 IDE, and AT commands were used to control the WiFi module (Figures 1 and 2). The total cost of the developed prototype was around $45.7 (Table 1).

Table 1. The total cost of the developed soil moisture monitoring system (US$ in 2019).

Component	Units	Unit Cost ($)	Subtotal ($)	Total ($)
STC89C52RC	1	1.12	1.12	
ADS1115	1	2.73	2.73	
ESP8266-12E	1	1.79	1.79	
SKU:SEN0193	4	7.24	28.96	
SHT30	1	3.98	3.98	
LCD1602	1	2.12	2.12	
Other components			5.00	
				45.7

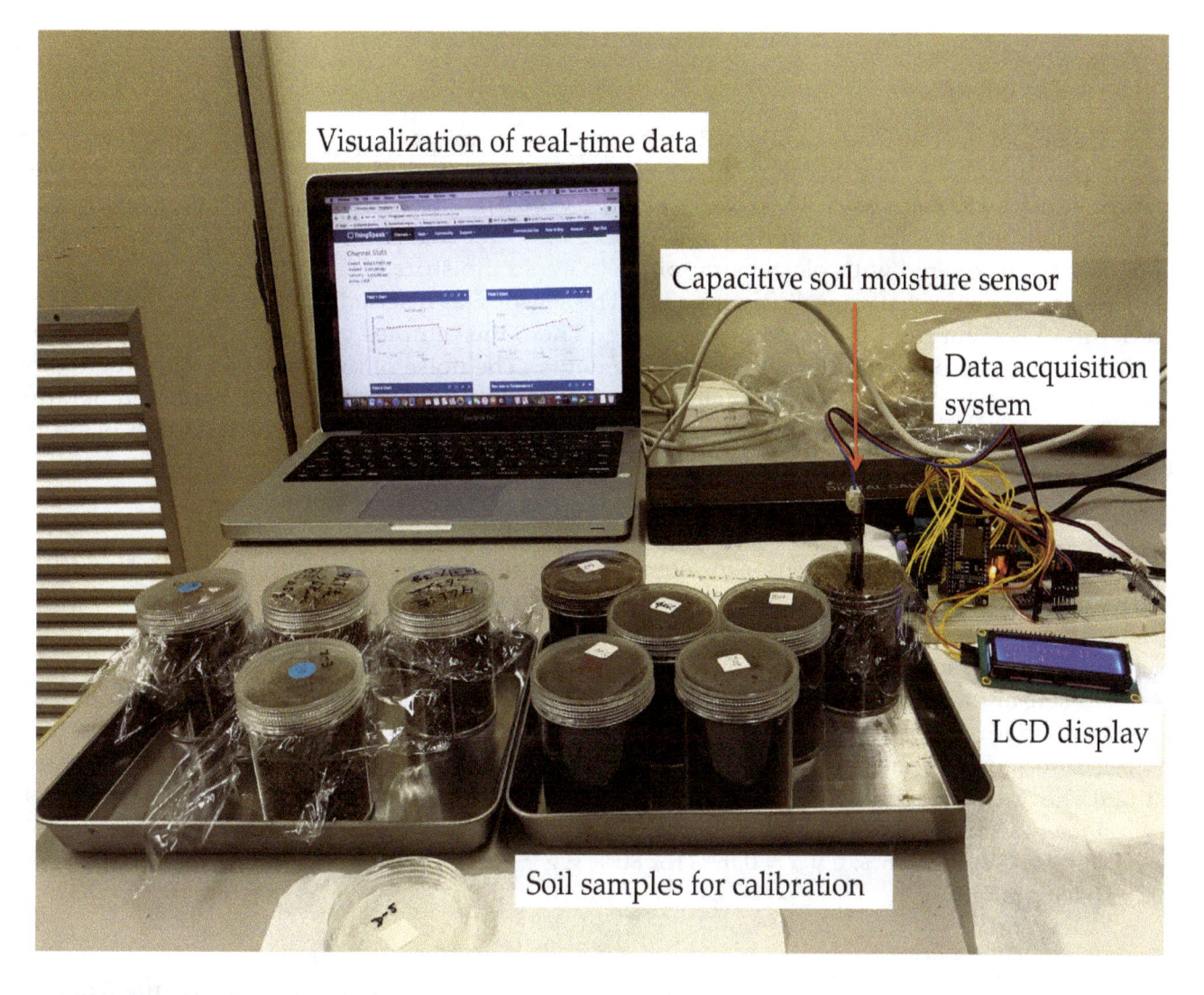

Figure 1. Acquisition and visualization of real-time data during the soil sample calibration process.

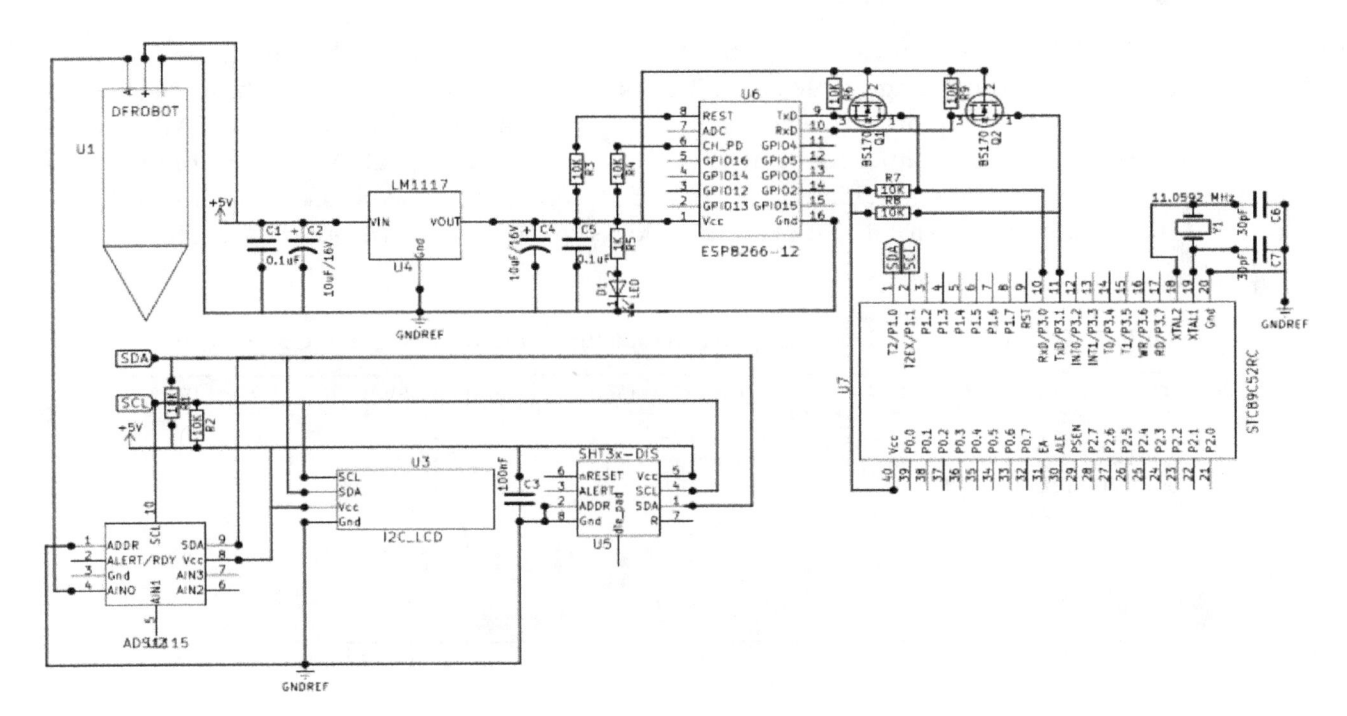

Figure 2. Circuit diagram of the data acquisition system.

2.2. Soil Sample Preparation

Soil samples with different soil moisture contents were prepared by adjusting the soil moisture content gravimetrically. The moisture-adjusted samples were packed into polycarbonate containers (average volume 110 cm^3 (Area: 18.08 cm^2 × Height: 6.1 cm)) with an average dry bulk density, ρ_d, of 0.6 g cm^{-3} for organic-rich gardening soil and 0.3 g cm^{-3} for the laboratory soil mixture. Then the containers were covered with plastic wrappings to avoid moisture loss during the calibration, and the samples were stored after sealing with a lid. The sensor was inserted to the center of the soil core at a depth of 5–6 cm from the upper soil surface. Then, the samples were let achieve equilibrium for 15 to 20 min, before performing the measurements. The noise of the data, i.e., data fluctuation, was very limited during the measurements. All experiments were performed at room temperature (25 °C). The raw counts of the samples for each gravimetric water content were stored using the Thingspeak platform.

2.3. Sensor-to-Sensor Variability Study

Prior to the calibration, the sensor-to-sensor reading variability (raw counts) of the SKU:SEN0193 sensors for predetermined soil moisture contents was evaluated (Table 2). The sensor-to-sensor variability study was performed only for the organic-rich soil. Samples with two different soil moisture contents (40 and 80%, g g^{-1} of organic-rich soil), in duplicate, and water were used to test sensor-to-sensor variability. Repeated measurements (twenty replications of sensor measurements) were obtained for every duplicated soil sample with four SKU:SEN0193 sensors (S_1, S_2, S_3, and S_4). The sensor-to-sensor variability was analyzed using a one-way Analysis of Variance (ANOVA) statistical test [26]. Further, to estimate the measurement noise of the SKU:SEN0193 sensors, the coefficient of variance (CV) of raw counts was determined. The CV values thus obtained were $CV_{S1} = 0.05\%$, $CV_{S2} = 0.10\%$, $CV_{S3} = 0.06\%$, $CV_{S4} = 0.09\%$ for 80%, g g^{-1} sample, i.e., an organic-rich soil sample in a field capacity range.

2.4. Calibration of the Sensor

Considering the sensor-to-sensor variability, the samples were measured repeatedly using two selected SKU:SEN0193 sensors to obtain a calibration function. The soil-specific calibration was performed using the organic-rich gardening soil and the laboratory soil mixture. The moisture contents of the tested soil and soil mixture were adjusted gravimetrically using distilled water. Moisture-adjusted samples were prepared with gravimetric water content of 0 (oven dried), 10, 20, 30, 40, 50, 60, 70, 80, 90, and 100 g g^{-1}. The moisture-adjusted samples were kept in closed plastic bags for 24 h for equilibration, and the gravimetric water content was verified by evaluating the moisture content of each sample by oven-drying at 105 °C [27]. The raw counts for the samples of the two different materials with each gravimetric water content were stored using the Thingspeak platform.

2.5. Validation of the Developed Calibration Function

The developed calibration function was validated by measuring organic-rich soil samples with different moisture contents: oven-dried samples, air-dried samples, samples with a moisture content of up to 100%, and saturated samples (>100%, g g^{-1}). The samples with different moisture contents were packed into polycarbonate containers. The soil water content of the samples was measured by the SKU:SEN0193 and SM-200 sensors. These values were confirmed gravimetrically by oven-drying. Similarly, data validation was performed in the selected values of the field capacity using samples with soil water contents of 60, 65, 70, 75, and 80% g g^{-1}. A soil-specific calibration was performed for the SM-200 sensor in accordance with its user manual [28]. The calculated coefficients a$_0$ and a$_1$, which conveniently parameterize the dielectric properties of soils, were 1.3 and 6.1, respectively, for the organic-rich soil (Equations (1)–(3)).

$$\sqrt{\varepsilon} = a_0 + a_1 \theta \tag{1}$$

$$a_0 = \sqrt{\varepsilon_{dry\_soil}} \tag{2}$$

$$\sqrt{\varepsilon} = 1.0 + 16.103V - 38.725V^2 + 60.881V^3 - 46.032V^4 + 13.536V^5 \tag{3}$$

where ε is the dielectric permittivity of the soil, θ (cm^3 cm^{-3}) is the volumetric water content, and V is the SM-200 reading in Volts of the corresponding soil moisture content.

A polynomial conversion (Equation (4)) was performed to calculate the volumetric water content θ (cm^3 cm^{-3}) after soil-specific calibration for the organic-rich soil:

$$\theta = \frac{\left[1.0 + 16.103V - 38.725V^2 + 60.881V^3 - 46.032V^4 + 13.536V^5\right] - a_0}{a_1} \tag{4}$$

3. Results and Discussion

The data measured by the low-cost SKU:SEN0193 sensor and the temperature and humidity sensor were stored and visualized via an internet-based platform. The soil water contents predicted using the linear equation (which we developed by considering the raw counts of the sensor for bulk water and air with average values of 10,653 and 20,240, respectively) were inaccurate for both the organic-rich soil and the laboratory soil mixture. Hence, new calibration functions were developed by plotting the volumetric water content θ of the tested materials as a function of the raw count.

The volumetric water content θ (cm^3 cm^{-3}) of the samples can be obtained by the measured gravimetric water content w (g/g, %) of the samples, the dry bulk density of the material ρ_b (g cm^{-3}), and the density of water ρ_w (g cm^{-3}) as follows (Equation (5)) [29]:

$$\theta = w \frac{\rho_b}{\rho_w} \tag{5}$$

3.1. Results of the Sensor-to-Sensor Variability Study

The results of the ANOVA test at a 5% significance level for 20 replications of sensor response measurements (for each duplicated soil sample) of 4 SKU:SEN0193 sensors indicated that the mean sensor response was significantly different for the 4 SKU:SEN0193 sensors (see Table 2), demonstrating significant sensor-to-sensor variability.

Table 2. Results of the ANOVA test for 20 replications of sensor response measurements of 4 SKU:SEN0193 sensors for 3 different conditions.

Condition	Consideration	Df *	Sum of Squares (Raw counts)	Mean Square (Raw counts)	F Value
40%, g g^{-1}	Sensor-to-sensor variability	3	14888295	4962765	30091
	Noise	76	12535	165	
80%, g g^{-1}	Sensor-to-sensor variability	3	33893919	11297973	9813
	Noise	76	87499	1151	
Water	Sensor-to-sensor variability	3	7610824	2536941	1121
	Noise	76	171977	2263	

* df: degrees of freedom.

3.2. Calibration of the SKU:SEN0193 Sensor for Organic-Rich Soil and Laboratory Soil Mixture

The sensor-to-sensor variability study suggested that the responses of the tested four sensors were different, and therefore, their accuracy should be improved by using a sensor-specific calibration model for each sensor [29]. Considering the sensor-to-sensor variability, the samples were measured repeatedly using two selected SKU:SEN0193 sensors to obtain a calibration function.

The average raw count of the two SKU:SEN0193 sensors was obtained for each moisture-adjusted sample. Soil moisture, temperature, and humidity data were recorded at intervals of 1 min. After the equilibrium, the data were recorded over 15–20 min for each sample (n = 15–20). The averaged raw count as a function of volumetric water content, θ, was plotted in Figure 3. Here, we show the results for three moisture categories for organic-rich soil: dry to moderately wet samples with volumetric water content around 0.0–0.26 cm^3 cm^{-3}, field capacity samples with volumetric water content around 0.34–0.50 cm^3 cm^{-3}, and saturated samples with volumetric water content around 0.62–0.74 cm^3 cm^{-3}. In addition, these data indicate that the sensor could distinguish the three different soil moisture levels of dry to moderately wet, field capacity, and saturated. Hence, the minimum and maximum values of the readily available soil moisture (permanent wilting percentage and field capacity) can be maintained using the SKU:SEN0193 sensor in organic-rich soil. Interestingly, the calibration curve obtained for the laboratory soil mixture did not show a clear difference among the three moisture categories. The laboratory soil mixture was prepared by mixing organic-rich soil and vermiculite. It has been reported that also high-cost sensors such as TDR do not accurately predict the soil moisture content of soil substitutes based on mineral media (vermiculite, tuff, perlite) [30]. Furthermore, it is known that less expensive capacitance moisture sensors operate in low frequencies and are thereby more sensitive to effects of soil textural variances and salinity [4,31]. Thus, the low sensitivity of the SKU:SEN0193 sensor for the soil moisture content of the laboratory soil mixture could be due to the high mineral content of the laboratory soil mixture.

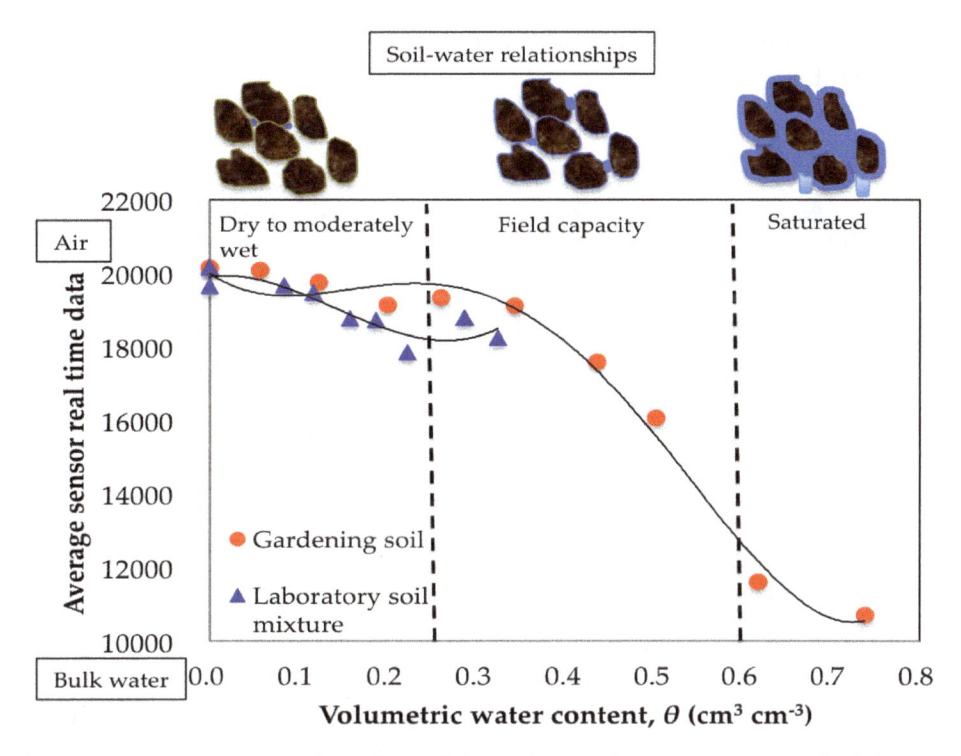

Figure 3. Average raw count as a function of the volumetric water content θ of the tested porous materials ($n = 20$). The raw count varies with the dielectric constants of the bulk water and air and ranges between 10,000 and 20,000 according to the sensor output.

3.3. Soil-Specific Calibration Function

The derived soil-specific calibration function for the organic-rich soil accurately predicted the soil moisture values of known samples, while that for the laboratory soil mixture did not. Therefore, further studies were carried out using organic-rich soils.

The derived calibration function (polynomial function) of the SKU:SEN0193 sensor for the organic-rich soil is ($R^2 = 0.98$):

$$\theta = 13.248 - 2.576 \times 10^{-3}\, raw + 1.726 \times 10^{-7}\, raw^2 - 3.839 \times 10^{-12}\, raw^3 \tag{6}$$

3.4. Data Validation with the Commercially Available SM-200 Sensor and the Low-Cost SKU:SEN0193 Capacitive Sensor

Soil samples were analyzed for their soil water content using the commercial available SM-200 sensor and the low-cost SKU:SEN0193 capacitive sensor. The volumetric water contents derived from the polynomial function of the SKU:SEN0193 sensor, the polynomial function of the SM-200 sensor for organic soil (manufacturer's function), and the soil-specific calibration (the function obtained from the measured coefficients) are plotted in Figure 4 as a function of the measured volumetric water content. The root-mean-square error (RMSE) was calculated for the soil water contents derived by the different calibration (polynomial) functions. The RMSE values of the SKU:SEN0193 sensor function, the SM-200 manufacturer's function, and the SM-200 soil-specific calibration function were 0.09, 0.07, and 0.06 $cm^3\ cm^{-3}$, respectively, for the samples with dry to saturated levels of soil water content. The soil water content at 0.45 $cm^3\ cm^{-3}$ (74% gravimetric water content) was predicted precisely by the derived polynomial functions of both sensors (Figure 4a). Hence, data prediction at the field capacity (a gravimetric water content of 60–80%) was performed by measuring the moisture-adjusted samples (Figure 4b). The RMSE was calculated as before. The RMSE values of the SKU:SEN0193 sensor function, the SM-200 manufacturer's function, and the SM-200 soil-specific calibration function were 0.05, 0.08, and 0.03 $cm^3\ cm^{-3}$, respectively, in the field capacity range. The data prediction ability of the sensor appeared similar to that of the SM-200 sensor (with soil-specific calibration) at a lower investment cost.

This study highlights the potential use of the low-cost SKU:SEN0193 capacitive moisture sensor for predicting the water content of soil.

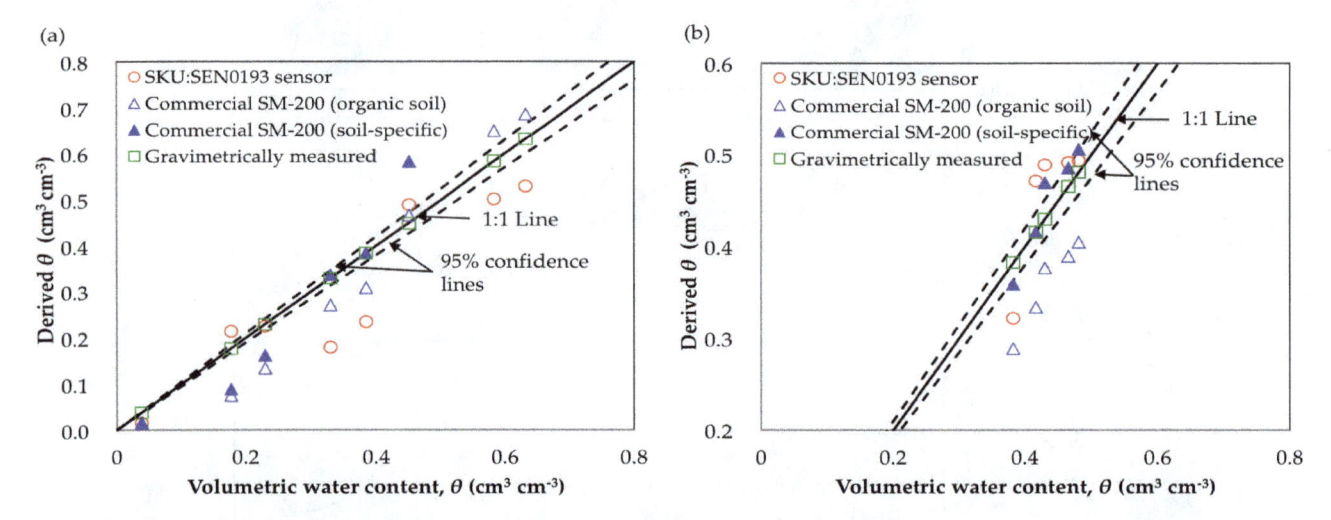

Figure 4. The samples' soil water contents θ derived from the polynomial function of the SKU:SEN0193 sensor, the polynomial function of the SM-200 sensor for organic soil (manufacturer's function), and the soil-specific calibration function as a function of volumetric water content: (**a**) in the range of soil water contents from dry to saturated, (**b**) in the range of the field capacity.

Prior to deploying the system in the field, it is necessary to evaluate the performance of the derived function for continuous operation. Thus, soil moisture loss from a sample in the field capacity range was evaluated using the derived polynomial function (gravimetric basis) (Figure 5). The experiment was performed for 12 h, from day to nighttime, under laboratory conditions. The room temperature and the relative humidity of the room ranged from 25 to 26 °C and 72 to 75%, respectively, during the measurements. The 3D scatter-plot in Figure 5 illustrates the soil moisture loss with time. It shows a gradual decrease in soil moisture during 12 h of measurements.

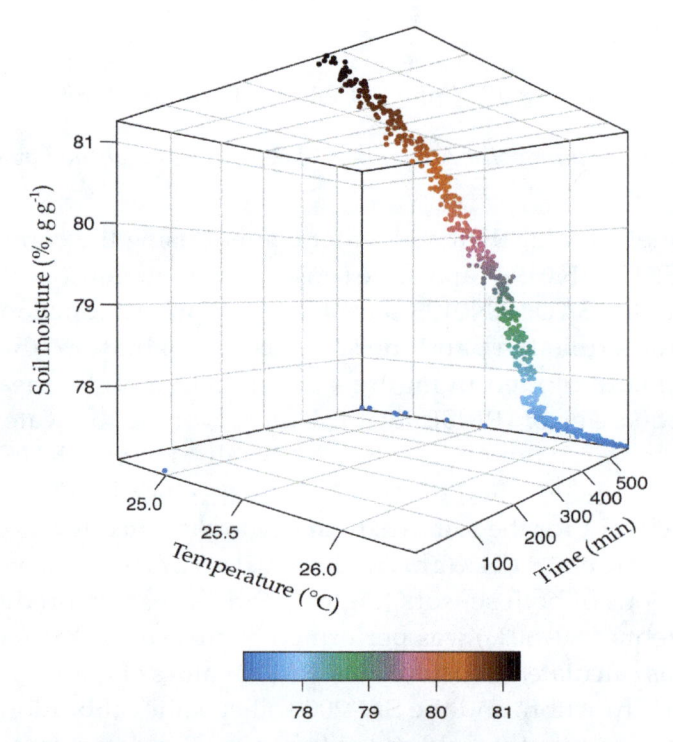

Figure 5. Soil moisture loss with time. The developed calibration function (gravimetric basis) was used to derive the soil moisture loss every minute (Equation (6)).

4. Conclusions

We evaluated the accuracy and reliability of the SKU:SEN0193 low-cost capacitive soil moisture sensor under laboratory conditions. The developed soil-specific calibration function for gardening soil performed satisfactorily during the sensor validation procedure for the prediction of soil water content. Furthermore, our data suggest that the soil-specific calibration function of the SKU:SEN0193 sensor can be used to predict the soil water contents in three different ranges of soil moisture. Hence, it can be used to maintain the minimum and maximum values of readily available soil moisture in indoor systems. In contrast, the SKU:SEN0193 capacitive soil moisture sensor did not perform acceptably for the laboratory soil mixture in predicting the soil moisture content. This result suggests that the accuracy of the sensor depends on the soil mixture constituents.

However, it is necessary to investigate the effects of soil temperature, bulk density of the soil profile, and salinity levels on the accuracy of the sensor measurements. Further studies should be undertaken to assess its behavior in a real working scenario under field conditions.

Author Contributions: Conceptualization, validation, and production of the final manuscript, E.A.A.D.N. and I.S.P.N.; Microcontroller program and prototype development, I.S.P.N.; Experiments for sensor calibration, E.A.A.D.N.; Supervision, revision and editing, T.F.

References

1. Kramer, P.J.; Boyer, J.S. Soil and water. In *Water Relations of Plants and Soils*; Academic Press: San Diego, CA, USA, 1995; pp. 84–114. Available online: http://udspace.udel.edu/handle/19716/2830. (accessed on 15 December 2018).
2. Zazueta, F.S.; Xin, J. *Soil Moisture Sensors*; Florida Cooperative Extension Service, Institute of Food and Agricultural Science; University of Florida: Gainesville, FL, USA, 1994.
3. Bogena, H.R.; Huisman, J.A.; Schilling, B.; Weuthen, A.; Vereecken, H. Effective calibration of low-cost water content sensors. *Sensors* **2017**, *17*, 208. [CrossRef]
4. Vaz, C.M.P.; Jones, S.; Meding, M.; Tuller, M. Evaluation of Standard Calibration Functions for Eight Electromagnetic Soil Moisture Sensors. *Vadose Zone J.* **2013**, *12*, 1–16. [CrossRef]
5. Terzic, E.; Terzic, J.; Nagarajah, R.; Alamgir, M. Capacitive sensing technology. In *Neural Network Approach to Fluid Quantity Measurement in Dynamic Environments*; Springer: London, UK, 2012.
6. Robbins, A.; Miller, W. *Circuit Analysis: Theory and Practice*; Delmar: Albany, NY, USA, 2000.
7. Fen-Chong, T.; Fabbri, A.; Guilbaud, J.; Coussy, O. Determination of liquid water content and dielectric constant in porous media by the capacitive method. *Comptes Rendus Mécanique* **2004**, *332*, 639–645. [CrossRef]
8. Kaatze, U. The dielectric properties of water in its different states of interaction. *J. Solution Chem.* **1997**, *26*, 1049–1112. [CrossRef]
9. Ganjegunte, G.K.; Sheng, Z.; Clark, J.A. Evaluating the accuracy of soil water sensors for irrigation scheduling to conserve freshwater. *Appl. Water Sci.* **2012**, *2*, 119–125. [CrossRef]
10. Parvin, N.; Degré, A. Soil-specific calibration of capacitance sensors considering clay content and bulk density. *Soil Res.* **2016**, *54*, 111–119. [CrossRef]
11. Payero, J.O.; Mirzakhani-Nafchi, A.; Khalilian, A.; Qiao, X.; Davis, R. Development of a low-cost Internet-of-Things (IOT) system for monitoring soil water potential using Watermark 200SS sensors. *Adv. Internet Things* **2017**, *7*, 71–86. [CrossRef]
12. Archer, N.A.L.; Rawlins, B.R.; Marchant, B.P.; Mackay, J.D.; Meldrum, P.I. Approaches to calibrate in-situ capacitance soil moisture sensors and some of their implications. *SOIL Discuss.* **2016**. [CrossRef]
13. Fares, A.; Awal, R.; Bayabil, H.K. Soil water content sensor response to organic matter content under laboratory conditions. *Sensors* **2016**, *16*, 1239. [CrossRef]
14. Mittelbach, H.; Lehner, I.; Seneviratne, S.I. Comparison of four soil moisture sensor types under field conditions in Switzerland. *J. Hydrol.* **2012**, *430–431*, 39–49. [CrossRef]
15. Ferrarezi, R.S.; Dove, S.K.; Van Iersel, M.W. An automated system for monitoring soil moisture and controlling irrigation using low-cost open-source microcontrollers. *HortTechnology* **2015** *25*, 110–118. [CrossRef]

16. Cardenas-Lailhacar, B.; Dukes, M. Effect of temperature and salinity on the precision and accuracy of landscape irrigation soil moisture sensor systems. *J. Irrig. Drain. Eng.* **2015**, *141*, 7. [CrossRef]
17. Inoue, M.; Ould Ahmed, B.A.; Saito, T.; Irshad, M.; Uzoma, K.C. Comparison of three dielectric moisture sensors for measurement of water in saline sandy soil. *Soil Use Manag.* **2008**, *24*, 156–162. [CrossRef]
18. Nemali, K.S.; Montesano, F.; Dove, S.K.; Van Iersel, M.W. Calibration and performance of moisture sensors in soilless substrates: ECH2O and Theta probes. *Sci. Hortic.* **2007**, *112*, 227–234. [CrossRef]
19. Kojima, Y.; Shigeta, R.; Miyamoto, N.; Shirahama, Y.; Nishioka, K.; Mizoguchi, M.; Kawahara, Y. Low-cost soil moisture profile probe using thin-film capacitors and a capacitive touch sensor. *Sensors* **2016**, *16*, 1292. [CrossRef] [PubMed]
20. Puértolas, J.; Alcobendas, R.; Alarcón, J.J.; Dodd, I.C. Long-distance abscisic acid signalling under different vertical soil moisture gradients depends on bulk root water potential and average soil water content in the root zone. *Plant Cell Environ.* **2013**, *36*, 1465–1475. [CrossRef]
21. Kodešová, R.; Kodeš, V.; MRáz, A. Comparison of two sensors ECH2O EC-5 and SM200 for measuring soil water content. *Soil Water Res.* **2011**, *6*, 102–110. [CrossRef]
22. ASTM D 2974. *Standard Test Methods for Moisture, Ash, and Organic Matter of Peat and Organic Soils*; ASTM International: West Conshohocken, PA, USA, 2007.
23. Ding, S.; Zhang, B.; Qin, F. Arabidopsis RZFP34/CHYR1, a ubiquitin E3 ligase, regulates stomatal movement and drought tolerance via SnRK2.6-mediated phosphorylation. *Plant Cell.* **2015**, *27*, 3228–3244. [CrossRef] [PubMed]
24. Ma, Q.; Xia, Z.; Cai, Z.; Li, L.; Cheng, Y.; Liu, J.; Nian, H. GmWRKY16 enhances drought and salt tolerance through an ABA-mediated pathway in *Arabidopsis thaliana*. *Front Plant Sci.* **2019**, *9*, 1979. [CrossRef]
25. Quarles, S.D. How to implement I2C serial communication using Intel MCS-51 microcontrollers; 1993. Intel Corporation: P.O. Box 7641, Mt. Prospect, IL 60056-7641. Available online: http://electro8051.free.fr/I2C/27231901.pdf. (accessed on 8 March 2018).
26. Rosenbaum, U.; Huisman, J.A.; Weuthen, A.; Vereecken, H.; Bogena, H.R. Sensor-to-sensor variability of the ECH2O, EC-5, TE, and 5TE sensors in dialectic liquids. *Vadose Zone J.* **2010**, *9*, 181–186. [CrossRef]
27. ASTM D 2216. *Standard Test Method for Laboratory Determination of Water (Moisture) Content of Soil, Rock, and Soil-Aggregated Mixture*; ASTM International: West Conshohocken, PA, USA, 2010.
28. Delta-T. *Devices Ltd. User Manual for the SM200 Soil Moisture Sensor*; SM200-UM-1; 1 May 2006; Delta-T. Devices Ltd.: Cambridge, UK, 2006.
29. Hillel, D. *Introduction to Environmental Soil Physics*; Academic press, Elsevier Science: San Diego, CA, USA, 2004; pp. 14–15. Available online: https://dewagumay.files.wordpress.com/2011/12/environmental-soil-physics.pdf (accessed on 16 December 2018).
30. Da Silva, F.F.; Wallach, R.; Polak, A.; Chen, Y. Measuring water content of soil substitutes with time-DOMAIN reflectometry (TDR). *J. Am. Soc. Hortic. Sci.* **1998**, *123*, 734–737. [CrossRef]
31. Kizito, F.; Campbell, C.S.; Campbell, G.S.; Cobos, D.R.; Teare, B.L.; Carter, B.; Hopmans, J.W. Frequency, electrical conductivity and temperature analysis of a low-cost capacitance soil moisture sensor. *J. Hydrol.* **2008**, *352*, 367–378. [CrossRef]

PERMISSIONS

All chapters in this book were first published in MDPI; hereby published with permission under the Creative Commons Attribution License or equivalent. Every chapter published in this book has been scrutinized by our experts. Their significance has been extensively debated. The topics covered herein carry significant findings which will fuel the growth of the discipline. They may even be implemented as practical applications or may be referred to as a beginning point for another development.

The contributors of this book come from diverse backgrounds, making this book a truly international effort. This book will bring forth new frontiers with its revolutionizing research information and detailed analysis of the nascent developments around the world.

We would like to thank all the contributing authors for lending their expertise to make the book truly unique. They have played a crucial role in the development of this book. Without their invaluable contributions this book wouldn't have been possible. They have made vital efforts to compile up to date information on the varied aspects of this subject to make this book a valuable addition to the collection of many professionals and students.

This book was conceptualized with the vision of imparting up-to-date information and advanced data in this field. To ensure the same, a matchless editorial board was set up. Every individual on the board went through rigorous rounds of assessment to prove their worth. After which they invested a large part of their time researching and compiling the most relevant data for our readers.

The editorial board has been involved in producing this book since its inception. They have spent rigorous hours researching and exploring the diverse topics which have resulted in the successful publishing of this book. They have passed on their knowledge of decades through this book. To expedite this challenging task, the publisher supported the team at every step. A small team of assistant editors was also appointed to further simplify the editing procedure and attain best results for the readers.

Apart from the editorial board, the designing team has also invested a significant amount of their time in understanding the subject and creating the most relevant covers. They scrutinized every image to scout for the most suitable representation of the subject and create an appropriate cover for the book.

The publishing team has been an ardent support to the editorial, designing and production team. Their endless efforts to recruit the best for this project, has resulted in the accomplishment of this book. They are a veteran in the field of academics and their pool of knowledge is as vast as their experience in printing. Their expertise and guidance has proved useful at every step. Their uncompromising quality standards have made this book an exceptional effort. Their encouragement from time to time has been an inspiration for everyone.

The publisher and the editorial board hope that this book will prove to be a valuable piece of knowledge for researchers, students, practitioners and scholars across the globe.

LIST OF CONTRIBUTORS

Rodrigo Gonçalves Trevisan, Natanael Santana Vilanova Júnior, Mateus Tonini Eitelwein and José Paulo Molin
Precision Agriculture Laboratory, Department of Biosystems Engineering, Luiz de Queiroz College of Agriculture, University of São Paulo, Piracicaba 13418-900, São Paulo, Brazil

Marta Aranguren, Ander Castellón and Ana Aizpurua
NEIKER-Basque Institute for Agricultural Research and Development, Department of Plant Production and Protection, Berreaga 1, 48160 Derio, Biscay, Spain

David Reiser, Oliver Bumann, Jörg Morhard and Hans W. Griepentrog
Institute of Agricultural Engineering, Hohenheim University, Stuttgart 70599, Germany

El-Sayed Sehsah
Department of Agricultural Engeenering, Faculty of Agriculture, Kafrelsheikh University, Kafrelsheikh 33516, Egypt

Pei Wang, Mingxiong Ou, Chen Gong and Weidong Jia
Key Laboratory of Modern Agricultural Equipment and Technology, Ministry of Education of PRC, Jiangsu University, Zhenjiang 212300, China
Key Laboratory of Plant Protection Engineering, Ministry of Agriculture and Rural Affairs of PRC, Jiangsu University, Zhenjiang 212300, China

Wei Yu
Key Laboratory of Plant Protection Engineering, Ministry of Agriculture and Rural Affairs of PRC, Jiangsu University, Zhenjiang 212300, China

Rui Pitarma, João Crisóstomo and Maria Eduarda Ferreira
Research Unit for Inland Development, Polytechnic of Guarda, Avenida Francisco Sá Carneiro 50, 6300-559 Guarda, Portugal

Fernando Palacios, Maria P. Diago and Javier Tardaguila
Televitis Research Group, University of La Rioja, 26006 Logroño (La Rioja), Spain
Instituto de Ciencias de la Vid y del Vino, University of La Rioja, CSIC, Gobierno de La Rioja, 26007 Logroño, Spain

Christoph W. Zecha, Johanna Link and Wilhelm Claupein
Department of Agronomy (340a), Institute of Crop Science, University of Hohenheim, Fruwirthstraße 23, 70599 Stuttgart, Germany

Gerassimos G. Peteinatos
Department of Weed Science (360b), Institute of Phytomedicine, University of Hohenheim, Otto-Sander-Straße 5, 70599 Stuttgart, Germany

João Valente, Rodrigo Almeida and Lammert Kooistra
Laboratory of Geo-information Science and Remote Sensing, Wageningen University & Research, 6708 PB Wageningen, The Netherlands

Simon Munder, Dimitrios Argyropoulos and Joachim Müller
Institute of Agricultural Engineering, Universität Hohenheim, Garbenstrasse 9, 70599 Stuttgart, Germany

Lili Yao, Qing Wang, Jinbo Yang, Yu Zhang, Yan Zhu, Weixing Cao and Jun Ni
National Engineering and Technology Center for Information Agriculture, Key Laboratory for Crop System Analysis and Decision Making, Ministry of Agriculture, Jiangsu Key Laboratory for Information Agriculture, Nanjing Agricultural University, Nanjing 210095, Jiangsu, China

Michael Koutsiaras, Vasilios Psiroukis and Spyros Fountas
Department of Natural Resources Management & Agricultural Engineering, Agricultural University of Athens, Iera Odos 75, 11855 Athens, Greece

Georgios Bourodimos
Department of Agricultural Engineering, Institute of Soil and Water Resources, Hellenic Agricultural Organization "DEMETER", Democratias 61, 13561 Aghii Anargiri Attikis, Greece

Athanasios Balafoutis
Institute for Bio-Economy & Agri-Technology, Centre of Research & Technology Hellas, Dimarchou Georgiadou 118, 38221 Volos, Greece

Sigfredo Fuentes, Eden Jane Tongsona and Claudia Gonzalez Viejo
School of Agriculture and Food, Faculty of Veterinary and Agricultural Sciences, The University of Melbourne, Parkville, VIC 3010, Australia

List of Contributors

Roberta De Bei, Renata Ristic, Stephen Tyerman and Kerry Wilkinson
School of Agriculture, Food andWine, The University of Adelaide, PMB 1, Glen Osmond, SA 5064, Australia

Riccardo Rainato, Luigi Sartori and Francesco Marinello
Department TESAF, University of Padova, viale dell'Università, 16, I-35020 Legnaro (PD), Italy

Ahmed Kayad
Department TESAF, University of Padova, viale dell'Università, 16, I-35020 Legnaro (PD), Italy
Agricultural Engineering Research Institute (AEnRI), Agricultural Research Centre, Giza 12619, Egypt

Lorenzo Picco
Department TESAF, University of Padova, viale dell'Università, 16, I-35020 Legnaro (PD), Italy
Faculty of Engineering, Universidad Austral de Chile, Campus Miraflores, Valdivia 5090000, Chile
Universidad Austral de Chile, RINA–Natural and Anthropogenic Risks Research Center, Campus Miraflores, Valdivia 5090000, Chile

Gurjinder S. Baath and Hardeep Singh
Department of Plant and Soil Sciences, Oklahoma State University, 371 Agricultural Hall, Stillwater, OK 74078, USA

Harpinder K. Baath and Johnson P. Thomas
Department of Computer Science, Oklahoma State University, 219 MSCS, Stillwater, OK 74078, USA

Prasanna H. Gowda
USDA-ARS, Southeast Area Branch, 114 Experiment Station Road, Stoneville, MS 38776, USA

Brian K. Northup and Srinivas C. Rao
USDA-ARS, Grazinglands Research Laboratory, 7207 W. Cheyenne St., El Reno, OK 73036, USA

Ekanayaka Achchillage Ayesha Dilrukshi Nagahage, Isura Sumeda Priyadarshana Nagahage and Takeshi Fujino
Graduate School of Science and Engineering, Saitama University, 255 Shimo-Okubo, Sakura-ku, Saitama 338-8570, Japan

Index

A
Aggregation, 73, 140
Aggregation Index, 140

D
Data Mining, 93, 103, 205, 213
Data Transmission, 187
Datum, 193
Decision Making, 50, 89, 136, 176, 188
Dependent Variable, 98-99
Diagnostic Accuracy, 33
Digital Signal, 43, 53, 141

E
Electrical Conductivity, 224
Embedded Computer, 41-42
Embedded Systems, 132
Energy Balance, 186
Ethylene, 105-120
Extrapolation, 60, 128

F
Feedback, 43
Field Of View, 42, 136, 139-140, 145, 149, 151, 186-187
Food Packaging, 198
Food Security, 155

G
Gas Chromatography, 106
Genetic Algorithm, 93
Geographic Information System, 16, 97, 104
Global Navigation Satellite Systems, 39
Global Positioning System, 75
Global Warming, 59, 175
Glycine, 200, 202

H
Heat Transfer, 61-62, 65
Heterogeneity, 54

I
Image Processing, 51, 74, 76
Imaging System, 74
Industrial Applicability, 74
Information Systems, 71

Infrared Spectroscopy, 152, 175, 187, 189-190, 200-201, 213-214
Ingredient, 10, 54
Inner Product, 203-204
Internet Of Things, 187, 198, 217

K
Kjeldahl Method, 148
Kjeldahl's Method, 20

L
Labor Costs, 40
Lichens, 69-70
Light-emitting Diodes, 95
Limiting Factor, 123
Linear Regression, 55-57, 97-99, 134, 204
Liquid Chromatography, 176, 189
Liquid Crystal Display, 53, 216

M
Machine Learning, 49, 73-74, 79, 82, 89, 91, 101, 104, 175-177, 179-180, 182, 184-188, 190, 200, 202, 204-205, 208, 212-213
Mass Spectrometry, 176, 189
Metabolism, 152
Methodologies, 51, 70
Microcontroller, 53, 56, 216-217, 223
Morphology, 73, 76, 80, 184
Morphometry, 177, 181
Motor Control, 43

N
Northern Hemisphere, 65

R
Resolution, 7, 12, 42, 59-61, 71, 75, 95-96, 100, 119, 140, 153, 177, 180, 186-187
Resource Management, 40
Root-mean-square Error, 215, 221

S
Salinity, 220, 223-224
Sampling Rate, 160
Semiconductor, 75
Senescence, 101, 106
Serial Communication, 224
Shear Stress, 45, 47

Skeletonization, 50
Soil Erosion, 191
Spatial Data, 16
Spatial Information, 90
Spectral Signature, 2
Spectroscopy, 137, 151-152, 175-176, 180, 184, 187-190, 200-201, 213-214
Standard Deviation, 95, 181-182, 195, 203
Standardization, 134, 170
Support Vector Machine, 79, 82, 85, 200, 202, 204, 206-209, 212
Sustainability, 39, 58-59, 70, 171, 173

T

Tandem Mass Spectrometry, 176, 189

Taxonomy, 36, 91
Thermal Conductivity, 62-63
Thermogravimetric Analysis, 124
Titration, 36
Transgenic, 58

U

Ubiquitin, 224

V

Variables, 34, 77, 97-98, 104-105, 121, 127, 168, 170, 173, 187, 196, 203-204
Viscosity, 156